EVERY THING MUST GO

Every Thing Must Go argues that the only kind of metaphysics that can contribute to objective knowledge is one based specifically on contemporary science as it really is, and not on philosophers' a priori intuitions, common sense, or simplifications of science. In addition to showing how recent metaphysics has drifted away from connection with all other serious scholarly inquiry as a result of not heeding this restriction, Ladyman and Ross demonstrate how to build a metaphysics compatible with current fundamental physics ('ontic structural realism'), which, when combined with their metaphysics of the special sciences ('rainforest realism'), can be used to unify physics with the other sciences without reducing these sciences to physics itself. Taking science metaphysically seriously, they argue, means that metaphysicians must abandon the picture of the world as composed of self-subsistent individual objects, and the paradigm of causation as the collision of such objects.

Every Thing Must Go also assesses the role of information theory and complex systems theory in attempts to explain the relationship between the special sciences and physics, treading a middle road between the grand synthesis of thermodynamics and information, and eliminativism about information. The consequences of the authors' metaphysical theory for central issues in the philosophy of science are explored, including the implications for the realism vs. empiricism debate, the role of causation in scientific explanations, the nature of causation and laws, the status of abstract and virtual objects, and the objective reality of natural kinds.

James Ladyman is Professor of Philosophy at the University of Bristol.

Don Ross is Professor of Philosophy and Professor of Economics at the University of Alabama at Birmingham, and Professor of Economics at the University of Cape Town.

Every Thing Must Go

Metaphysics Naturalized

JAMES LADYMAN
and
DON ROSS

with
DAVID SPURRETT
and
JOHN COLLIER

OXFORD
UNIVERSITY PRESS

Great Clarendon Street, Oxford OX2 6DP

Oxford University Press is a department of the University of Oxford.
It furthers the University's objective of excellence in research, scholarship,
and education by publishing worldwide in

Oxford New York

Auckland Cape Town Dar es Salaam Hong Kong Karachi
Kuala Lumpur Madrid Melbourne Mexico City Nairobi
New Delhi Shanghai Taipei Toronto

With offices in

Argentina Austria Brazil Chile Czech Republic France Greece
Guatemala Hungary Italy Japan Poland Portugal Singapore
South Korea Switzerland Thailand Turkey Ukraine Vietnam

Oxford is a registered trade mark of Oxford University Press
in the UK and in certain other countries

Published in the United States
by Oxford University Press Inc., New York

© James Ladyman, Don Ross, David Spurrett, and John Collier 2007

The moral rights of the authors have been asserted
Database right Oxford University Press (maker)

First published 2007
First published in paperback 2009

All rights reserved. No part of this publication may be reproduced,
stored in a retrieval system, or transmitted, in any form or by any means,
without the prior permission in writing of Oxford University Press,
or as expressly permitted by law, or under terms agreed with the appropriate
reprographics rights organization. Enquiries concerning reproduction
outside the scope of the above should be sent to the Rights Department,
Oxford University Press, at the address above

You must not circulate this book in any other binding or cover
and you must impose the same condition on any acquirer

British Library Cataloguing in Publication Data

Data available

Library of Congress Cataloging in Publication Data

Data available

Typeset by Laserwords Private Limited, Chennai, India
Printed in Great Britain
on acid-free paper by the
MPG Books Group, Bodmin and King's Lynn

ISBN 978–0–19–927619–6(Hbk.);   978–0–19–957309–7 (Pbk.)

1 3 5 7 9 10 8 6 4 2

in the case of metaphysics we find this situation: through the form of its works it pretends to be something that it is not

Rudolf Carnap

ant is a mass term for anteaters

Daniel Dennett

For Caragh and Nelleke

Preface

This is a polemical book. One of its main contentions is that contemporary analytic metaphysics, a professional activity engaged in by some extremely intelligent and morally serious people, fails to qualify as part of the enlightened pursuit of objective truth, and should be discontinued. We think it is impossible to argue for a point like this without provoking some anger. Suggesting that a group of highly trained professionals have been wasting their talents—and, worse, sowing systematic confusion about the nature of the world, and how to find out about it—isn't something one can do in an entirely generous way. Let us therefore stress that we wrote this book not in a spirit of hostility towards philosophy or our fellow philosophers, but rather the opposite. We care a great deal about philosophy, and are therefore distressed when we see its reputation harmed by its engagement with projects and styles of reasoning we believe bring it into disrepute, especially among scientists. We recognize that we may be regarded as a bit rough on some other philosophers, but our targets are people with considerable influence rather than novitiates. We think the current degree of dominance of analytic metaphysics within philosophy is detrimental to the health of the subject, and make no apologies for trying to counter it.

Lest the reader suppose on the basis of the above remarks that the whole book is a campaign of aggressive destruction, we emphasize that they describe only part of the first chapter. Having argued there that analytic metaphysics as it is now practised is irrelevant to our metaphysical project, we spend the rest of the book attempting to encourage truly naturalistic metaphysics by example. For reasons related to our naturalism, and stated early in the text, we expect that our particular positive account of the nature of the world will be deemed mainly or perhaps even entirely incorrect by future philosophers who will know future science. This is likely to be the fate of any generalizations of wide scope based on limited empirical observations. But we hope the kind of metaphysic we construct here—one motivated by currently pursued, specific scientific hypotheses, and having as its sole aim to bring these hypotheses advanced by the various special sciences together into a comprehensive world-view—will go on being constructed by others for as long as science itself is carried out.

The structure of the book is as follows. Chapter 1, as indicated above, is partly destructive in aim. It is intended to persuade the reader that standard analytic metaphysics (or 'neo-scholastic' metaphysics as we call it) contributes nothing to human knowledge and, where it has any impact at all, systematically misrepresents the relative significance of what we do know on the basis of science. We go on to explain and defend our particular form of naturalism, and our view of the relationship between physics and the rest of science. This is the most accessible chapter of the book, and it is possible that some readers, who are more interested in philosophers' attitudes towards science than in metaphysics *per se*, may want to read only it.

Chapter 2, by contrast, will seem to most philosophers to be the most conventional part of the book. Here we assemble foundations for the metaphysics to come on the basis of a particular position, 'ontic structural realism' (OSR), that we defend as the best synthesis of several decades of reflection and argument in the philosophy of science on the debate between empiricists and scientific realists, particularly in the light of the history of science and the relationships between successful theories and their successors. This provides the conceptual framework within which we then consider relationships among contemporary theories in different sciences so as to construct a unified world picture.

Chapters 3 and 4 constitute the heart of our positive naturalistic metaphysics, so we hope the reader will indulge their considerable length and complexity. In Chapter 3 we ask which among possible unifying principles are motivated by advanced physical theory as we now find it. We furthermore show how OSR is motivated directly by this physical theory, thus exemplifying our principle from Chapter 1 that the input for philosophizing must come from science. In Chapter 4 we inquire as to how the general image of reality suggested by contemporary (fundamental) physics can be reconciled—composed into one world with—the many special sciences that appear to have quite different theoretical structures and commitments. On the basis of this investigation we propose a theory of ontology—of what there is—that we call Rainforest Realism (RR) because of the relatively lush ontology it propounds.

In Chapter 5 we show how our naturalistic metaphysics, which consists in the combination of OSR and RR, achieves consilience among a wide variety of facts about the sciences, including the ways in which they investigate and understand causal claims, the status of scientific laws, and the principles of classification and arrangements of reality into types that scientists use as they discover and refine predictive and explanatory generalizations. The combined position, our positive naturalistic metaphysics, is called 'Information-Theoretic Structural Realism' (ITSR), for reasons that will become evident. In Chapter 5 the reader who struggled through the often dense material in Chapters 3 and 4 will receive her payoff, as she sees the progress that ITSR permits us to make on a range of major preoccupations in the philosophy of science.

Finally, the brief Chapter 6 orients our metaphysic in the context of work by other philosophers that is closest in positive content to ours. We first ask how our general account differs from that of Kant, since strong affinities will have occurred to some readers. A similar question, and for similar reasons, is asked about our relationship to the philosophy of Daniel Dennett. We next consider traditional points of issue between realism and empiricism, showing that we side with realists on some, with empiricists on others, and that we reject the underlying terms of debate on still others. At this point we explain why it is appropriate to regard our view as a kind of 'neo-positivism', despite our allowing for the significance of a minimalist positive metaphysics where the positivists insisted on none. Finally, before concluding, we note the arguments of an ultimately anti-naturalist theistic philosopher who argues for supernaturalism on the grounds that if one doesn't adopt it one will be forced to a theory exactly like the one we give in this book. The theist considers this the basis for a conclusion by reductio; we agree, but make the opposite judgement about what is absurd.

This book began as a collaboration among four authors. As the project developed, the two of us (Ladyman and Ross) found our thinking shifting in a more empiricist direction than any of us had started from, and under the impetus of this we increasingly drove the project in a direction of our own. The other two original authors, however, David Spurrett and John Collier, were so important as consultants and researchers throughout the project that they are acknowledged on the cover and title page. They also wrote parts of three chapters, as is indicated in the Contents.

Our next greatest debt of thanks goes to Nelleke Bak, who checked the entire manuscript for consistency (while also formatting our references). This is obviously more important in a co-authored work than is normally the case for a philosophy manuscript. This is a long book that states many propositions; without Nelleke's diligence and acumen, we would have produced a book that more than once proclaimed P and $\sim$P, and would thus have implicitly announced every proposition altogether.

The following colleagues and friends read our first draft and commented trenchantly and constructively on it in detail: Jimmy Doyle, Katherine Hawley, Chris Honey, Harold Kincaid, Ausonio Marras, Alex Rosenberg, Emma Ruttkamp, and David Wallace, and an anonymous reader for Oxford University Press. The book is very much better than it was thanks to their assistance, though we repeat the standard mantra of authors that the remaining errors were made at our insistence.

For their long-standing influence on our ideas, for discussion of many of the issues we address, for their encouragement of the project, and for comments on specific parts of the text, we thank Dan Dennett and Bas van Fraassen. In a similar vein, Collier acknowledges Cliff Hooker and Kai Neilson. The other great intellectual debt that we owe is to Steven French for the breadth and depth of his work on the subject matter of Chapters 2 and 3 and its profound influence on Ladyman. Much of our articulation and defence of OSR is derived from previously published joint work by French and Ladyman. We are also grateful to the following people for discussions and other help concerning some or all of the issues we address: Alexander Bird, Harvey Brown, Jeremy Butterfield, Michael Esfeld, Hannes Leitgeb, Samir Okasha, Oliver Pooley, Simon Saunders, and Finn Spicer.

We had institutional support for which we are grateful. Ladyman enjoyed the support of a Philip Leverhulme Prize that enabled a trip to South Africa to work with Ross to finalize the manuscript. Ross enjoyed an appointment as a Benjamin Meeker Visiting Professor at the University of Bristol in 2004 that facilitated the main collaboration with Ladyman. Ross also received substantial travel and other research financing from the Center for Ethics and Values in the Sciences at the University of Alabama at Birmingham. He also enjoyed support from the University Research Committee of the University of Cape Town.

We have presented our work in progress at meetings in Bristol, Sydney, Halifax, Oxford, Pietermaritzburg, and St Johns, and extend our thanks accordingly. This project was conceived and sustained at the annual International Philosophy of Science Conference held each year at the Inter-University Centre in the wondrous World Heritage Site of Dubrovnik, Croatia. We heartily thank the Directors of the workshop and the people of Dubrovnik for their repeated hospitality.

JL & DR
Cape Town, South Africa, August 2006

Contents

1. **In Defence of Scientism** — 1
   Don Ross, James Ladyman, and David Spurrett
   1.1 Naturalistic metaphysics — 1
   1.2 Neo-scholastic metaphysics — 7
   1.3 The principle of naturalistic closure — 27
   1.4 The primacy of physics — 38
   1.5 Unity of science and reductionism — 45
   1.6 Fundamental and other levels — 53
   1.7 Stances, norms, and doctrines — 57

2. **Scientific Realism, Constructive Empiricism, and Structuralism** — 66
   James Ladyman and Don Ross
   2.1 Scientific realism — 68
   2.2 Theory change — 83
   2.3 Structuralism — 93
   2.4 What is structural realism? — 122
   2.5 Conclusion — 129

3. **Ontic Structural Realism and the Philosophy of Physics** — 130
   James Ladyman and Don Ross
   3.1 Identity and individuality in quantum mechanics — 132
   3.2 Individuality and spacetime physics — 141
   3.3 Objectivity and invariance — 145
   3.4 The metaphysics of relations — 148
   3.5 Objections to ontic structuralism — 154
   3.6 Mathematical structure and physical structure — 159
   3.7 Further reflections on physics — 161
   3.8 Conclusion — 189

4. Rainforest Realism and the Unity of Science — 190
Don Ross, James Ladyman, and John Collier

    4.1 Special sciences and disunity hypotheses — 190
    4.2 Dennett on real patterns — 196
    4.3 Concepts of information in physics and metaphysics — 210
    4.4 Rainforest realism — 220
    4.5 Fundamental physics and special science — 238

5. Causation in a Structural World — 258
Don Ross, James Ladyman, and David Spurrett

    5.1 Russell's naturalistic rejection of causation — 258
    5.2 Philosophers and folk on causation — 266
    5.3 Causes in science — 269
    5.4 Letting science hold trumps — 274
    5.5 Laws in fundamental physics and the special sciences — 281
    5.6 Real patterns, types, and natural kinds — 290

6. Conclusion—Philosophy Enough — 298
Don Ross and James Ladyman

    6.1 Why isn't this Dennett? Why isn't it Kant? — 298
    6.2 A reductio — 300
    6.3 Neo-positivism — 303

References — 311
Index — 339

1
In Defence of Scientism

Don Ross, James Ladyman, and David Spurrett

1.1 NATURALISTIC METAPHYSICS

The aim of this book is to defend a radically naturalistic metaphysics. By this we mean a metaphysics that is motivated exclusively by attempts to unify hypotheses and theories that are taken seriously by contemporary science. For reasons to be explained, we take the view that no alternative kind of metaphysics can be regarded as a legitimate part of our collective attempt to model the structure of objective reality.

One of our most distinguished predecessors in this attitude is Wilfrid Sellars. He expressed a naturalistic conception of soundly motivated metaphysics when he said that the philosopher's aim should be 'knowing one's way around with respect to the subject matters of all the special [scientific] disciplines' and 'building bridges' between them (1962, 35). It might of course be wondered whether or why science has any role for non-specialist bridge-builders. The argument that there is a useful such role must, for the naturalist, be by way of demonstration. We aim to provide such a demonstration later in the book (Chapters 4 and 5), when we turn to positive claims and build specific bridges. First, however, there is work of a destructive sort that needs to be done.

There is a rich tradition of naturalistic metaphysics in Western philosophy. Competing strongly with this tradition—often within the body of a single philosopher's work—is a tradition which aims at domesticating scientific discoveries so as to render them compatible with intuitive or 'folk' pictures of structural composition and causation. Such domestication is typically presented as providing 'understanding'. This usage may be appropriate given one everyday sense of 'understanding' as 'rendering more familiar'. However, we are interested here in a sense of 'understanding' that is perhaps better characterized by the word 'explanation', where an explanation must be true (at least in its most general claims). We argue that a given metaphysic's achievement of domestication furnishes no evidence at all that the metaphysic in question is true, and thus no reason for believing that it explains anything.

Quine (1969), in arguing for the naturalization of epistemology, claimed that the evolutionary processes that designed people should have endowed us with cognition that reliably tracks truth, on the grounds that believing truth is in general more conducive to fitness than believing falsehood. This is an empirical hypothesis, and it may well be a sound one. However, it does not imply that our everyday or habitual intuitions and cognition are likely to track truths reliably across all domains of inquiry. We believe it to be probable that human intelligence, and the collective representational technologies (especially public languages) that constitute the basis for what is most biologically special about that intelligence (see Clark 1997, 2004, Ross forthcoming b), evolved mainly to enable us to navigate complex social coordination games (Byrne and Whiten 1988, 1997; Whiten and Byrne 1997; Ross 2005, forthcoming a). People are probably also relatively reliable barometers of the behavioural patterns of animals they get to spend time observing, at making navigational inferences in certain sorts of environments (but not in others), and at anticipating aspects of the trajectories of medium-sized objects moving at medium speeds. However, proficiency in inferring the large-scale and small-scale structure of our immediate environment, or any features of parts of the universe distant from our ancestral stomping grounds, was of no relevance to our ancestors' reproductive fitness. Hence, there is no reason to imagine that our habitual intuitions and inferential responses are well designed for science or for metaphysics.

Fortunately, people learned to represent the world and reason mathematically—that is, in a manner that enables us to abstract away from our familiar environment, to a degree that has increased over time as mathematics has developed—and this has allowed us to achieve scientific knowledge. Since this knowledge can be incorporated into unified pictures, we also can have some justified metaphysics. Based as it is on incomplete science, this metaphysics probably is not true. However, if it is at least motivated by our most careful science at time t, then it is the best metaphysics we can have at t. We will argue for a metaphysics consistent with and motivated by contemporary science by, in the first place, contrasting it with the kind of metaphysics that has arisen through the tradition of domestication. There are various ways of trying to discover the character of the latter. One way that is not very reliable (because too dependent on unverifiable conjectures about history) is to ask what metaphysical pictures might have worked well for our early hominid ancestors. Two better ways infer our habitual tendencies in metaphysical reasoning from assumptions encoded in natural languages, and from the efforts of our most important philosophers when they work in the spirit of domestication, as most do some of the time and some do all of the time.

Lakoff and Johnson (1980) and Lakoff (1987) have done pioneering work at uncovering habitual metaphysics, at least among English speakers, by the linguistic route. They document the extent to which the deep metaphors of English, which govern everyday inferences made in that language, are structured according to an implicit doctrine of 'containment'.[1] On this doctrine, the world is a kind of container bearing objects that change location and properties over time. These objects cause things to happen by interacting directly with one another. Prototypically, they move each other about by banging into one another. At least as important to the general picture, they themselves are containers in turn, and their properties and causal dispositions are to be explained by the properties and dispositions of the objects they contain (and which are often taken to comprise them entirely).

Though we are closely informed about the deep metaphorical structure of only a tiny proportion of human languages, it seems likely that the structure described above is reflected in most of at least the Western ones, including the ancient Western ones. It may thus be thought unsurprising that the earliest known Western philosophy puzzled itself over ways in which the apparently endless regress of containment might be stopped. Thales suggested that everything is ultimately 'made of' water. Other early philosophers suggested air and fire instead. Popular consensus eventually settled down, for a long time, on the idea that all corporeal things are made of mixtures of four basic elements—earth, water, air, and fire—with differences among kinds of things to be explained by reference to differences in elemental proportions. Then science replaced the four elements with tiny, indivisible, ultimate particles—first early modern corpuscles, then atoms, then systems of subatomic particles still often conceived in popular imagination as sub-microscopic solar systems, whose 'stars' and 'orbiting planets' are supposedly the new ultimate constituents from which everything is composed.

The modern mereology has in some ways strengthened the commitments of the containment metaphor. Aristotle, famously, had a complex and multi-faceted concept of causation. Causation to the modern domesticating metaphysician is, by contrast, typically identified with what Aristotle called 'efficient causation'. A characteristic of efficient causation, in the context of the containment metaphor, is that it is imagined to 'flow' always from 'inside out'. Thus the ultimate constituents of the world that halt the regress of containment are also taken to be the ultimate bearers of causal powers, which somehow support and determine

---

[1] Here are some of their examples of everyday English phrases that encode the implicit metaphysic of containment: There was a lot of good running *in* the race; *Halfway into the race* I ran out of energy; How did Jerry *get out of* washing the windows?; *Outside of* washing the windows, what else did you do?; He's *immersed in* washing the windows right now; We're *out of* trouble now; I'm slowly getting *into* shape; He *fell into* a depression (Lakoff and Johnson 1980, 31–2).

the whole edifice of (often complex) causal relations that constitute the domain of observable dynamics.

The metaphysics of domestication tends to consist of attempts to render pieces of contemporary science—and, at least as often, simplified, mythical interpretations of contemporary science—into terms that can be made sense of by reference to the containment metaphor. That is, it seeks to account for the world as 'made of' myriad 'little things' in roughly the way that (some) walls are made of bricks. Unlike bricks in walls, however, the little things are often in motion. Their causal powers are usually understood as manifest in the effects they have on each other when they collide. Thus the causal structure of the world is decomposed by domesticating metaphysics into reverberating networks of what we will call 'microbangings'—the types of ultimate causal relations that prevail amongst the basic types of little things, whatever exactly those turn out to be. Metaphysicians, especially recently, are heavily preoccupied with the search for 'genuine causal oomph', particularly in relation to what they perceive to be the competition between different levels of reality.[2] We will argue that this is profoundly unscientific, and we will reject both the conception of causation and levels of reality upon which it is based.

We will argue in this book that, in general, the domesticating metaphysics finds no basis in contemporary science. Some successful science, and some reasonable metaphysics, were done in the past on the basis of it. However, the attempt to domesticate twenty-first-century science by reference to homely images of little particles that have much in common with seventeenth- and eighteenth-century mechanistic and materialist metaphysics is forlorn. There are, we will argue, no little things and no microbangings. Causation does not, in general, flow from the insides of containers to their outsides. The world is in no interesting ways like a wall made of bricks in motion (that somehow manages not to fall apart), or, in the more sophisticated extension of the metaphor dominant since modern science, like a chamber enclosing the molecules of a gas. Indeed, it is no longer helpful to conceive of either the world, or particular systems of the world that we study in partial isolation, as 'made of' anything at all.[3]

As we will discuss in Chapter 5, the containment metaphor and its accompanying ontology of little things and microbangings has more problems than its mere failure to follow from science. It cannot be defended by someone on the grounds that psychological repose and cultural familiarity are values that might be defended against the objective truth. This is because the increasing heights of abstraction in representation achieved by science over the past century have now carried its investigations so far beyond the reaches of our ancestral habitation that the containment metaphor can no longer be applied to the scientific image

---

[2] We take it that 'causal oomph' is a synonym of 'biff' (Armstrong 2004). We discuss this in 5.2.
[3] We do not intend here to impugn the accounts of composition that are ubiquitous in the special sciences. Rather our target is the metaphysical idea of composition discussed further in 1.2.3.

without doing at least as much violence to everyday intuitions as does our denial of the metaphor. This emerges most clearly in the struggles of professional domesticators—that is, (some) philosophers. Much recent metaphysics, in trying to save a version of the habitual picture that has already been transformed by half-digested science, ends up committed to claims that are as least as shocking to common sense as anything we will urge. For example, Trenton Merricks (2001) is led to deny the existence of tables and chairs because he thinks physics tells us that they decompose without residue into atoms, and he denies that baseballs can break windows because he thinks that windows must be broken by particular atomic constituents of a baseball, thus rendering the effects of the ball as a whole causally otiose. (We return to Merrick's work in 1.2.3 below.)

According to the account we will give, science tells us many surprising things, but it does not impugn the everyday status of objects like tables and baseballs. These are, we will argue, aspects of the world with sufficient cohesion at our scale that a group of cognitive systems with practically motivated interest in tracking them would sort them into types for book-keeping purposes. They are indeed not the sorts of objects that physics itself will directly track as types; but this is a special instance of the more general fact that physics, according to us, does not model the world in terms of types of objects in the first place.

We can imagine some readers worrying that our whole effort here will rely on a premise to the effect that scientific objectivity is all that matters, or that if it comes into conflict with our desire to feel at home in our own 'Lebenswelt' then it is the second that must always give way. We depend on no such premise. People who wish to explore the ways in which the habitual or intuitive anthropological conceptual space is structured are invited to explore social phenomenology. We can say 'go in peace' to Heideggerians, noting that it was entirely appropriate that Heidegger did not attempt to base any elements of his philosophy on science, and focused on hammers—things that are constituted as objects by situated, practical activity—rather than atoms—things that are supposed by realists to have their status as objects independently of our purposes—when he reflected on objects. We, however, are interested in objective truth rather than philosophical anthropology. Our quarrel will be with philosophers who claim to share this interest, but then fail properly to pay attention to our basic source of information about objective reality.

There is another set of philosophers who are broadly naturalistic in the sense that, like us, they allow science priority over domesticating conceptual analysis, but whose perspective fits uncomfortably with our understanding of metaphysics as consisting in unification of science. These are philosophers such as Cartwright (1999) and Dupré (1993) who argue, on the basis of reflections on the ontology implied by science, that the world is not metaphysically unified. The objection will naturally be raised for us that, by our definition of naturalistic metaphysics, generation of arguments for disunity is not naturalistic metaphysical inquiry. We indeed claim that if the world were fundamentally disunified, then discovery of this would be tantamount to discovering that there is no metaphysical work to be done: objective inquiry would start and stop with the separate investigations of the mutually unconnected special sciences. By 'fundamentally disunified' we refer to the idea that there is no overarching understanding of the world to be had; the best account of reality we could establish would include regions or parts to which no generalizations applied. Pressed by Lipton (2001), Cartwright (2002) seems to endorse this. However, she admits that she does so (in preference to non-fundamental disunity) not because 'the evidence is ... compelling either way' (2002, 273) but for the sake of aesthetic considerations which find expression in the poetry of Gerald Manley Hopkins. Like Hopkins, Cartwright is a lover of 'all things counter, original, spare, strange' (ibid). That is a striking motivation to be sure, but it is clearly not a naturalistic one. Similarly, although Dupré's arguments are sometimes naturalistic, at least as often they are in service of domestication. He frequently defends specific disunity hypotheses on the grounds that they are politically or ethically preferable to unifying ('imperialistic') ones. (See especially Dupré 2001, and Ross 2005, chs. 1 and 9.) The urge to try to make the world as described by science safe for someone's current political and moral preferences may even be the main implicit motivation for most efforts at domestication.

We must admit that the hypothesis that metaphysics is possible is itself a metaphysical hypothesis, but this is a purely semantic point, arising simply from the fact that in the normal arrangement of domains of inquiry, there is no named level of abstraction beyond the metaphysical. (By contrast, wondering whether physics is unified is metaphysics, wondering whether chemistry is unified is physics, and so on.) Our substantive claim is that the worthwhile work to be done by naturalistic metaphysics consists in seeking unification, but this is not based on an analysis of 'metaphysics'. Let us just stipulate, then, that inquiry into the possibility or impossibility of metaphysics is 'metametaphysics'. Then naturalistic metametaphysics, we hold, should be based on naturalistic metaphysics, which should in turn be based on science. At various points in the discussion to come, we engage with naturalistic advocates of disunity as we do with fellow naturalists in general. When we do, we will avoid pedantically announcing ourselves as shifting into metametaphysics. Otherwise we will assume that naturalistic metaphysics is possible, and that we are successfully doing some of it here.

Jonathan Lowe (2002) has two arguments against naturalized metaphysics:

(i) 'to the extent that a wholly naturalistic and evolutionary conception of human beings seems to threaten the very possibility of metaphysical knowledge, it equally threatens the very possibility of *scientific* knowledge' (6). Since natural selection cannot explain how natural scientific knowledge is possible, the fact that it cannot explain how metaphysical knowledge is possible gives us no reason to suppose that such knowledge is not possible.

(ii) Naturalism depends upon metaphysical assumptions.

In response, we maintain that even if one granted the tendentious claim that natural selection cannot explain how natural scientific knowledge is possible, we have plenty of good reasons for thinking that we do have such knowledge. On the other hand, we have no good reasons for thinking that a priori metaphysical knowledge is possible. With respect to Lowe's second claim, it is enough to point out that even if naturalism depends on metaphysical assumptions, the naturalist can argue that the metaphysical assumptions in question are vindicated by the success of science, by contrast with the metaphysical assumptions on which autonomous metaphysics is based which are not vindicated by the success of metaphysics since it can claim no such success.

1.2 NEO-SCHOLASTIC METAPHYSICS

In this section we describe the philosophical environment that motivates our project. Our core complaint is that during the decades since the fall of logical empiricism, much of what is regarded as 'the metaphysics literature' has proceeded without proper regard for science. The picture is complicated, however, by the fact that much activity in what is classified as philosophy of science is also metaphysics, and most of this work is scientifically well informed. This book is an exercise in metaphysics done as naturalistic philosophy of science because we think that no other sort of metaphysics counts as inquiry into the objective nature of the world. In this and the following few sections, we aim to show why, despite the fact that our book is about metaphysics, almost all of our discussion from Chapter 2 onward will engage with problems and disputes emanating from the philosophy of science and from science itself.

As long as science enjoys significant prestige there will be attempts to pass off as science ideological pursuits, such as intelligent design 'theory' (*sic*) and 'hermeneutic economics' (Addleson 1997), and attempts to challenge or undermine the epistemic credentials of science.[4] We have nothing to add to the contributions of those who have criticized these attempts.[5] Though we follow the logical positivists and empiricists in concerning ourselves with the 'demarcation problem', our concern here is not with populist pseudo-science. It is instead with a sophisticated cousin of pseudo-science, pseudo-naturalist philosophy, especially as this occurs in metaphysics. Espousal of 'naturalism' is widespread in philosophy, but explicit criteria for being consistently naturalist are rare. In 1.3 below we provide a new formulation of the naturalist credo. First, in the

---

[4] For example, Dupré (2001), Harris (2005).
[5] Kitcher (1982) and Pennock (2000) are outstanding examples.

present section we sketch some of the historical background to the emergence of 'neo-scholastic' metaphysics, and in the next section we argue against it.[6]

Around a century ago, Bertrand Russell rejected the dominant philosophical idealism of his day, and most of the principles of speculative metaphysics that had supported it. In its place he proposed and helped to develop an approach to epistemology and metaphysics based on the logical analysis of claims justified by empirical experience, particularly by empirical science. The logical positivists briefly carried Russell's programme to extremes, believing that by reduction of all empirically significant statements to reports about sense-data they could do away with metaphysics altogether. In this ambition they of course failed—not least because one of their central working concepts, that of a 'sense-datum', is itself a scientifically unsupported one. Their understanding of 'empirical significance' in terms of the verificationist theory of meaning was likewise a piece of metaphysics they did not derive from science. Although many philosophers in the twentieth century regarded metaphysics as a relic of earlier ages, it never ceased to be done, even by those who intended to avoid it.

Though positivism and its successor, logical empiricism, died as serious philosophical options, they expired gradually. The most persuasive criticisms of logical empiricism were given by logical empiricists themselves (for example, Hempel 1950), and the basic ideas behind Quine's famous (1951) work of destruction were anticipated by prior insights of, among others, Carnap (see Creath 1991). Nevertheless, for at least twenty years after the original core commitments of positivism had all been surrendered—so, into the 1970s—philosophers often conveyed roughly the following message to their students:

> The technical aims of positivism and logical empiricism—to show how all meaningful discourse can be reduced to, or at least rigorously justified by reference to, reports of observations regimented for communication and inference by formal linguistic conventions—have been shown to be unachievable. Nevertheless, the positivists, following the lead of Hume and Russell before them, introduced into (non-continental) philosophy a profound respect for empirical science and its pre-eminence in all inquiry that continues to be the basis of the philosophical project. We may no longer believe in the verificationist theory of meaning, in the myth of the given, or in the analytic–synthetic distinction. Nevertheless, it is in the spirit of the positivists that we can say, with Quine, 'philosophy of science is philosophy enough'. (Quine 1953[1966], 151)[7]

Of course, if the positivists are wrong in fact, however right they might have been in aim and spirit, then metaphysics can't be regarded as impossible or

---

[6] The term 'neo-scholastic' metaphysics is fairly widely used among philosophers of science. We owe our usage of it to Ross and Spurrett (2004) but it is so apt that we suspect it has been independently invented on many occasions.

[7] This imaginary quotation distils what Ross was taught by an almost unanimous consensus among his graduate school professors in the 1980s.

foolish on the basis of their arguments. A key breakthrough in the rehabilitation of metaphysics came in the mid-1970s, when Kripke's *Naming and Necessity* (1973), and a series of papers on meaning in science by Hilary Putnam (gathered in Putnam 1975b), convinced many philosophers to believe in both metaphysical reference relations, and in the mind-independent reality of the objects of successful scientific theories.[8] Thanks to Putnam and David Lewis, trips to possible worlds became standard instruments of philosophical argument.

Initially, this sort of metaphysics could be indulged in guiltlessly by philosophers who admired the positivists, because it was profoundly respectful of science. Indeed, in providing philosophers with a way of regarding leading scientific theories as literally true—rather than just instrumentally useful or descriptively adequate to experience—it seemed to pay science even deeper tribute than positivism had done. In its early days the metaphysical turn was partly inspired by interpretative problems about meaning continuity arising from the philosophical history of science championed by Kuhn. Such history became a core part of philosophy's subject matter; and the metaphysics of essences, natural kinds, and rigid designation gave philosophers a means of avoiding the relativist path that was bound to end in the tears of sociology. Indeed, some philosophers (especially followers of a particular interpretation of the later Wittgenstein) followed social scientists in regarding the reborn metaphysics as 'scientistic' (Sorrell 1991). We wish this charge were better justified than it is.

The revival of metaphysics after the implosion of logical positivism was accompanied by the ascendancy of naturalism in philosophy, and so it seemed obvious to many that metaphysics ought not to be 'revisionary' but 'descriptive' (in Peter Strawson's terminology, 1959). That is, rather than metaphysicians using rational intuition to work out exactly how the absolute comes to self-consciousness, they ought instead to turn to science and concentrate on explicating the deep structural claims about the nature of reality implicit in our best theories. So, for example, Special Relativity ought to dictate the metaphysics of time, quantum physics the metaphysics of substance, and chemistry and evolutionary biology the metaphysics of natural kinds. However, careful work by various philosophers of science has shown us that this task is not straightforward because science, usually and perhaps always, underdetermines the metaphysical answers we are seeking. (See French 1998, 93). Many people have taken this in their stride and set about exploring the various options that are available. Much excellent work has resulted.[9] However, there has also been another result of the recognition that science doesn't wear metaphysics on its sleeve, namely the resurgence of the kind of metaphysics that floats entirely free of science. Initially granting themselves

---

[8] Fodor (2004) also sees the Kripke and Putnam arguments for referential realism as the moment when analytical philosophy broke free of its Quinean moorings, on the basis of wishful thinking rather than sound argument.

[9] See Butterfield (2006) for recent exemplary work.

permission to do a bit of metaphysics that seemed closely tied to, perhaps even important to, the success of the scientific project, increasing numbers of philosophers lost their positivistic spirit. The result has been the rise to dominance of projects in analytic metaphysics that have almost nothing to do with (actual) science. Hence there are now, once again, esoteric debates about substance, universals, identity, time, properties, and so on, which make little or no reference to science, and worse, which seem to presuppose that science must be irrelevant to their resolution. They are based on prioritizing armchair intuitions about the nature of the universe over scientific discoveries. Attaching epistemic significance to metaphysical intuitions is anti-naturalist for two reasons. First, it requires ignoring the fact that science, especially physics, has shown us that the universe is very strange to our inherited conception of what it is like. Second, it requires ignoring central implications of evolutionary theory, and of the cognitive and behavioural sciences, concerning the nature of our minds.

1.2.1 Intuitions and common sense in metaphysics

The idea that intuitions are guides to truth, and that they constitute the basic data for philosophy, is of course part of the Platonic and Cartesian rationalist tradition.[10] However, we have grounds that Plato and Descartes lacked for thinking that much of what people find intuitive is not innate, but is rather a developmental and educational achievement. What counts as intuitive depends partly on our ontogenetic cognitive makeup and partly on culturally specific learning. Intuitions are the basis for, and are reinforced and modified by, everyday practical heuristics for getting around in the world under various resource (including time) pressures, and navigating social games; they are not cognitive gadgets designed to produce systematically worthwhile guidance in either science or metaphysics. In light of the dependence of intuitions on species, cultural, and individual learning histories, we should expect developmental and cultural variation in what is taken to be intuitive, and this is just what we find. In the case of judgements about causes, for example, Morris et al. (1995) report that Chinese and American subjects differed with respect to how they spontaneously allocated causal responsibility to agents versus environmental factors. Given that the 'common sense' of many contemporary philosophers is shaped and supplemented by ideas from classical physics, the locus of most metaphysical discussions is an image of the world that sits unhappily between the manifest image and an out of date scientific image.[11]

---

[10] DePaul and Ramsey (1998) contains a number of papers assessing the epistemic status of intuition in philosophy though none of them are primarily concerned with the role of intuition in analytic metaphysics.

[11] It is ironic that the most prominent defender of antirealism about scientific knowledge, namely Bas van Fraassen, is also one of the fiercest contemporary critics of speculative metaphysics not least because it has nothing to do with science (see especially his 2002).

While contemporary physics has become even more removed from common sense than classical physics, we also have other reasons to doubt that our common sense image of the world is an appropriate basis for metaphysical theorizing. Evolution has endowed us with a generic theory or model of the physical world. This is evident from experiments with very young children, who display surprise and increased attention when physical objects fail to behave in standard ways. In particular, they expect ordinary macroscopic objects to persist over time, and not to be subject to fusion or fission (Spelke et al. 1995). For example, if a ball moves behind a screen and then two balls emerge from the other side, or vice versa, infants are astonished. We have been equipped with a conception of the nature of physical objects which has been transformed into a foundational metaphysics of individuals, and a combinatorial and compositional conception of reality that is so deeply embedded in philosophy that it is shared as a system of 'obvious' presuppositions by metaphysicians who otherwise disagree profoundly.

This metaphysics was well suited to the corpuscularian natural philosophy of Descartes, Boyle, Gassendi, and Locke. Indeed, the primary qualities of matter which became the ontological basis of the mechanical philosophy are largely properties which form part of the manifest image of the world bequeathed to us by our natural history. That natural history has been a parochial one, in the sense that we occupy a very restricted domain of space and time. We experience events that last from around a tenth of a second to years. Collective historical memory may expand that to centuries, but no longer. Similarly, spatial scales of a millimetre to a few thousand miles are all that have concerned us until recently. Yet science has made us aware of how limited our natural perspective is. Protons, for example, have an effective diameter of around 10^{-15} m, while the diameter of the visible universe is more than 10^{19} times the radius of the Earth. The age of the universe is supposed to be of the order of 10 billion years. Even more homely sciences such as geology require us to adopt time scales that make all of human history seem like a vanishingly brief event.

As Lewis Wolpert (1992) chronicles, modern science has consistently shown us that extrapolating our pinched perspective across unfamiliar scales, magnitudes, and spatial and temporal distances misleads us profoundly. Casual inspection and measurement along scales we are used to suggest that we live in a Euclidean space; General Relativity says that we do not. Most people, Wolpert reports, are astounded to be told that there are more molecules in a glass of water than there are glasses of water in the oceans, and more cells in one human finger than there are people in the world (ibid. 5). Inability to grasp intuitively the vast time scales on which natural selection works is almost certainly crucial to the success of creationists in perpetuating foolish controversies about evolution (Kitcher 1982). The problems stemming from unfamiliar measurement scales are just the tip of an iceberg of divergences between everyday expectations and scientific findings. No one's intuitions, in advance of the relevant science, told them that white light would turn out to have compound structure, that combustion primarily involves something being taken up rather than given off (Wolpert 1992, 4), that birds are the only living descendants of dinosaurs, or that Australia is presently on its way to a collision with Alaska. As Wolpert notes, science typically explains the familiar in terms of the unfamiliar. Thus he rightly says that 'both the ideas that science generates and the way in which science is carried out are entirely counter-intuitive and against common sense—by which I mean that scientific ideas cannot be acquired by simple inspection of phenomena and that they are very often outside everyday experience' (ibid. 1). He later strengthens the point: 'I would almost contend that if something fits with common sense it almost certainly isn't science' (ibid. 11). B. F. Skinner characteristically avoids all waffling on the issue: 'What, after all, have we to show for non-scientific or pre-scientific good judgment, or common sense, or the insights gained through personal experience? It is science or nothing' (Skinner 1971, 152–3).

However, in exact reversal of this attitude metaphysicians place great emphasis on preserving common sense and intuitions. Michael Loux and Dean Zimmerman explain the methodology of metaphysics as follows: 'One metaphysical system is superior to another in scope in so far as it allows for the statement of satisfactory philosophical theories on more subjects—theories that preserve, in the face of puzzle and apparent contradiction, most of what we take ourselves to know' (2003, 5). Here is a conception of metaphysics according to which its function is to reassure the metaphysician that what they already believe is true. Yet philosophers are often completely deluded when they claim that some intuition or other belongs to common sense. Not only are genuine common-sense intuitions the product of cultural learning, but philosophers who have spent years customizing their cognition with recondite concepts and philosophical technology, as well as habituating themselves to interpreting the world in terms of specific philosophical theories, do not share as many intuitions with the folk as they usually suppose. What metaphysicians take themselves to know by intuition is independent of the latest scientific knowledge and is culturally specific.

Lewis famously advocated a metaphysical methodology based on subjecting rival hypotheses to a cost–benefit analysis. Usually there are two kinds of cost associated with accepting a metaphysical thesis. The first is accepting some kind of entity into one's ontology, for example, abstracta, possibilia, or a relation of primitive resemblance. The second is relinquishing some intuitions, for example, the intuition that causes antedate their effects, that dispositions reduce to categorical bases, or that facts about identity over time supervene on facts about instants of time. It is taken for granted that abandoning intuitions should be regarded as a cost rather than a benefit. By contrast, as naturalists we are not concerned with preserving intuitions at all, and argue for the wholescale abandonment of those associated with the image of the world as composed of little things, and indeed of the more basic intuition that there must be something of which the world is made.

There are many examples of metaphysicians arguing against theories by pointing to unintuitive consequences, or comparing theories on the basis of the quantity and quality of the intuitions with which they conflict. Indeed, proceeding this way is more or less standard. Often, what is described as intuitive or counterintuitive is recondite. For example, L. A. Paul (2004, 171) discusses the substance theory that makes the de re modal properties of objects primitive consequences of their falling under the sortals that they do: 'A statue is essentially statue shaped because it falls under the statue-sort, so cannot persist through remoulding into a pot' (171). This view apparently has 'intuitive appeal', but sadly, 'any counterintuitive consequences of the view are difficult to explain or make palatable'. The substance theory implies that two numerically distinct objects such as a lump of bronze and a statue can share their matter and their region, but this 'is radically counterintuitive, for it seems to contradict our usual way of thinking about material objects as individuated by their matter and region' (172). Such ways of thinking are not 'usual' except among metaphysicians and we do not share them.

Paul says '[I]t seems, at least prima facie, that modal properties should supervene on the nonmodal properties shared by the statue and the lump' (172). This is the kind of claim that is regularly made in the metaphysics literature. We have no idea whether it is true, and we reject the idea that such claims can be used as data for metaphysical theorizing. Paul summarizes the problem for the advocate of substance theory as follows: 'This leaves him in the unfortunate position of being able to marshal strong and plausible commonsense intuitions to support his view but of being unable to accommodate these intuitions in a philosophically respectable way' (172). So according to Paul, metaphysics proceeds by attempts to construct theories that are intuitive, commonsensical, palatable, and philosophically respectable. The criteria of adequacy for metaphysical systems have clearly come apart from anything to do with the truth. Rather they are internal and peculiar to philosophy, they are semi-aesthetic, and they have more in common with the virtues of story-writing than with science.

The reliance on intuitions in metaphysics often involves describing a situation that is intuitively possible and then concluding something important about the actual world from the 'existence' of this possible world. For example, Sider tells us that it is clearly possible to imagine a world consisting of matter that is infinitely divisible: 'Surely there is a gunk world in which some gunk is shaped into a giant sphere' (1993, 286).[12] Surely? Lowe tells us that 'it seems that an individual material sphere could exist as a solitary occupant of space' (2003b, 79). He goes on to claim that while it may not be 'causally possible' for an individual organism to be an isolated existent, it is nonetheless 'metaphysically possible'. This is justified on grounds that 'two different individuals cannot both individuate, or help to

---

[12] The term 'gunk' was established by Lewis (1991).

individuate, each other. This is because individuation in the metaphysical sense is a determination relation ... As such, individuation is an *explanatory* relation' (93, his emphasis) in the metaphysical sense of 'explanatory'. He goes on: 'Certainly, it seems that any satisfactory ontology will have to include self-individuating elements, the only question being which entities have this status—space-time points, bare particulars, tropes, and individual substances all being among the possible candidates' (93). Certainly? None of these 'obvious' elements of reality (including the pseudoscientific 'space-time points') are known to either everyday intuition or science.

In the course of this brief survey we have encountered a number of *sui generis* versions of ordinary and/or scientific notions, namely, individuation, determination, explanation, and possibility. With these and other inventions, metaphysicians have constructed a hermitically sealed world in which they can autonomously study their own special subject matter.[13] We return to individuation in depth in Chapter 3 where we reject on scientific grounds the idea that ontology depends on self-individuating elements. For now, we merely note that the candidates for such elements that Lowe identifies are all pure philosophical constructions. As Glymour puts it: 'the philosopher faces the dragons in the labyrinth of metaphysics armed only with words and a good imagination' (1999, 458).[14]

There are ways in which intuitions *could* be useful. As discussed by Dennett (2005, 31–2), the artificial intelligence researcher Patrick Hayes (1979) thought at one time that the best method for trying to simulate our minds in computers or robots was to try to discern, by introspection, our 'naïve physics' (people's behaviourally manifest theory of how the physical world works), axiomatize this theory, and then implement it as the inference engine of a working android. This work, though it proved to be harder than Hayes expected and he did not finish it, is of some interest and value. Of course, Hayes did not imagine that naïve physics corresponded to *true* physics. As Dennett notes, philosophers who speculatively elaborate on intuitions *might*, if they were sophisticated in the way that Hayes was, be interpreted as doing introspective anthropology.[15] Obviously, this would not be metaphysics—the attempt to discover general *truths* about the objective world. However, what Dennett goes on to say about his critical target, the activity of neo-scholastic philosophers of mind, can with equal justice be said about many metaphysicians:

---

[13] Another example is the composition relation as studied by metaphysicians that we discuss in 1.2.3.

[14] Of course, as Glymour mentions, metaphysicians usually also use logic and set theory to formulate their theories. From our perspective this does not confer any extra epistemic status on their activity, but it may bamboozle the outsider or the student into supposing that the activity has much in common with mathematics and science.

[15] In fact, introspective anthropology is done all the time, and most people regard it as highly valuable. Its expert practitioners are mainly writers of fiction. We do not recommend turning their job over to philosophers.

They have proceeded as if the deliverances of their brute intuitions were not just *axiomatic-for-the-sake-of-the-project* but *true*, and, moreover, somehow inviolable... One vivid ... sign of this is the curious reversal of the epithet 'counterintuitive' among philosophers of mind. In most sciences, there are few things more prized than a counterintuitive result. It shows something surprising and forces us to reconsider our often tacit assumptions. In philosophy of mind a counterintuitive 'result' (for example, a mind-boggling implication of somebody's 'theory' of perception, memory, consciousness or whatever) is typically taken as tantamount to a refutation. This affection for one's current intuitions ... installs deep conservatism in the methods of philosophers. (Dennett 2005, 34)

As Dennett then says, methodological conservatism is not invariably bad policy—science certainly implements it, though not with respect to intuitions. However, as we noted in 1.1, neo-scholastic metaphysicians cannot even defend their project by falling back on a general defence of conservatism, since in the service of defending one intuition they frequently outrage other ones; we gave the example of Merricks's (2001) denial that tables and chairs exist and that baseballs can break windows.

We do not deny that intuitions in one sense of the term are important to science. It is frequently said of, for example, a good physicist that he or she has sound physical intuition. Economists routinely praise one another's 'economic intuitions'—and routinely break the bad news to struggling graduate students that they lack such intuitions. However, the meaning of 'intuitions' in these uses differs sharply from the metaphysician's. The physicist and economist refer to the experienced practitioner's trained ability to see at a glance how their abstract theoretical structure probably—in advance of essential careful checking—maps onto a problem space. Intuitions in this sense have nothing to do with deliverances of putative untrained common sense. Furthermore, even the intuitions of the greatest scientist are regarded by other scientists as heuristically and not evidentially valuable. By contrast, for neo-scholastic metaphysicians intuitive judgements are typically all that ever passes for evidence.[16]

1.2.2 A priori metaphysics

Representing the resurgent voice of the analytic metaphysicians here again is Lowe: 'metaphysics goes deeper than any merely empirical science, even physics, because it provides the very framework within which such sciences are conceived and related to one another' (2002, vi). According to him the universally applicable concepts that metaphysics studies include those of identity, necessity, causation,

---

[16] Much of our critique of the role of intuitions in metaphysics applies to other areas of philosophy. Weatherson (2003) argues against the weight that has been given to intuitions in epistemology post-Gettier. He defends the traditional conceptual analysis of knowledge against Gettier cases on this basis. From our perspective, the role of intuitions in that analysis is just as suspect as their role in undermining it.

space and time. Metaphysics must say what these concepts are and then address fundamental questions involving them such as whether causes can have earlier effects. Metaphysics' other main job according to Lowe, is to systematize the relations among fundamental metaphysical categories such as things, events, properties, and so on. We might reasonably ask how we could proceed with these tasks. Lowe follows Frank Jackson (1998) and many others in advocating the familiar methodology of reflecting on our concepts (conceptual analysis). But why should we think that the products of this sort of activity reveal anything about the deep structure of reality, rather than merely telling us about how some philosophers, or perhaps some larger reference class of people, think about and categorize reality? Even those fully committed to a conception of metaphysics as the discovery of synthetic a priori truths shy away from invoking a special faculty of rational intuition that delivers such knowledge; rather they usually just get on with their metaphysical projects and leave the matter of explaining the epistemology of metaphysics for another occasion. Ted Sider defends this strategy by pointing out that lack of an epistemological foundation for science and mathematics does not prevent practitioners from getting on with the business of advancing the state of knowledge in these domains (2001, xv). However, the appropriateness of this comparison is precisely what is at issue here. Mathematics and science have undoubtedly borne fruits of great value; a priori metaphysics has achieved nothing remotely comparable, if it has achieved anything at all.[17]

According to Lowe, it is the job of metaphysics to tell us what is possible, but he concedes that which of the possible fundamental structures of reality exist can be answered only with empirical evidence. In his (2006) he outlines a view of the goal of metaphysics that we endorse:

[R]eality is one and truth indivisible. Each special science aims at truth, seeking to portray accurately some part of reality. But the various portrayals of different parts of reality must, if they are all to be true, fit together to make a portrait which can be true of reality as a whole. No special science can arrogate to itself the task of rendering mutually consistent the various partial portraits: that task can alone belong to an overarching science of being, that is to ontology. (4)

However, we differ with Lowe on how this task is to be accomplished, because we deny that a priori inquiry can reveal what is metaphysically possible. Philosophers have often regarded as impossible states of affairs that science has come to entertain. For example, metaphysicians confidently pronounced that non-Euclidean geometry is impossible as a model of physical space, that it is impossible that there not be deterministic causation, that non-absolute time is impossible, and so on. Physicists learned to be comfortable with each of these

---

[17] The extent to which the epistemology of science and mathematics is a mystery is overstated in our view. We take it that philosophy of science has significantly illuminated the nature and basis of scientific inference and justification.

ideas, along with others that confound the expectations of common sense more profoundly.

1.2.3 Pseudo-scientific metaphysics

As a consequence of the recognition that neither deductive nor inductive logic can fully account for scientific knowledge of unobservables, most scientific realists admit that fallible appeals to the explanatory power of claims about unobservable causes of observable phenomena are our only source of epistemic access to the former. Some metaphysicians have realized that they can imitate science by treating their kind of inquiry as the search for explanations too, albeit in a different domain.[18] Taking the familiar explanatory virtues of unity, simplicity, non-ad hocness, and so on, they can now argue with each other about whose particular metaphysical package scores highest on some loosely weighted vector of these virtues and requires the fewest unexplained explainers. On the basis of such reasoning, metaphysics is now often regarded as if it were a kind of autonomous special science, with its explananda furnished by the other sciences.[19]

There are three ways in which analytic metaphysicians who rhetorically emulate science sometimes or often fail to follow through on their naturalistic pretence:

(1) They ignore science even though it seems to be relevant.

(2) They use outdated or domesticated science rather than our best contemporary science.

(3) They take themselves to be able to proceed a priori in the investigation of matters upon which they claim science does not bear.

It is rare to find metaphysicians defending (1), and arguing that if science and metaphysics seem to conflict the latter may trump the former, but here is a breathtaking declaration of philosophical arrogance from Peter Geach: ' ... "at the same time" belongs not to a special science but to logic. Our practical grasp of this logic is not to be called into question on account of recondite physics ... A physicist who casts doubt upon it is sawing off the branch he sits upon' (1972, 304). Note that the 'recondite' physics we are being advised not to take metaphysically seriously is Special Relativity.[20] In a similar vein, Ned Markosian (2005) is happy to defend mereological atomism, the thesis that there is a 'bottom level' to reality, composed of simples, despite his concession that the empirical evidence does not support this. There are, he thinks, good a priori

---

[18] For an explicit defence of the use of IBE in metaphysics by analogy with the use of IBE in science see Swoyer (1983).

[19] See, for example, Bealer (1987).

[20] Michael Tooley defends his theory of time by proposing an alternative to Special Relativity (1997, ch. 11) in which there is absolute simultaneity. He does this on metaphysical grounds, though he claims support from the phenomena of quantum entanglement. We return to these issues in 3.7.1.

grounds for believing it. Physicists do not believe there are such things as good a priori grounds for holding beliefs about the constitution of the physical world, and we suggest that only a foolhardy philosopher should be willing to quarrel with them on the basis of his or her hunches.

However, it is much more common for metaphysicians to ignore science without acknowledging or defending the idea of doing so. Alyssa Ney notes that 'very few of those philosophers who call themselves physicalists spend any time worrying about what physicists are actually up to' (forthcoming a, 1). A good example of this is Markosian again, who defines physical objects as all and only those that have spatial locations, and physicalism as the view that all objects are physical objects. Leaving aside the worry that this will allow irreducible mental properties provided they are spatially located (as Ney points out, forthcoming a, 15), Markosian's proposal would condemn most of the entities posited by fundamental physics to the status of non-physical including the universe itself.

In general in the philosophy of mind, in debates about perception, semantic content, and self-knowledge, it is common to proceed without paying any attention to science. Consider, for example, Jaegwon Kim's (1998) *Mind in a Physical World*. Despite its commitment to physicalism, it has no index entry for 'physics' and not a single work of physics appears in the list of references. Kim's argument, however, depends on non-trivial assumptions about how the physical world is structured. One example is the definition of a 'micro-based property' which involves the bearer 'being completely decomposable into nonoverlapping proper parts' (1998, 84). This assumption does much work in Kim's argument—being used, *inter alia*, to help provide a criterion for what is physical, and driving parts of his response to the charge, an attempted *reductio*, that his 'causal exclusion' argument against functionalism generalizes to all non-fundamental science.[21] As well as ignoring physics, Kim, and much of the metaphysical philosophy of mind of which he is a prominent exponent, ignores most of the interesting questions about the mind that scientists investigate.[22]

It is not always straightforward to spot cases of (2) and (3), because many contemporary metaphysicians explicitly claim to be naturalists and to be taking the scientific image as their explanandum. Indeed, many populate their discussions with scientific examples. In particular, they make recourse to examples from physics, for the obvious reason that, as the scientific discipline that makes the most general claims about the universe, physics most closely approaches metaphysics in scope. Unfortunately metaphysicians seem implicitly to assume that (a) non-actual physics can be used as part of the explananda for metaphysics,

---

[21] For more on this charge, and rejoinders to Kim's response, see Ross and Spurrett (2004), Marras (2000), Bontly (2001), and Kim (2005).

[22] For an eloquent critique of Kim on these grounds see Glymour (1999). Like Glymour we hold Kim's work in the highest regard as an exemplar of its kind. Our intention is to draw attention to the fact that something has gone awry when the very best philosophical work in a given domain is so estranged from science.

and that (b) this is acceptable because whatever the actual details of mature physics, they will somehow be able to 'dock' with the non-actual physics in question at some level of abstraction or generality, so that philosophers need not worry about or even pay attention to those details. For an explicit defence of (b) here is Frank Jackson:

> it is reasonable to assume that physical science, despite its known inadequacies, has advanced sufficiently for us to be confident of the *kinds* of properties and relations that are needed to give a complete account of non-sentient reality. They will be broadly of a kind with those that appear in current physical science ... (1998, 7)

We reject both (a) and (b).

There are several kinds of uses of non-actual physics regularly encountered in metaphysics. Among them are appeals to obsolete features of classical physics, and reliance on intuitions or common-sense conceptions of the material world. For example, Lewis explicitly states that his doctrine of 'Humean supervenience' is based on a model of the world in which the fundamental physical properties of a 'world like ours' are local, 'perfectly natural intrinsic properties of points, or of point-sized occupants of points' and declares himself that this picture is 'inspired by classical physics' (1999, 226). Van Inwagen (1990) assumes for the purposes of his metaphysical argument the truth of atomism, understood as the view that all (material) things ultimately decompose into mereological atoms, where a mereological atom lacks proper parts. None of these assumptions, on which are based arguments of considerable attention in the metaphysics literature, finds any basis in contemporary science. Kim's micro-based properties, completely decomposable into non-overlapping proper parts, and Lewis's 'intrinsic properties of points, or of point-sized occupants of points' both fall foul of the non-separability of quantum states, something that has been a well-established part of microphysics for generations. It is still hard to improve upon Schrödinger's formulation, in which the term 'entanglement' was coined:

> When two systems, of which we know the states by their respective representatives, enter into temporary physical interaction due to known forces between them, and when after a time of mutual influence the systems separate again, then they can no longer be described in the same way as before, viz. by endowing each of them with a representative of its own. I would not call that *one* but rather *the* characteristic trait of quantum mechanics, the one that enforces its entire departure from classical lines of thought. By the interaction the two representatives [the quantum states] have become entangled. (1935, 555)[23]

It is also hardly news that the conflict between quantum mechanics and general relativity comes to the fore when considering very small scales, especially points. (This is one of the motivations for string theory, which does not postulate infinitesimal objects or processes.) So Lewis's world of 'perfectly natural intrinsic properties of points, or of point-sized occupants of points' seems highly unlikely

---

[23] We return to this issue in Chapter 3.

to be the actual one. Van Inwagen's Democritean image of a world mereologically composed of simple atoms corresponds to it even less; this image has no more in common with reality as physics describes it than does the ancient cosmology of four elements and perfect celestial spheres. Yet Van Inwagen does not market his work as history of (early modern) philosophy; it is supposed to be contemporary metaphysics.

Consider also Lewis's discussion of the distinction between internal and external relations in his (1986). He asks us at one point to 'consider a (classical) hydrogen atom, which consists of an electron orbiting a proton at a certain distance' (62). There are not, nor were there ever, any 'classical hydrogen atoms'. At the same time that physicists came to believe in protons, they also became aware that the laws of classical mechanics could not apply to electrons orbiting them. Indeed the notion of an electronic *orbit* has about as much relation to the common-sense notion of an orbit as the mathematical notion of compactness has to the everyday notion of compactness, which is to say hardly any. Lewis thus encourages his readers to think that his metaphysics is addressed to the scientific image of the world rather than the manifest one, but he gives the game away because 'classical' here means nothing other than 'commonsensical'. Note that we are not arguing that what Lewis goes on to do with his account of internal and external relations is affected one way or the other by how he chooses to introduce the distinction; he could of course have used another example. Our point is that the rhetorical effect of his fictitious example is to suggest that his metaphysics has something to do with science when it does not.

When it comes to debates about the nature of matter in contemporary metaphysics it tends to be assumed that there are two possibilities: either there are atoms in the sense of partless particles, or there is 'gunk' in the sense of matter whose every part has proper parts (infinitely divisible matter).[24] This debate is essentially being conducted in the same terms as it was by the pre-Socratic philosophers among whom the atomists were represented by Democritus and the gunkists by Anaxagoras. In early modern philosophy Boyle, Locke, and Gassendi lined up for atomism against gunkists Descartes and Leibniz. It is preposterous that in spite of the developments in the scientific understanding of matter that have occurred since then, contemporary metaphysicians blithely continue to suppose that the dichotomy between atoms and gunk remains relevant, and that it can be addressed a priori. Precisely what physics has taught us is that matter in the sense of extended stuff is an emergent phenomenon that has no counterpart in fundamental ontology. Both the atoms in the void and the plenum conceptions of the world are attempts to engage in metaphysical theorizing on the basis of extending the manifest image. That metaphysicians continue to regard the world

---

[24] This debate becomes particularly baroque in the hands of Daniel Nolan (2004) who considers whether the infinite divisibility in question is denumerable or higher-order.

as a spatial manifold comprising material objects that must either have smallest spatial parts or be made of infinitely divisible matter is symptomatic of their failure to escape the confines of the domestic realm.

Similar points are pertinent to the debate about composition among analytic metaphysicians. A good part of most of the special sciences concerns the particular kinds of composition relevant to their respective domains. For example, biologists concern themselves with how cells compose multicellular organisms, economists with how individual markets compose economies, chemists with how oxygen and hydrogen compose water, and so on. Metaphysicians do not dirty their hands with such details but seek instead to understand something more fundamental, namely the general composition relation itself. But why suppose that there is any such thing? It is supposed to be the relation that obtains between parts of any whole, but the wholes mentioned above are hugely disparate and the composition relations studied by the special sciences are *sui generis*. We have no reason to believe that an abstract composition relation is anything other than an entrenched philosophical fetish.[25]

Composition in real science as opposed to metaphysics is usually a dynamic and complex feature that is much more interesting than its metaphysical counterpart. Consider, for example, the notion of composition at work in economics. Economic models are typically models of 'systems', which are taken to participate in larger systems. However, the relations between systems and sub-systems are not compositional in the philosopher's sense because they are model-relative. A system is distinguished by reference to variables that can be treated as endogenous, that is, as having their values co-determined as a set *given* some simultaneous choice of another set of variables as exogenous. Economists freely admit that interesting phenomena typically admit of multiple parsings along different endogenous/exogenous boundaries for varying predictive and explanatory purposes. In general, though economists are mainly concerned, most of the time, to discover which variables are 'control levers' for which others, their theoretical structure finds no use for the kind of rigid distinction between causal relations and compositional relations that neo-scholastic metaphysicians assume as fundamental. The case of composition in the physical sciences is similar. Water, for example, is composed by oxygen and hydrogen in various polymeric forms, such as $(H_2O)_2$, $(H_2O)_3$, and so on, that are constantly forming, dissipating, and reforming over short time periods in such a way as to give rise to the familiar properties of the macroscopic kind water.[26] The usual philosophical identity claim 'water is H_2O' ignores a fascinating and complex scientific account that is still not complete.

---

[25] Cf. Paul's mention of the 'primitive relation of fusing, already a part of standard ontology' (2004, 173). Again 'fusion' in the metaphysician's sense has nothing to do with real composition, and the 'standard' ontology appealed to here is standard, if at all, only among metaphysicians.

[26] See Van Brakel (1986).

A key general grievance we raise against traditional metaphysicians is that in continuously constructing simplistic caricatures of science, they render it substantially less interesting than it really is. Donald Davidson (1970) and Jerry Fodor (1974) both take it as true of physics that it discovers causal laws that take the form of exceptionless generalizations relating atomic events.[27] This is notwithstanding the fact that thus understood physical laws have numerous counterexamples, a point made forcefully by Russell (1913) while arguing that the laws physicists do produce are laws of functional interdependence, not statements of regularities. As we discuss in Chapter 3, Davidson's and Fodor's picture finds even less support from contemporary fundamental physics than it did from the physics known to Russell.

Next consider Merricks's *Objects and Persons* (2001), which, like van Inwagen (1990), defends the radical view that there are no statues, rocks, tables, stars, or chairs—only 'elementary particles' and people. The general idea here is that, given mereological atomism, the things additively composed out of atoms without residue are metaphysically redundant. (Merricks thinks that people are not composed of atoms, though their bodies are.) This is apparently supposed to be naturalistic, since Merricks declares that he has 'in mind here the atoms of physics, not Democritus' (2001, 3), and also that what he says about these atoms should be considered as 'placeholders for claims about *whatever microscopic entities are actually down there*' (2001, 3, emphasis added). That is, no matter what physics does, Merricks is confident that it will deliver atoms of the sort he requires for his arguments.

In a symposium on Merricks's book, Lowe (2003a) objects that it is 'hubristic' for philosophers to dictate to physicists about what is real in their domain of study, and suggests that physicists might be 'mystified and irritated' by Merricks's line of argument. (We imagine them laughing, in the unlikely event that they notice at all.) Merricks (2003b, 727) responds by claiming that Lowe's 'invocation of Physics is a red herring' because Merricks's metaphysic doesn't depend on him having any knowledge about disciplines that study what he says doesn't exist, and because all the relevant experts need to secure their authority in a given domain, including physics, is 'nucleons and electrons (or more fundamental entities) arranged' so as to present like (for example) helium. So, according to Merricks, as a metaphysician he is entitled to take as a premise for his arguments a claim about (what he takes to be) a matter of physical fact (that the world decomposes into atoms). On the other hand, the metaphysician apparently need not know anything about physics in order to make assertions about whether physicists are ontologically confused. As we discuss in detail in Chapter 3, none of the main contending theories in fundamental physics give the slightest encouragement to Merricks's conviction that the world is mereologically composed of any little

---

[27] Davidson (1970) actually suggests, but without explaining, that the 'exceptionless' criterion could be 'relaxed'. (See 5.5 and also Glymour 1999.)

things at all. But the point is that Merricks doesn't think this matters. All that does matter for metaphysics, it seems, is that people who know just a bit of superficial science are comfortable with thinking about a world made out of ultimate little things and collisions amongst them.

Crawford Elder (2004) is a metaphysician who attempts to restrain his fellow philosophers from deriving scientifically hilarious conclusions that amount to *reductio*s of their intuitions. However, Elder thinks that the way to do this is to show that the hilarious conclusions in question don't follow in the fantasy world of ultimate little things and microbangings. (See, for instance, the example Elder considers in which 'microparticles' composing 'host objects' 'hurtle' at one another in 'a microphysical mêlée' (94–6). The world, it seems, is like a pinball machine, though Elder doesn't mention flashing lights or funny sound effects.) All of his counter-arguments also depend on similar intuitions about this imaginary world, and are similarly irrelevant to what the actual world is like. Furthermore, Elder often says things that make it doubtful that he is merely donning the pretence of his opponents' bad assumptions for the sake of argument. 'With rare and strange exceptions,' he pronounces at one point, in what seems to be *propria persona*, 'we suppose that extended objects of any kind cannot simultaneously occupy two discontinuous spatial regions' (15). Physics knows nothing of the class of 'extended objects'; and the physical objects that occupy two or more discontinuous spatial regions are basic and ubiquitous.

What we mean to draw attention to here is Elder's implicit assertion that classical objects are the standard case, while entangled objects are exotic. Elsewhere, Elder trots out what he considers 'the most scientifically grounded picture' of alteration in fundamental physical composition that an opposing metaphysician might appeal to. This turns out to involve, once again, 'the subatomic microparticles that future physics will discover to be the truly fundamental building blocks of the physical world—"physical simples"' (51). Elder doesn't commit himself to believing in this picture; but on the basis of what evidence does he consider it to be the 'most scientifically grounded' one? That other philosophers tirelessly entertain it? Eventually he drops his careful agnostic guard about what he takes physics to hold: 'It really is true that each individual microphysical movement, in the complex microphysical event that the physicalist identifies as shadowing the cause in a typical special-science transaction, causes some other microphysical movement' (108). If this is indeed 'really true' then, as we will see in later chapters, Elder knows it on the basis of something other than science.

Finally, we exhibit David Armstrong defining metaphysical naturalism as the doctrine that everything that exists is in space and time, despite the fact that contemporary physics takes very seriously the idea that spacetime itself is emergent from some more fundamental structure (Armstrong 1983). Metaphysical naturalism, of all things. Note that all of these examples are, aside from ignoring science, models of professional philosophy, being clearly written, carefully argued, and responsive to the objections of those with opposing views. They are all centrally placed in the literature. Mainstream contemporary analytic metaphysics has, like the nineteenth-century metaphysics against which Russell revolted, become almost entirely a priori. Metaphysics informed by real physics is much less common.

In 1.1 we announced our resistance to the 'domestication' of science. It would be easy to get almost any contemporary philosopher to agree that domestication is discreditable if the home for which someone tries to make science tame is a populist environment. Consider, for example, the minor industry that seeks to make sense of quantum mechanics by analogies with Eastern mysticism. This is obviously, in an intellectual context much less rigorous than that of professional philosophy, an attempt to domesticate physics by explaining it in terms of things that common sense thinks it comprehends. Few philosophers will regard the gauzy analogies found in this genre as being of the slightest metaphysical interest. Yet are quantum processes any more like those described by Newtonian physics than they are like the temporal and spatial dislocations imagined by mystics, which ground the popular comparisons? People who know almost no formal physics are encouraged by populists to find quantum mechanics less wild by comparing it to varieties of disembodiment. Logically, this is little different from philosophers encouraging people who know a bit of physics to make quantum accounts seem less bizarre by comparing them to what they learned in A-level chemistry.[28] We might thus say that whereas naturalistic metaphysics ought to be a branch of the philosophy of science, much metaphysics that pays lip-service to naturalism is really philosophy of A-level chemistry.

One response to what we've said so far would demand a justification for our evident commitment to the view that philosophy of A-level chemistry is a bad thing at all. Science itself, after all, makes use of flatly non-actual scenarios and notions, including frictionless planes, perfectly elastic collisions, ideal gasses, etc. As Hüttemann (2004, 20) argues, un-instantiated laws can be established in science, and consequently bear explanatory weight. For such laws to be established, we need reasons for thinking that the closer conditions get to some (possibly unattainable) limit, the more the behaviour of a system approximates an ideal indicated by the un-instantiated law. One of Hüttemann's examples concerns a law specifying the specific heat of a sample of lithium fluoride crystal. The law in question supposes a crystal entirely devoid of impurities. Even without ever having an example of such a crystal, we can rank the behaviour of samples we do have with respect to the extent of their impurities and the degree to which they conform to the law, and thereby justify thinking that the law holds in the limiting case.

---

[28] Non-British readers may be unfamiliar with this. We refer to the idea of a scientific education that gets as far as the solar system model of atomic structure and no further.

However, examples of use of non-actual science by philosophers differ from scientists' uses of idealizations in two crucial ways. First, scientifically motivated justifications for the non-actual physics, along the lines just sketched in Hüttemann's example, are not offered by the philosophers. Second, it is typically the case that whole non-actual *worlds*, such as 'Newtonian worlds', are discussed by neo-scholastic metaphysicians, rather than ideal tendencies that may be partially manifest in more realistic settings. Scientific idealizations and approximations are usually accompanied by explicit statements of the contexts in which they are appropriate and/or the degrees of freedom for which they are accurate. Scientific idealization and approximation is about local not global verisimilitude.[29]

Remembering that metaphysics is supposed to be about the general structure of reality, let us ask ourselves what worth there could be in philosophical thought experiments that begin with the metaphysician declaring, for example, that she will be considering a 'Newtonian world' for simplicity (or some other reason). The actual world cannot be Newtonian in this sense. Newton himself realized as much when he noted that more than gravity would be needed to hold objects together. In fact, it is dubious to suppose that Newtonian mechanics could be true in the required sense and matter still exist, since we have learned that the only matter with which we have any acquaintance is not governed by Newton's laws. Furthermore, Newtonian mechanics always left unanswered questions about, for example, the cause and propagation of gravitational forces, why inertial and gravitational mass seem to be equal in magnitude for all objects, whether energy is conserved, and also about the nature of space and time. Newtonian physics was a research programme that attempted to extend an incomplete theory, which ultimately led to the recognition that it was after all a false theory. This is why the presupposition that philosophers' bogus physics can serve as a placeholder for whatever real physics turns out to say cannot be invoked in this case. Even if there could be Newtonian worlds, there are no grounds for thinking that whatever answers we get to metaphysical questions by considering thought experiments about them tell us anything about how things are in the actual world.

Refusal to take seriously the implications of living in a world that has turned out not to be Newtonian is also exemplified when philosophers imagine that the strange features of quantum physics can be contained. So it is often claimed that although quantum mechanics seems to imply indeterminism and single-case probabilities, these can be confined to the microscopic level. Plainly, however, if there is indeterminism among quantum events and there is any coupling of them to macroscopic events, as there surely is, then the indeterminism will infect the macroscopic. For a homely example that suffices to make the point, imagine a physicist deciding that she'll go for lunch after exactly so many clicks of the Geiger counter. It also seems that quantum entanglement contributes

---

[29] See Wallace (2001).

to many macroscopically observable properties of things, like their specific heat capacities. (Entanglement flatly contradicts Kim's proposal that everything is exhaustively structured by micro and 'micro-based' properties, as defined earlier, *and* destabilizes the whole neo-scholastic basis for distinguishing between micro-level and macro-level properties.) Philosophers simply obfuscate in pretending that the macroscopic world could be just like it is even if there wasn't all that quantum weirdness, as if the latter was an add-on at the level of the very small and not a fundamental aspect of the world. The metaphysician may claim that in the absence of a solution to the measurement problem quantum mechanics cannot teach us any metaphysical lessons. However, we know from Bell's theorem that any empirically adequate successor to quantum mechanics will have to violate local realism and hence some part or other of the 'common-sense' intuitions of metaphysicians.[30]

Metaphysicians surely know that contemporary physics is hugely more complicated and less intuitively comprehensible than either classical physics or toy worlds based on features of classical physics. Most, however, resist the obvious lesson that any attempt to learn about the deep structure of reality from thought experiments involving domesticated physics is forlorn. If it really doesn't matter that classical physics is false then we might as well do our metaphysical theorizing on the basis of Aristotelian or Cartesian physics. But then the absurdity would be patent. Nobody who assumed an Aristotelian distinction between forced and natural motion, and then declared that key parts of what she said about the world were to be understood as placeholders for 'whatever story about proper places and fundamental substances physics eventually says are real', would be taken seriously. Yet metaphysicians considering possible worlds consisting of only a few particles are as likely to arrive at deep truths about the universe on that basis as if they considered a world consisting of, for example, only a single unactualized potentiality.

We cannot go back to anti-metaphysical positivism. This book is not hostile to metaphysics; indeed, it is an exercise in metaphysics. However, we think that the kind of intellectual atmosphere that led Hume, and later Russell, the Vienna Circle, and Reichenbach, to denounce whole leading branches of the philosophy of their times as scholastic have arisen again. It seems, inductively, that such moments recur endemically in the discipline. We suppose this happens mainly because philosophers inevitably spend most of their time arguing with one another, until they forget that there is anyone else around or any important source of opinions besides rational arguments. When philosophy becomes institutionally solipsistic, however, it risks making itself intellectually and culturally irrelevant. No scientist has any reason to be interested in most of the conversation that now goes on under the rubric of metaphysics. We are dismayed that a large part of our profession deserves to be ignored by those who actually interrogate nature in the

---

[30] Again, see Chapter 3.

field and the lab. Fortunately, in philosophy's previous episodes of detachment from empirical inquiry, the Humes, Russells, and Carnaps have turned up when needed to save the enterprise from itself. While not claiming a mantle of their magnitude, we are here embarked on a mission of disciplinary rescue in their spirit.

In setting out upon this mission, we require some criteria for determining when purportedly naturalist metaphysics has descended into philosophy of A-Level chemistry, or some other variety of pseudo-naturalism. Simply transforming the presuppositions we identify into negative commandments won't do the job. Furthermore, we think that contemporary science provides evidence for some *positive* metaphysical claims and theses. Following all our critical remarks in this section, we can only move on to sketch these after first stating the conditions under which we think that positive metaphysics can be appropriately motivated.

1.3 THE PRINCIPLE OF NATURALISTIC CLOSURE

In this section we explicitly formulate the fully naturalistic principles that constitute the ground rules for our project. The discussion in the section above indicates the sort of metaphysics we eschew by citing instances. Socrates would remind us that this can't in itself constitute a substantive claim. We require some proper principle which distinguishes what we regard as useful from useless metaphysics.

Note that in stating this as our aim, we immediately distance ourselves from the positivists and align our attitude more closely with that of Peirce and pragmatism. As Putnam (1995) reminds us, both the positivists and the pragmatists sought to demarcate the scientific from the unscientific by use of verificationist principles. However, Putnam emphasizes that 'for the positivists, the whole idea was that the verification principle should *exclude* metaphysics ... while for the pragmatists the idea was that it should *apply to* metaphysics, so that metaphysics might become a responsible and significant enterprise' (293, his emphasis).

Why should radical methodological naturalists suppose that there is any 'responsible and significant' job for metaphysics to do? Our answer is that one of the important things we want from science is a relatively unified picture of the world. We do not assert this as a primitive norm. Rather, we claim, with Friedman (1974) and Kitcher (1981), that it is exemplified in the actual history of science. Scientists are reluctant to pose or to accept hypotheses that are stranded from the otherwise connected body of scientific beliefs. This is rational, reflecting the fact that a stranded hypothesis represents a mystery, and therefore calls out for scientific work aimed at eliminating it. It also reflects the fact that an important source of justification for a hypothesis is its standing in reciprocal explanatory relationships—networked consilience relationships—with other hypotheses (see Thagard 1992). (Good hypotheses are also, of course, expected to have at least some relatively direct independent evidence in their favour.) However, evaluating the global consilience network is not a task assigned to any particular science, partly because important efficiency considerations recommend disciplinary specialization. Metaphysics, as we will understand it here, is the enterprise of critically elucidating consilience networks across the sciences.

The reader will have noticed that this justification of metaphysics, and indeed its identification of the nature of justifiable metaphysics, is pragmatic in character. Our appeal to pragmatism here is in turn based on one meta-methodological and one epistemological claim that we endorse. The meta-methodological claim is that there is no such thing as 'scientific method', by which we mean: no particular set of positive rules for reasoning that all and only scientists do or should follow. There are of course many observed prohibitions (for example, 'Do not induct on samples known to be selected in unrepresentative ways' and 'Do not invent data'), but these apply to all sound reasoning, not to distinctively 'scientific' reasoning. Thus science is, according to us, demarcated from non-science solely by institutional norms: requirements for rigorous peer review before claims may be deposited in 'serious' registers of scientific belief, requirements governing representational rigour with respect to both theoretical claims and accounts of observations and experiments, and so on. We do not suppose that these norms are arbitrary or products of path-dependent historical factors. They are justified by the fact that individual human beings are poorly prepared by evolution to control complex inductive reasoning across domains that did not pose survival problems for our ancestors. We can, however, achieve significant epistemological feats by collaborating and by creating strong institutional filters on errors. This point gives rise in turn to the epistemological claim mentioned at the beginning of the paragraph. Since science just *is* our set of institutional error filters for the job of discovering the objective character of the world—that and no more but also that *and no less*— science respects no domain restrictions and will admit no epistemological rivals (such as natural theology or purely speculative metaphysics).[31] With respect to anything that is a putative fact about the world, scientific institutional processes are absolutely and exclusively authoritative.

Still following Putnam (1995), we can now note two other respects in which our pragmatist attitude to metaphysics resembles Peirce's rather than the positivists'. First, with Peirce we emphasize that scientific (and useful metaphysical) reasoning is a community enterprise and is not, except on rare occasions such as the achievements of Darwin and Einstein, reliably supported by feats of individual reasoning—let alone consequences of reflection on intuitions.

---

[31] Thus the very popular idea, recently championed by Gould (1999), that religion and science provide complementary accounts of different domains of reality must be rejected except where—as Gould sometimes implies—some particular religion, or religion in general, is interpreted as making no factual claims. Any fact any religion purports to establish will, if there is any evidence for it at all, be a target for scientific explanation.

Second, we stress that what sound metaphysics should be connected with are substantial bodies of scientific results taken together, rather than individual claims taken one at a time. (Positivists, of course, discovered this second point themselves: this was the path that led through Carnap to Quine.) These points are connected to one another by the following claim: individuals are blessed with no epistemological anchor points, neither uninterpreted sense-data nor reliable hunches about what 'stands to reason'. The epistemic supremacy of science rests on repeated iteration of institutional error filters.

In the previous section we rejected the idea that scientifically disconnected metaphysics should step into the breach wherever science has in principle nothing to say. We follow Peirce in endorsing a non-positivist version of verificationism—a version that is universally respected by the institutional practices of science. This verificationism consists in two claims. First, no hypothesis that the approximately consensual current scientific picture declares to be beyond our capacity to investigate should be taken seriously. Second, any metaphysical hypothesis that is to be taken seriously should have some identifiable bearing on the relationship between at least two relatively *specific* hypotheses that are either regarded as confirmed by institutionally *bona fide* current science or are regarded as motivated and in principle confirmable by such science.

With respect to the first aspect of this verificationism, let us be clear that 'capacity' is to be read in a strong modal sense. In saying that something is beyond our capacity to investigate we do not just mean that it's beyond our *practical* capacity—because we would have to last too long as a species, or travel too far or too fast or use a probe no one now has any idea how to build. We refer instead to parts of reality from which science itself tells us information cannot, in principle, be extracted for receipt in our region of spacetime or in regions of spacetime to which we or our instruments could in principle go. Suppose that the Big Bang is a singular boundary across which no information can be recovered from the other side.[32] Then, if someone were to say that 'The Big Bang was caused by Elvis', this would count, according to our principle, as a pointless speculation. There is no evidence against it—but only for the trivial reason that no evidence could bear on it at all. We take it that the claim about Elvis is obviously uninteresting. Our point here, however, is that it is not just uninteresting because it is silly and unmotivated. Claims that the Big Bang was caused by God, or by the action of a black hole, would be uninteresting, in the imagined circumstances, in exactly the same sense. Unlike the claim about Elvis, they may have evident psychological motivations. Typical motivations for the claim about God are various and often complex, though they almost invariably include instances of attempted domestication of science (in our sense). The imagined claim about the black hole is more obviously

---

[32] We do not assert that this is so. This question was regarded as closed until a few years ago, but has recently become controversial again.

domesticating. Why, if the Big Bang were truly a singularity in the sense of being an information boundary, would a black hole be any more likely to be involved in its generation than Elvis or God or an infinity of other things someone could imagine? Black holes seem plausible only because they are appropriately scaled entities for interacting with very large cosmic events on *our* side of the boundary. The explanation is (relatively) attractive only because it is familiar. An aspect of leaving science undomesticated is recognizing that it itself may tell us that there are questions we absolutely cannot answer because any attempted answer is as probable as any other. This does not imply that we should look to an institution other than science to answer such questions; we should in these cases forget about the questions.[33]

It should be clear from what has just been said that our verificationism, unlike that of the logical positivists, is not a claim about meaning. The statement 'The Big Bang was caused by Elvis' is perfectly meaningful in all reasonable senses of the term. When we call the statement 'pointless' we intend nothing technical. We mean only that asking it can make no contribution to objective inquiry. (It might, of course, make a contribution to comedy or art.)

The second claim stated as constitutive of our verificationism is, so far, ambiguous in three ways that require sorting out.

First, 'identifiable bearing on' is a weasel phrase, open to multiple readings. Let us thus make its intended meaning more precise. Naturalism requires that, since scientific institutions are the instruments by which we investigate objective reality, their outputs should motivate all claims about this reality, including metaphysical ones. We have stated our view that the point of metaphysics is to articulate and assess global consilience relations across bodies of scientifically generated beliefs. Thus one naturalist constraint on metaphysics might be expressed as follows:

Any new metaphysical claim that is to be taken seriously should be motivated by, and only by, the service it would perform, if true, in showing how two or more specific scientific hypotheses jointly explain more than the sum of what is explained by the two hypotheses taken separately, where a 'scientific hypothesis' is understood as an hypothesis that is taken seriously by institutionally *bona fide* current science.

This proposal clearly calls for an account of explanation that allows us to make clear sense of the idea that someone could have 'more' explanation given one structure of beliefs than another. Here, we rely on a substantial body of work on scientific explanation by Philip Kitcher (1981, 1989). Kitcher's basic idea is that

---

[33] Science might posit the existence of a region of spacetime that is absolutely inaccessible in the sense that we can obtain no information bearing on any of its properties other than whatever relations with other regions licensed inferring its existence in the first place. Then any hypotheses about these other properties would be pointless metaphysics.

unification of science consists in maximizing the ratio of kinds of phenomena we can explain to the number of kinds of causal processes[34] we cite in the explanations. We make progress toward such maximization every time we show that two or more phenomena are explained by a common *argument pattern*. An argument pattern is a kind of template for generating explanations of new phenomena on the basis of structural similarity between the causal networks that produced them and causal networks that produced other, already explained phenomena. Lest the idea be thought too vague, let us quote Kitcher's official introduction of argument patterns in full:

> A *schematic sentence* is an expression obtained by replacing [at least] some... nonlogical expressions occurring in a sentence with dummy letters. Thus, starting with the sentence 'Organisms homozygous for the sickling allele develop sickle-cell anaemia,' we can generate a number of schematic sentences: for example, 'Organisms homozygous for A develop P' and 'For all x, if x is O and A then x is P'... A set of *filling instructions* for a schematic sentence is a set of directions for replacing the dummy letters of the schematic sentence, such that for each dummy letter, there is a direction that tells us how it should be replaced. For the schematic sentence 'Organisms homozygous for the sickling allele develop sickle-cell anaemia,' the filling instructions might specify that A be replaced by the name of an allele and P by the name of a phenotypic trait. A *schematic argument* is a sequence of schematic sentences. A *classification* for a schematic argument is a set of statements describing the inferential characteristics of the schematic argument: it tells us which terms of the sequence are to be regarded as premises, which are inferred from which, what rules of inference are used, and so forth. Finally, a *general argument pattern* is a triple consisting of a schematic argument, a set of sets of filling instructions, and a classification for the schematic argument. (Kitcher 1989, 432)

Kitcher goes on to exemplify application of the argument-pattern concept by constructing the argument patterns for three large-scale scientific theories: classical genetics, Darwinian selection theory, and Dalton's theory of the chemical bond. Ross (2005, 377–8) constructs the argument pattern for neoclassical microeconomics. We have a unified world-view to the extent that we use a smaller rather than a larger number of argument patterns in science, and to the extent that what get used as schematic sentences in these argument patterns are themselves derived from other non-ad hoc argument patterns.

It will be noted that this account of unification is given in terms of the expressions of propositional descriptions of scientific discoveries and generalizations. To this extent it might appear to be positivistic in one of the senses we've rejected. However, our claim (and Kitcher's claim) is not that science consists of argument patterns, but that our being able to describe our scientific knowledge in terms of a compact set of argument patterns reflects our (collectively) knowing how to use a compact set of *problem-solving strategies*—ways

---

[34] Later (in Chapter 5) we will justify substituting 'information transmitting processes' for 'causal processes' in any use of the concept of causation that is, like Kitcher's, general and metaphysical rather than parochial to a special science.

of designing experiments and other measurement procedures—when confronted with new phenomena. The body of direct scientific knowledge consists of problem-solving strategies, and our ability to communicate this knowledge by means of a compact set of argument patterns is diagnostic of unified science. (Beginning in Chapter 2, we will assimilate schematic arguments to what we will call *structures*.)

In case this is thought to be too abstract to be put to operational use, note that Paul Thagard (1992 and elsewhere) has constructed connectionist learning networks that successfully predict features of scientific theory on the basis of inducting argument patterns from exemplars. It is true that the argument for Kitcher's version of unification cannot be derived as an *analytical* principle that is independent of all instances; the claim is that it applies to what scientists regard as prize-worthy achievements.

A second ambiguous idea that occurs in our second verificationist claim requires attention. What does 'specific' mean in the phrase 'specific scientific hypothesis'?

The history of verificationism warns us of both the importance and the difficulty of this issue. The positivists, in trying to arrive at their principle of empirical significance, struggled in vain with the problem of finding a criterion strict enough to exclude unscientific speculation, but liberal enough to avoid ruling out any claims that might matter to scientists (Hempel 1965, ch. 4). If one allows that any claim that itself is relevant to science is an empirically significant scientific hypothesis, then it becomes trivially easy to make almost any metaphysical hypothesis come out as scientifically relevant. Suppose one granted that 'All objects people can see without instruments are larger than atoms' is a significant scientific hypothesis. Almost all of the metaphysics we instanced as scientifically irrelevant a few pages ago can be demonstrably connected, by putatively explanatory relations, to statements at this level of generality. To have teeth, a naturalistic restriction on metaphysics must block the justification of metaphysical hypotheses only by reference to such purely generic and qualitative truths. But 'specific' is vague. Suppose we tried to make it precise enough to define an exact line between overly generic scientific hypotheses and the kind which we think should motivate metaphysical ideas? In that case, the history of positivism warns us, we would inevitably invite counterexamples derived from the fact that there is no canonical, pre-definable *kind* of statement of hypotheses that scientists regard as important. An attempt to *analyse* 'specific' in a way precise enough to block trivial justification of most metaphysics would inevitably become an arbitrary ban on many metaphysical hypotheses that important groups of scientists do find interesting.

The positivists made a fundamental mistake in seeking a principle for demarcating scientific from unscientific speculation by logical and semantic analysis. The problem with this approach is that the analytic generalizations used for demarcation must either be regarded as a priori or as tautologies, in which case they cannot be thought to be derived from science. A naturalistic demarcation principle should be based on reference to criteria that are empirically observed to regulate the practices of science. The principle at which we aim should not be trotted out as an analysis of the concept of 'sound metaphysics', and should not govern kinds of linguistic entities or propositions. What we should attempt to articulate is a heuristic principle, something that reminds us which sorts of critical questions should be addressed to philosophers when they offer metaphysical proposals. It must be clear enough in its intended force to rule out clear cases of neo-scholastic metaphysics, but it need not be something that could be applied algorithmically.

This might seem to be special pleading on our part. It might be thought that, after thundering about our intention to scourge the land of bad metaphysics, we've now just admitted that we can't even exactly say what bad metaphysics is. However, while our point above is obviously 'pleading' of a sort—don't ask us to do the impossible, please—our pragmatist framework of assumptions saves it from being ad hoc. We demarcate good science—around lines which are inevitably fuzzy near the boundary—by reference to institutional factors, not to directly epistemological ones. (Again, this reference is indirectly epistemological, and not irreducibly sociological, if the institutional factors that make science epistemologically superior themselves admit of epistemological justification, as they do.) This in turn implies that our principle must have the status of a normative heuristic, not that of a logical analysis. It also suggests a strategy for rendering the requirement of 'specificity' less vague. We can do this not by reference to representational (syntactic or semantic) properties of hypotheses themselves, but by reference to well-understood norms of scientific practice that are identified empirically.

Here, then, is such a norm. Almost all successful participants in '*bona fide* institutional science'—on which we will say more below—learn in graduate school, or soon after, which sorts of hypotheses one can*not* propose as the targets of investigation in a grant proposal to a 'serious' foundation or funding agency with non-zero prospects of success. Of course, there will be many hypotheses with respect to which judgements will differ as to how far above zero their funding prospects should be estimated to be. And some work that many or even most people would judge to be unimportant or silly actually gets proposed and funded. However, any physicist would agree that a study aimed at testing the claim that 'All objects people can see without instruments are larger than atoms' would not be worth writing up for funding.[35] The problem is not, of course, that the claim is unscientific, let alone empirically insignificant. The problem is that the claim is generic rather than specific.

---

[35] We invite the reader to imagine a physicist, fresh from reading Nolan (2004), writing a grant proposal to investigate the idea that the universe is made of hypergunk (see n. 23 above).

It is fine for our purposes if our verificationist principle errs on the side of permissiveness, allowing, in principle, 'importance' to be bestowed on some metaphysical claims motivated by slightly eccentric or dull specific scientific hypotheses. The principle is intended as a negative heuristic that blocks neo-scholastic metaphysics. This sort of metaphysics is, we suggested in the previous section, partly distinguished by substituting the philosophy of A-level chemistry for the philosophy of actual science when it tries to link itself to science and thus justify its conception of itself as naturalistic. The kinds of generic—often true—principles that comprise A-level chemistry are just the sorts of things that are not targets of investigation in projects that earn research grants.

Again, we stress that specificity is not a feature that is necessarily correlated with empirical significance as the positivists intended that idea. The generic claims that often motivate neo-scholastic metaphysics have no special epistemological features in common. Some of them are clear truths (as is the case with our example); others are folk approximations to the truth that are so approximate that the best semantic designation for them is: false. This just serves to emphasize that a consistent naturalist should aim only at a kind of demarcation principle quite different from the positivists' criterion of cognitive significance. We seek a principle, referenced to the institutional factors that make science epistemically superior, for distinguishing well-motivated from ill-motivated metaphysical proposals; we do not seek a principle for separating sense from nonsense.

Some philosophers will worry that if we index specificity to fundable research, we relativize the motivators for metaphysical revision to particular moments in the history of science. Others will raise a closely related worry that, in declaring that generic scientific hypotheses should not motivate metaphysical revisions, we are ignoring the fact that generic hypotheses do sometimes get refuted or revised by scientific progress—and that when they do, this is typically the basis for our most important metaphysical adjustments. This would indeed be a decisive objection if someone tried to operationalize empirical significance by reference to fundable activity. Again, however, that is not what we propose to do; fundability is simply being suggested as a proxy indicator (in the economist's sense) of what is likely to be scientifically interesting. Hypotheses we would regard as unacceptably generic (when used as the basis for motivating metaphysical innovations) are indeed sometimes overturned. But when they fall, they do so under pressure from accumulated results based on investigations of more specific phenomena with which the generic hypothesis in question is inferentially related. Typically, the accumulated results in question amount to substantial sets of data. Thus the same work investigating specific hypotheses that motivated the rejection of the generic hypothesis would be expected also to motivate portentous metaphysical adjustments. On the issue of temporal relativity, we see no direct objection to indexing the naturalistic constraints on metaphysical hypotheses to historically adjustable norms. After all, the point of such a constraint is to require metaphysicians to be motivated by what the scientific communities with which they are contemporary find significant; we should not demand that metaphysicians be more prescient than the scientists of their times.

One consequence of naturalism that cannot be avoided is that if our current scientific image of the world changes much, as we suppose it will, then it will then turn out that the best current metaphysics is substantially wrong. Neither we nor anyone else can do better than articulate the best metaphysical picture the current evidence suggests in attempting to sketch the image of the world that science presents to us now. We look forward to future science proving us wrong but hope that philosophers armed with current science will have trouble doing so. Note that to the extent that metaphysics is closely motivated by science, we should expect to make progress in metaphysics iff we can expect to make progress in science. In Chapter 2 we indicate at length why we hold fallibilism about science to be compatible with optimism about epistemic progress in science. This argument carries directly over to scientifically motivated metaphysics.

There are two reasons why a specific hypothesis might not be deemed suitable for investigation that should not impugn that hypothesis's possible relevance to metaphysics. The first is that the hypothesis in question might already be regarded as confirmed. This in itself presents no issue for our principle as stated, since that principle allows that confirmed hypotheses are potential motivators of sound metaphysics. But in light of what we admitted above about temporal relativity, there seems to be a problem. Suppose that a hypothesis was confirmed at a time when scientific judgement as to what was sufficiently specific to warrant investigation was different—presumably less strict—than at present. The only way to avoid the absurd conclusion that sound metaphysics becomes unsound simply because science becomes more specialized is to stipulate that every hypothesis that was actually a target of *direct* investigation by recognizably institutionalized science (so, including the Royal Society but not including Plato's Academy) is a potential motivator of sound metaphysics. We do not think there was ever a time or place after the scientific revolution when institutionally bona fide scientific resources were devoted to directly investigating claims of a level of generality that justified or would justify neo-scholastic metaphysics.

The second reason why a hypothesis that is a potential motivator of sound metaphysics might not be considered as a subject of direct scientific investigation is that no one can think of a practical experiment or measurement. Some interesting microphysical hypothesis, for example, might be testable only using an accelerator that is too expensive to build. We of course want to distinguish hypotheses that are non-investigable in this sense from hypotheses that are non-investigable because information that would bear on them is in principle unobtainable by any observer. Let us therefore say simply that a specific scientific hypothesis is one that *would* be deemed suitable for direct investigation given the absence of any constraints resulting only from engineering, physiological, or economic restrictions or their combination.

A remaining locus of ambiguity in our version of verificationism is its appeal to 'institutionally *bona fide*' science and scientific research funding bodies. In general, we are happy to leave this open to the rational judgements of observers of institutional processes. What we importantly wish to exclude that will not be obvious from anything said so far, however, are research projects that are primarily motivated by anthropocentric (for example, purely engineering driven) ambitions, as opposed to ambitions anchored around attempts to determine the objective structures in nature. As philosophers, we naturally owe a detailed account of what we take this to mean, and providing such an account, and a justification for it, is another of the primary objectives of the book as a whole. In Chapter 4 we will articulate a theory of what it is for a scientific theory to be taken to describe a part of objective reality—a 'real pattern', as we will say (following Dennett 1991a). It will follow from this analysis that for a pattern to be real—for the object of a scientific theory or other description to be deemed an aspect of objective reality—it must be such that a community of inquirers who wished to maximize their stock of true beliefs would continue to be motivated to track the pattern notwithstanding any shifts in practical, commercial, or ideological preferences that are not justified by new evidence bearing on the epistemic redundancy or non-redundancy of the pattern. Some activity appropriately called 'scientific' because it is governed by the institutional error-filtering processes characteristic of science—for example, some research done in medical, engineering, law, and public policy schools and institutes—does not aim at objectivity in this sense. That is just to say that such activity sometimes deliberately tolerates pursuit of objectively redundant facts for the sake of the practical utility of certain representations by people aiming at real-time solutions to problems arising from their non-epistemic preferences. Engineers, for example, sometimes study refinements of generalizations from classical physics that are strictly false according to contemporary physics. Metaphysics should not be motivated by such activity.

In excluding anthropocentrically motivated investigations as relevant motivators of metaphysical hypotheses, we also rule out ideologically driven research. Suppose, for example, that some self-styled 'creation scientists' sought and obtained funding from one of their own dedicated foundations to pursue a physical hypothesis that, if true, would comport with their belief in a very young Earth. We would of course wish to exclude this as a motivator of relevant metaphysics. The fact that these 'scientists' would have to seek their funding from non-standard sources is, by our lights, precisely the most reliable indicator that their activity should not be taken seriously by the metaphysician—more reliable, in particular, than any specific analysis of the specific arguments and assumptions cited in the motivations for the project.[36] We thus take this hypothetical example as casting further illumination on the value of indexing serious

---

[36] This does not imply, of course, that such criticisms are not often valuable in their own right.

science to epistemological factors by way of mediating institutional factors as proxies rather than by directly epistemological criteria. To reiterate: we assume that the institutions of modern science are more reliable epistemic filters than are any criteria that could be identified by philosophical analysis and written down. Note that we do not derive this belief from any wider belief about the reliability of evolved human institutions in general. Most of those—governments, political parties, churches, firms, NGOs, ethnic associations, families, etc.—are hardly epistemically reliable at all. Our grounding assumption is that the specific institutional processes of science have inductively established peculiar epistemic reliability.

One more general idea underlying our naturalistic constraint on metaphysics remains to be expressed. In the next section of this chapter, we will argue for a principle we call the *Primacy of Physics Constraint* (PPC). This articulates the sense in which evidence acceptable to naturalists confers epistemic priority on physics over other sciences. In Chapter 5 we will elaborate on the details about how physics constrains other sciences. It will turn out that what is most importantly different about physics—or, at least, about the part of physics we will call 'fundamental'—is that it has wider scope, in a sense we will make precise, than other sciences. Metaphysics, as the project of unifying the scientific world-view, shares the maximum scope of fundamental physics, in the same precise sense.

It follows from this view—which is part of the content of our metaphysical theory rather than part of its motivation—that a hypothesis that unified specific hypotheses from sciences other than fundamental physics, but unified them with no specific hypotheses from fundamental physics, would not be a metaphysical hypothesis. It would instead be a hypothesis of a special science of wider scope than those it partially unified. Again, it is premature to go into the details of this in advance of the analysis to be given in Chapters 4 and 5. For now we simply assert that although specific hypotheses from any non-anthropocentric scientific inquiry may be motivating premises for naturalistic metaphysical hypotheses, at least one specific hypothesis that the metaphysical hypothesis in question unifies with others must be derived from fundamental physics.

Based on these reflections, here is a refined formulation of a naturalist constraint on metaphysical hypotheses that we will henceforth refer to as the *Principle of Naturalistic Closure* (PNC):

Any new metaphysical claim that is to be taken seriously at time t should be motivated by, and only by, the service it would perform, if true, in showing how two or more specific scientific hypotheses, at least one of which is drawn from fundamental physics, jointly explain more than the sum of what is explained by the two hypotheses taken separately, where this is interpreted by reference to the following terminological stipulations:

Stipulation: A 'scientific hypothesis' is understood as an hypothesis that is taken seriously by institutionally bona fide science at t.

Stipulation: A 'specific scientific hypothesis' is one that has been directly investigated and confirmed by institutionally *bona fide* scientific activity prior to t or is one that might be investigated at or after t, in the absence of constraints resulting from engineering, physiological, or economic restrictions or their combination, as the primary object of attempted verification, falsification, or quantitative refinement, where this activity is part of an objective research project fundable by a *bona fide* scientific research funding body.

Stipulation: An 'objective research project' has the primary purpose of establishing objective facts about nature that would, if accepted on the basis of the project, be expected to continue to be accepted by inquirers aiming to maximize their stock of true beliefs, notwithstanding shifts in the inquirers' practical, commercial, or ideological preferences.

The PNC as thus formulated has so far been motivated only in a provisional way. At many points in the argument to come, we will flag further considerations that we take to enhance its justification.

1.4 THE PRIMACY OF PHYSICS

There is a methodological rule observed in the history of recent science to the effect that practitioners of special sciences at any time are discouraged from suggesting generalizations or causal relationships that violate the broad consensus in physics at that time, while physicists need not worry reciprocally about coherence with the state of the special sciences. We call this (so far roughly formulated) rule the 'Primacy of Physics Constraint' or PPC. Here we consider two questions: (1) How does the PPC relate to the more standard commitments of avowed 'physicalists'? (2) Why is the PPC observed in science? We conclude this section with a more precise formulation of the PCC.

Physicalism is usually defined as the view that everything is 'in some sense' physical, or sometimes that everything supervenes on the physical. Papineau (2001, 3) says that 'everything is physically constituted'. Hellman and Thompson's Ontological Physicalism is the view that everything is 'exhausted' by mathematical-physical entities (1975, 553–4). Another possibility, also canvassed in Hellman and Thompson (557), is to interpret the core notion of physicalism as that of 'one realm of facts *determining* another' (our italics), where causal priority, sufficiency, or necessity would be among the possible kinds of determination. We deny all these claims. We also deny the local supervenience of the mental on the physical, the token identity of mental states and physical states, the existence of a hierarchy of 'levels of reality', and the claim that all causation is physical causation.[37] Nonetheless, as Kim says, there is no consensus about how physicalism is to be formulated (2005, 33), and the PPC is compatible with some weak definitions of physicalism such as this one also due to Hellman and Thompson:

> Mathematical physics, as the most basic and comprehensive of the sciences, occupies a special position with respect to the over-all scientific framework. In its loosest sense physicalism is a recognition of this special position. (Hellman and Thompson 1975, 551)

Certainly, the PPC provides no comfort to dualists or emergentists. Physicalism, in common with the PPC, is committed to a generic *asymmetry*: special sciences do not relate to physics the way that it relates to them.

The view articulated and defended here clearly accords physics a special status. The Principle of Naturalistic Closure stated above requires that for a metaphysical claim to be taken seriously it must relate to at least one specific scientific hypothesis of fundamental physics. The notion of a real pattern, central to our approach to the special sciences, laid out in Chapter 4 requires projectibility by a physically possible device. The fact that we take seriously the notion that there are special sciences at all, that is sciences which are of restricted scope compared to physics, also shows that physics is more than one science among many for us.

We argue below that science—its current state and its history—supports the primacy of physics and physicalism in the loosest sense. Physicalism is generally regarded, at least by most physicalists, as a naturalist position that is motivated by science.[38] Yet, with a few recent exceptions (for example, Papineau 2001, Melnyk 2003), physicalists rarely offer direct arguments for physicalism using premises drawn from science itself. The debates in which physicalists do engage, including defending physicalism by dealing with various objections to it, are striking for the near total absence of reference to current scientific theories or results. Much of the contemporary debate over physicalism concerns variations on the knowledge argument (paradigmatically concerning what Mary, the colour-perception-deprived yet cognitively omnipotent colour scientist, could

---

[37] There is much debate about how to define the notion of 'physical' in this context following Hempel's dilemma (1965). Ontological physicalism is (almost certainly) false if it refers to existing physics for its ontology, for it can safely be assumed that present-day physics will be superseded by a more advanced physics of the future that will posit an ontology different in at least some respects from that of the former. But it is also (certainly) trivial (today) if it defines the physical as that which will be posited by a future, completed fundamental physics. Wilson (2006), following Kim (1996), Spurrett and Papineau (1999), Papineau (2001, 12), Crook and Gillett (2001), and Loewer (2001, 40), argues that physical entities ought to be characterized as those that are treated by fundamental physics and that are not fundamentally mental. This approach severs the link between physics and physicalism. Other recent discussions of how to formulate physicalism include Dowell (2006), Markosian (2000), Montero (2006), and Ney (forthcoming a). We do not concern ourselves with this problem because we do not defend ontological physicalism.

[38] One exception is Jackson (1998) who seems to regard physicalism as true a priori.

or could not know about colour (Jackson 1986)) and reflections on the putative possibility of zombies, inverted spectra, and other exotica utterly unrepresented in the literature of cognitive psychology and cognitive neuroscience. Earlier, one would more likely find debate over physicalism expressed through discussion of worries over epiphenomenal ectoplasm (Horgan 1982), or worlds physically differing from our own only in the position of one ammonium molecule, but at which there are no mental properties at all (Kim 1993). A striking feature of these debates in at least some versions of each of them is that no facts accessible to (third-person) science bear on whether the scenarios in question are actual or not: there are no epiphenomenal ectoplasm detectors, zombies are identical to us as far as any third-person investigation can tell, and worlds at which there are no mental properties at all pose the same problem on a larger scale. This dislocation from what could be discovered empirically is odd in debate over a position ostensibly motivated by naturalism.

Since at least the 1970s a strong tendency in much debate over physicalism has been to argue that epistemological commitments, especially to some form of theory reduction, are unsupportable and inessential to physicalism. Physicalism is generally taken to express an ontological rather than an epistemological or methodological claim, but not universally: Oppenheim and Putnam's (1958) well-known account of a reductive form of physicalism involved explanation of all facts by reference to theories concerning the behaviour of elementary particles. (Note that they didn't call their position 'physicalism'.) Much more recently Kim has said that physicalism is a thesis that any 'phenomenon in the world can be physically explained if it can be explained at all' (2005, 149–50).

There is a tension between the goal of providing a naturalist defence of physicalism, and that of making physicalism an ontological thesis but not an epistemological one. A naturalist defence must use evidence, and such evidence will consist in a catalogue of explanatory—that is, epistemological—successes. In asserting the thesis that everything is physical the ontological physicalist singles out a particular science, physics, as having a special role in ontology. For example, Philip Pettit says the world 'contains just what a true complete physics would say it contains' (1993, 213). Thus understood physicalism is in tension with the naturalism that supposedly motivates all forms of physicalism. That is, a responsible naturalist who defers to science as it stands in matters of belief formation will find herself ontologically committed to all sorts of entities and properties that aren't straightforwardly physical, in the sense of being studied as such by physicists.[39] (Dupré 1993 uses this as the basis for a

---

[39] In referring to ontological commitment we are assuming the naturalist to be a scientific realist. Although it doesn't matter for the specific arguments advanced in this discussion, we later, especially in Chapter 2, defend a specific version of ontic structural realism rather than standard scientific realism.

naturalist rejection of ontological physicalism.) Consider, for example, markets, fixed action patterns, mating displays, episodic memories, evolutionarily stable strategies, and phonemes. Powerful explanations and successful predictions have been produced by sciences that aren't physics and which refer to such entities. They are good ammunition for an epistemic success argument in favour of naturalism and against ontological physicalism.

Hüttemann and Papineau (2005) suggest that there are two main forms of ontological physicalism. One, part–whole physicalism, holds that everything real is in some sense made out of or is exhausted by basic constituents that are themselves physical. The other, supervenience or levels physicalism, holds that the putatively non-physical is nonetheless dependent on the physical.

There have been prominent part–whole physicalists, including Oppenheim and Putnam (1958), and Pettit (1993).[40] Since major parts of the remainder of this book, especially Chapter 3, are a sustained argument against central presuppositions of part–whole physicalism, we'll say little about it here. As Hütteman and Papineau point out, using a simple example concerning a classical mechanical system comprising three sub-systems, the view that the properties and behaviour of macroscopic entities is asymmetrically determined by their micro-constituents and the laws governing micro-activity has little to recommend it compared to a view in which 'parts and wholes mutually determine each other'. As we see later, attention to contemporary physics makes matters much, worse for the part–whole physicalist.

For a significant part of its history, especially in the wake of Davidson's influential paper on 'Mental Events' (1970) ontological physicalism has been widely understood as the thesis that the (putatively) non-physical, including the mental, the biological, and so forth, *supervenes* on the physical. There has been much discussion of different formulations of the supervenience relation.[41] The general idea is that one set of (for example, mental) properties supervenes on another (for example, physical or biological) set if, roughly, something cannot change with respect to its supervening properties without undergoing some change with respect to its subvening (or base) properties. Hütteman and Papineau's distinction is not always drawn, and some accounts of the supervenience relationship simultaneously suggest part–whole physicalism. An example is Lewis's discussion of the dot-matrix picture, the global properties of which depend upon the specific arrangement of a grid of dots (Lewis 1986, 14).

There is another form of ontological physicalism, namely Andrew Melnyk's (2003) realization physicalism. Realization, according to Melnyk, is a relation

---

[40] We return to the former in the next section. Pettit's microphysicalism is the claim that '[T]here are microphysical entities that constitute everything and microphysical regularities govern everything' (1993, 214–16). As Ney (forthcoming a, 13) points out, and as we discuss in Chapter 3, this is refuted by quantum mechanics since entangled states are not constituted by the entities that enter into them. We also return to microphysicalism in 1.6.
[41] See McLaughlin and Bennett (2005).

between tokens, in which the tokening of one type meeting certain conditions (the realizer) guarantees the tokening of another, functional type. A functional token is *physically* realized if the 'associated condition' for its tokening is met in virtue of the distribution of physical tokens and the holding of physical laws. So physical realizationism is the view that:

(R) Every property instance is *either* an instance of a physical property *or* a physically realized instance of some functional property; every object is *either* an object of some physical object kind *or* a physically realized object of some functional object kind; every event is *either* an event of some physical event kind *or* a physically realized event of some functional event kind.

Melnyk's realizationism isn't an identity theory since the realization relation requires neither token nor type identities, and although it entails some forms of global supervenience, it isn't really a supervenience theory either. In the next section we discuss different versions of reductionism and indicate why we don't endorse Melnyk's physicalism, even though its sparse commitments and bracing naturalism make it broadly PNC-compatible.

We now explain why we accord physics a special status. Most of the evidence for the primacy of physics was discovered in the nineteenth and early twentieth centuries. The developments in question involved attempts to test for the presence or absence of candidate non-physical forces or other influences that might affect the chances of some physical facts, as well as more general extensions of physical theory. That non-physical forces were on the agenda is partly explicable by reference to empirical advances in the life sciences and chemistry that added to the list of explananda phenomena that mechanists were initially unable to explain. Newton's introduction of forces was emulated with enthusiasm, and specific forces were proposed in several areas, including forces of attraction and repulsion for electrostatics, magnetism, and the cohesion of bodies; forces of irritability and sensibility to account for perception; forces to explain fermentation, the origin of micro-organisms, and chemical bonding. Some physical forces *were* found. None of the non-physical ones were.

At the same time, scientists made progress in unifying physical forces, and the physical treatment of force, work, and energy. Key steps here included Faraday's research on electromagnetic induction which also showed the unity of apparently different sorts of electricity, whether electrostatic, induced, or from batteries; Joule's research on the quantitative equivalence between heat and mechanical work; and Helmholtz' work on deriving the principle of conservation of the sum of kinetic and potential energy from rational mechanics, and relating this principle to the work of Joule. For a time, chemistry was a striking counterexample to this trend. Although various chemical regularities had been discovered, there was no contender for an explanation of chemical bonding in terms of more fundamental physical processes, and the possibility that there were as yet unknown chemical forces was recognized by leading scientists. Broad referred to chemistry as the 'most plausible' candidate for an 'example of emergent behaviour' (Broad 1925: 65), and Mill had chemistry in mind in his earlier treatment of composition of forces, which was a key source for British emergentism. However, following a series of advances by Thomson, Rutherford, and others, Bohr successfully constructed, first, a dynamical model of the hydrogen atom, then of heavier atoms, and finally aspects of the structure of the periodic table (Pais 1991, 146–52). A key measure of his success was deriving the hitherto descriptive Balmer formula for the emission spectra of hydrogen and some other simple elements from his model. A physical theory of chemical bonding had been developed, and while it did not apply readily to all molecules, or indeed all atoms, and certainly did not herald the theoretical reduction of chemistry to physics, it did dispose of the view that chemical phenomena involved distinct non-physical forces or forms of influence.[42]

This history has been widely taken to support two complementary arguments for the primacy of physics. The first argument is inductive: in the history of science a succession of specific hypotheses to the effect that irreducibly non-physical entities and processes fix the chances of physical outcomes have failed.

The second argument is also inductive. Over the history of science a succession of processes in living systems, and in the parts of some living systems dedicated to cognition, have come to be largely or entirely understood in physical terms, by which we mean in terms of the same quantities and laws as are invoked in physical theorizing about non-living systems. For example, the electrochemical functioning of neurons is understood partly in the language of insulators, resistors, and charge and density gradients, as is the operation of the ion pumps and ATP transport molecules that build the charge gradients, and supply the (physical) energy for the operation of the pumps. So it is not merely that anti-primacy-of-physics hypotheses have been rejected in the history of science, but that specifically physical hypotheses and explanations *have* been successful in their place.

Over this history physical theory itself has been unified and extended, for example in developing conservation principles, partly through the same experimental work concerning the conversion of different sorts of energy that supports the previous inductive arguments. Subsequently, for example, physical forces previously viewed as independent (electromagnetism and weak nuclear forces) have been given unified treatment in the Standard Model of particle physics, according to which the same electroweak interaction manifests in two different ways at low energies as a result of the photon having no rest mass. Consolidations and unifications of this sort are part of the reason for supposing that there is a coherent body of fundamental physical theory of sufficient scope and power that it is the only candidate for the 'most basic and comprehensive of the

---

[42] For arguments against reductionism about chemistry see Scerri and McIntyre (1997).

sciences'. In consequence, fundamental physics is regarded as having primacy over the rest of physics.

Most philosophers who consider themselves naturalists (Dupré and Cartwright being exceptions) regard what has just been said as uncontroversial. However, there is a frequent tendency to go on to use the primacy of fundamental physics as if classical physics is still the approximate content of fundamental physics. This, we contend, is the basic source of the widespread confusion of naturalism with the kind of ontological physicalism we reject. Classical physics was (at least in philosophers' simplifications) a physics of objects, collisions, and forces. When 'fundamental' physics is interpreted in these terms, as an account of the smallest constituents of matter and their interactions, it seems reasonable to many to think that everything decomposes into these constituents and that all causal relations among macroscopic entities are closed under descriptions of their interactions. In other words, they think that the history sketched above supports ontological physicalism. However, this is the philosophy of A-level chemistry at work. In Chapters 2 and 3 we will argue that contemporary physics motivates a metaphysics of *ontic structural realism* (Ladyman 1998, French and Ladyman 2003a and 2003b). This will yield an interpretation of fundamental physics that has nothing to do with putative tiny objects or their collisions. Our endorsing the primacy of fundamental physics should be read in light of this. To anticipate, fundamental physics for us denotes a set of mathematically specified structures without self-individuating objects, where any measurement taken anywhere in the universe is in part measurement of these structures. The elements of fundamental physics are not basic proper parts of all, or indeed of any, objects. (Nor is there any motivation for supposing that the fundamental structures describe gunk.) The primacy of fundamental physics as we intend it does not suggest ontological physicalism.[43]

We now give the more precise formulation of the PPC:

Special science hypotheses that conflict with fundamental physics, or such consensus as there is in fundamental physics, should be rejected for that reason alone. Fundamental physical hypotheses are not symmetrically hostage to the conclusions of the special sciences.

This, we claim, is a regulative principle in current science, and it should be respected by naturalistic metaphysicians. The first, descriptive, claim is reason for the second, normative, one.

---

[43] Note that the standard history is also taken to support the causal closure of the physical world or the completeness of physics (Papineau 1993, Spurrett and Papineau 1999). This principle is not entailed by the PPC since one might deny that there are causes in physics (perhaps persuaded by Russell's 1913 arguments, see 5.1–5.4), and so deny the causal closure of the physical while still defending the PPC.

Precisely because the PPC holds in science, one finds few direct invocations of it in scientific literature, because it would likely have been applied as a filter on the road to publication, rather than fought over in public. Nonetheless, there are places where one would expect to find it stated, including in textbooks where methodological precepts are passed on, and by those advocating radical or new ideas who are concerned that they not be mistaken for proponents of crazy ideas. We close this section with a few such instances.

Theories of agency and volition are especially likely to raise suspicion of dualism or other breaches of physical primacy. Near the opening of his (2001) account of the psychology of preference equivocation, addiction, and compulsion, Ainslie takes care to warn his readers that he requires that 'all explanations of behaviour should at least be consistent with what is known in the physical and biological sciences' (2001, 11). Kauffman's popular account of complexity theory argues that key phenomena associated with life are 'emergent'. He is careful, though, to emphasize that he does not mean anything 'mystical' by this, and explains that no 'vital force or extra substance is present in the emergent' system (1995, 24). Similarly, Stephanie Forrest (1991, 1) refers to emergent computing as having as its object 'computational models in which the behaviour of the entire system is in some sense more than the sum of its parts' and that in these systems 'interesting global behaviour *emerges* from many local interactions'. But then, in case this smacks of downward causation, she adds that 'the concept of emergent computation cannot contribute magical computational properties' (1991, 3). We suggest that the use of words like 'mystical' and 'magical' here is considered, and that it reflects the fact that rejection of the PPC amounts not merely to doing bad science, but to not doing science at all.

1.5 UNITY OF SCIENCE AND REDUCTIONISM

We have said that the *raison d'être* of a useful metaphysics is to show how the separately developed and justified pieces of science (at a given time) can be fitted together to compose a unified world-view. By 'pieces of science' we refer to any and all propositions[44] that motivate metaphysical reflections according to the PNC, including theoretical propositions of any degree of generality,[45] and representations of observations and experimental manipulations.

In philosophy of science, one finds two general strategies for unification. One is representational. It seeks formal or semantic entailment relations holding

---

[44] The reference to 'propositions' does not imply we think science is essentially linguistic. Other 'pieces of science' are embodied bits of procedural know-how among scientists that we assume can be modelled as representations of one sort or another if and as the need arises.

[45] Some scientific propositions will be sufficiently general as themselves to be metaphysical. Our notion of metaphysics is thus recursive, and requires no attempt to identify a boundary between metaphysical and scientific propositions.

amongst unifying propositions and the propositions they unify. Kitcher (for example, 1981, 1993) and Thagard (1992) are leading post-positivist proponents of this strategy. The other strategy, more frequent among philosophers who seek metaphysical rather than just epistemological unity, rests on some version or other of (usually ontological physicalist) reductionism. If there were one thing or some unified class of things that all of science could, via reductions between the parts of science, be supposed to be ultimately about, then science would be unified. Oppenheim and Putnam's (1958) famous argument that science was in the process of being unified was reductionistic and the particular form of reductionism they envisaged bottomed out at elementary physical particles.

We begin our discussion of the Unity of Science with this paper, even though no contemporary philosopher endorses much of what it says, because Oppenheim and Putnam's main premises are drawn from empirical science and it is therefore an exemplar of PNC-compatible philosophy.

For Oppenheim and Putnam the Unity of Science is a 'working hypothesis' which they argue commands more respect than the alternatives. Their discussion focuses on the Unity of Science, first, as 'an ideal state of science' and, second, to 'a pervasive *trend* within science, seeking the attainment of that ideal' (1958, 4). The specific notion of reduction that Oppenheim and Putnam use is developed from an account given by Kemeny and Oppenheim (1956). This so-called 'micro-reduction' is intended to provide a formulation of reductionism without commitment to the earlier (pre-1961, for example, Nagel 1949) programme of Nagel and others based on bridge principles or 'co-ordinating definitions'. The further developed version used by Oppenheim and Putnam says that, given two theories T_1 and T_2, T_2 can be said to be reduced to T_1 iff:

(1) The vocabulary of T_2 contains terms not in the vocabulary of T_1.
(2) Any observational data explainable by T_2 are explainable by T_1.
(3) T_1 is at least as well systematized as T_2. (1958, 5)

This rather weak construal of a reduction relation between theories (T) is applied by extension to *branches* of science (B) and the key requirement for being a micro-reduction here is that 'the branch B_1 deals with the parts of the objects dealt with by B_2' (1958, 6), which presupposes both a universe of discourse for each branch, and the part/whole notion (Pt).[46] Micro-reduction requires decomposition of entities (or objects in the universe of discourse) of B_2 into proper parts all within the universe of discourse of B_1. The micro-reduction relation is transitive, so micro-reductions could be cumulative. Oppenheim and Putnam note this and point out that two additional properties of micro-reductions (irreflexivity and asymmetry) can be derived from transitivity given only 'the (certainly true) empirical assumption that there does not exist an infinite descending chain of proper parts, i.e., a series of things $x_1, x_2, x_3 \ldots$ such that x_2 is a proper part of

---

[46] See also Hempel and Oppenheim (1953).

x_1, x_3 is a proper part of x_2, etc.' (1958, 7). (Note the assumption that atomism is certainly true.)

Oppenheim and Putnam construct a theory of the relationships between fundamental physics and the special sciences, and among the special sciences, by reference to this mereology. They reason that 'there must be several levels' of micro-reduction, and that 'scientific laws which apply to all things of a given level and to all combinations of those things also apply to all things of higher level. Thus the physicist, when he speaks about 'all physical objects,' is also speaking about living things—but not qua living things' (1958, 8). They also argue that induction on the recent history of science as of their writing shows a trend towards unification by micro-reduction across a cascade of six levels, which they identify as: (6) social groups; (5) multi-cellular living things; (4) cells; (3) molecules; (2) atoms; (1) elementary particles. The empirical case is not made by any careful demonstrations of inter-theoretic reductions satisfying the three criteria above. Rather, they claim that science 'directly' supports their proposed hierarchy of levels through a track record of successful decompositions of higher-level types into adjacent lower-level ones and syntheses of lower-level types into higher-level ones. They also claim 'indirect' empirical support from the fact that, according to them, science has consistently shown types at lower levels to be prior, ontogenetically and genealogically, to the types they compose.

A hodgepodge of considerations is responsible for the almost universal rejection of the package of theses that Oppenheim and Putnam promote. The view that science progresses mainly by reduction, or even the idea that this is an important question, has been fiercely contested since the work of Feyerabend and Kuhn. The atomism of Oppenheim and Putnam is now rejected by many philosophers. Oppenheim and Putnam clearly suppose that relations amongst constituents of compound systems are generally additive. Thanks to the aggressive championing of complexity theory in almost all disciplines (but especially in the social, behavioural, and biological sciences) by the Santa Fe Institute and its friends, this hunch has been displaced by a widespread emphasis on 'emergence' and inter-level feedback loops. Hence in many respects the inappropriateness of Oppenheim and Putnam's most crude background assumptions is not even controversial nowadays. Our most important reasons for rejecting the Oppenheim and Putnam programme are that we deny their atomism (see especially Chapter 3 below), and also maintain that Oppenheim and Putnam's supposition that two sciences (or 'branches' of science) could explain *the same* 'observational data' takes for granted a denial of what we call the scale relativity of ontology (see Chapter 4 below).

There are other sorts of reductionism worth distinguishing. Let 'micro-reductionism' denote any form of reductionism that includes a decomposition condition like one found in Oppenheim and Putnam. 'Nagelian reductions', on the other hand, are either reductions as envisaged in Nagel's own flagship account (Chapter 11 of Nagel 1961), or in later views developed as refinements or elaborations of Nagel's model. For Nagel, reduction is the (deductive) explanation of a theory by another theory. Nagel pointed out that typically, in real cases, in order for reduction to be effected 'additional assumptions' are needed. For example in his famous discussion of a reduction of a single law of thermodynamics (the Boyle–Charles law for ideal gasses, stating $pV = kT$ where p is pressure, V volume, k a constant, and T the temperature of a volume of gas) to statistical mechanics, Nagel argues that these additional assumptions include: supposing a sample of gas to consist in a large number of molecules; arbitrarily dividing the volume into many sub-volumes, and assigning independent probabilities of position and momentum in the sub-volumes to the molecules; treating the molecules as elastic spheres colliding with each other and the container; and treating collisions with the container as perfectly elastic. Nagel also proposed stipulations connecting the vocabularies of the two theories. For example he defines pressure as 'the average of the instantaneous momenta transferred from the molecules to the walls'). He then argued that it is 'possible to deduce that the pressure p is related in a very definite way to the mean kinetic energy E of the molecules, and that in fact $p = 2E/3V$, or $pV = 2E/3$' (Nagel 1961, 344). The latter equation shares 'pV' with the Boyle–Charles law, suggesting that 'the law could be deduced from the assumptions mentioned *if* the temperature was in some way related to the mean kinetic energy of the molecular motions'. Given all this, Nagel suggests that we could 'introduce the postulate' that $2E/3 = kT$ (Nagel 1961, 344–5).

As Marras (2005, 342) points out, the postulate that temperature is 'in some way related' to mean kinetic energy of molecules is a final step in the process of reduction, justified by preceding work showing that an *analogue*, or what some call an 'image' (Beckermann 1992; Bickle 1992, 1998), of one theory can be constructed in a supplemented version of another. Nagel rejects the suggestion that the additional assumptions are analytic—'no standard exposition of the kinetic theory of gasses pretends to establish the postulate by analyzing the meaning of the terms occurring in it' (1961, 355)—and says that there is no simple fact of the matter, independent of specifying the pragmatic context in which the question is raised, about whether instead the assumptions are conventions or empirical. Nagel argues against the 'unwitting doubletalk' of treating the terms related by additional assumptions as having been made 'identical by definition', because doing so would involve changing the meanings of the terms in one or both theories, so not reducing one original theory to the other (Nagel 1961, 357). So, for example, temperature is not a statistical property while mean kinetic energy is, and there are theoretically distinct ways of determining the mean within statistical mechanics (Feyerabend 1962; Nickles 1973, 193; Brittan 1970, 452; Yi 2003). Nagel allowed that the additional postulates often were empirical hypotheses 'asserting that the occurrence of the state of affairs signaled by a certain theoretical expression ... in the primary science is a sufficient (or necessary and sufficient) condition' for the occurrence of the corresponding state of affairs designated in some way by the 'secondary' science (Nagel 1961, 354).

Clearly, Nagelian reductions need not be micro-reductions. The conditions for each type of reduction are distinct (though compatible), and, part of the point of micro-reduction is to dispense with the 'bridge laws' requirement found in earlier proposals by Nagel. Conversely, there is no requirement in Nagel that the reducing science concern itself with the proper parts of the things studied by the reduced science. This is especially clear in the case of more 'homogenous' reductions (where the theories share 'descriptive terms') such as the derivation of some of Galileo's laws, or Kepler's, from Newton's, or those of Newton from the Special Theory of Relativity, given suitable additional assumptions and restrictions in each case.[47] Nobody supposes that these are micro-reductions.

Some purported refinements of Nagel's programme claim to 'remove' features that, as Marras (2005) points out, aren't present in the first place, such as refinements that downplay the importance of bridge laws (for example, Bickle 1998). It is important to note that if the multiple realization argument is correctly understood as aimed at the bridge laws requirement for theoretic reduction, and it is also the case that Nagelian reduction (refined or raw) doesn't require bridge laws, then the multiple realization argument may fail to undermine it. This is certainly the view of 'new wave' reductionists (for example, Bickle 1996).

Oppenheim and Putnam placed a bet on the direction of science. We bet differently. By contrast, Nagelian reduction makes no predictions about the direction of science. We reject micro-reductionism but not Nagelian reductionism, because we think that there are real examples of Nagelian reductions (though not of caricatures of Nagelian reductions involving bridge laws) that are significant contributions to science, and steps toward unification. There are, however, other arguments to the effect that Nagelian reductions are rare or non-existent and in some cases impossible.[48] This brings us to type reductionism.

Type reductionism (sometimes confused with Nagelian reductionism) is our name for the view that is the critical target of several classic anti-reductionist papers including Davidson (1970) and Fodor (1974). Fodor's paper invites confusion with its title—'Special sciences, or the disunity of science as a working hypothesis'—which suggests it is aimed against Oppenheim and Putnam's (1958)

---

[47] None of these reductions are perfect. Some need assumptions to derive one law that are incompatible with those used to derive another, or require assumptions which are known to be false (for example, that changing distance makes no difference to rate of fall, in order to 'derive' Galileo's law of fall from Newton's laws). See Sklar (1967), Brittan (1970), Nickles (1973), Batterman (2002, 18).

[48] Bickle (1992, 218) notes that Davidson's argument for the anomalism of the mental was also supposed to show the 'conceptual impossibility of reducing intentional psychology to a lower-level science'.

'Unity of Science as a working hypothesis'. Fodor's argument attacks bridge laws, which as noted above, are not involved in Oppenheim and Putnam's micro-reductions, and his argument ignores issues around mereological composition and atomism, which are so involved. In the discussions of Fodor (and many others in the voluminous literature that followed him) nearly all of the scientific detail found in Nagel is absent, and replaced by toy examples.

Fodor discusses reductionism and the multiple realization argument against it in terms of (natural) kinds and laws. So, if there is a true physical law $P1 \to P2$, and a true special science law $S1 \to S2$, the bridge laws will be biconditionals linking P1 and S1, and P2 and S2. It is exactly these biconditional laws that the multiple realization argument is supposed to show are either unlikely or impossible, but this is not Fodor's main reason for rejecting them. Fodor claims, as is supposedly 'obvious to the point of self-certification', that:

(a) interesting generalizations (e.g. counterfactual supporting generalizations) can often be made about events whose physical descriptions have nothing in common; (b) it is often the case that whether the physical descriptions of the events subsumed by such generalizations have anything in common is, in an obvious sense, entirely irrelevant to the truth of the generalizations, or to their interestingness, or to their degree of confirmation, or, indeed, to any of their epistemologically important properties; and (c) the special sciences are very much in the business of formulating generalizations of this kind. (Fodor 1974, 84)

This is over-stated. Physical descriptions of the tokens of at least many special science types often have a great deal (and certainly far more than 'nothing') in common. Questions of physical similarity aren't irrelevant to, for example, whether two animals are both vertebrates, or whether two different samples of sediment are clays or oozes. Indeed, many special science generalizations are directly about physical properties—consider the reported correlations between testicle size and mating strategy (for example, Møller 1994), or encephalization and group size in social primates (for example, Dunbar 1992), or the varying attractiveness to human females of the scents of more symmetrical men during ovulation (Thornhill and Gangestad 1999). Furthermore, even if Fodor's argument is devastating for the view that reduction can be achieved by biconditional bridge laws, it does no harm to prospects for Nagelian reductions.

What then does multiple realization show? Any real system or process features in a variety of different scientific generalizations, and multiple realization possibilities in one of them need not line up with those in others. A manufactured microphone in which vibrations move a metal coil positioned near a magnet, thereby inducing a current in a circuit, is one way of transducing sound into electricity. The hair cells in our ears, which bend in response to vibrations, twisting molecular gates at their bases, and releasing a cascade of processes involving ions at a synapse, is another. This is a credible instance of multiple realization. Like any, it has its limits: anyone can think of ways in which the two are not interchangeable because they don't both appear in all the same regularities.[49] As the criteria for instantiating a given regularity are made more demanding, or the number of regularities we seek to satisfy at once rises, the number of satisfiers will tend to drop. So a microphone makes a fine transducer, but it doesn't get well looked after by our kind of immune system, or get well built by our kind of diet. Contra Putnam, we couldn't be made of Swiss cheese and it does matter.[50]

This cross-classification of regularities at different scales is disastrous for the 'sundering' form of reduction defended by Kim (for example, 1998), in which relations of multiple realization require fragmenting special science kinds into as many sub-kinds (each a reductive identification of the sort we're resisting) as there are different realizers. Science does sometimes partition established kinds; for example, distinguishing different sorts of memory and separating whales from fish. But there is no good reason for supposing that this will generally happen when finding differences in some sense 'inside' the systems studied and plenty of reason for thinking the opposite. The fact that the relations between generalizations at different scales won't involve neat nesting means that no scale automatically gets to be the one against which others are fragmented. Good generalizations at any scale deserve the same scientific respect, consistent with the PPC.

Note finally that the functionalism promoted by Fodor (like Davidson's anomalism) is committed to the token identity of special science and physical events (or property instantiations, etc.). We've already indicated our endorsement of some Nagelian reasons for caution about endorsing identity claims between the referents of different theories. Later, in Chapter 4, we argue that ontology should be understood as scale-relative, presenting a further barrier to identifications. We note here that disputes over whether the identities are, or have to be, or can't be, event identities, thing identities, property identities, and so on, are all from our point of view distractions. On our structural realist metaphysic (see Chapters 2 and 3), neither things, nor properties, nor events turn out to be ontologically fundamental, meaning that by the lights of Fodor and others we hollow out (so to speak) the notion of a (natural) kind (see 5.6). Further, there is room in our view only for a very limited notion of material identity.

The upshot of all this is that while we are not type reductionists, and agree that the multiple realization argument tells against that doctrine, we are also not proponents of the sorts of non-reductive physicalism (involving token identity and/or local supervenience) paradigmatically associated with the rejection of type reductionism.

---

[49] The cross-classification of regularities at different scales being described here has sometimes been called 'multiple supervenience' (for example, Gasper 1992, Meyering 2000). The reasons for our wariness of endorsing even token identities prohibit our endorsing most supervenience theses, so we eschew that label. Our view (see Chapter 4) is very tolerant of cross-classifying regularities, though.
[50] 'We could be made of Swiss cheese and it wouldn't matter' (Putnam 1975b, 291).

Core-sense reductions are also sometimes called 'ontological reductions'.[51] Since the expression 'ontological reduction' is also used in other settings, we have adopted 'core-sense' reductionism from Melnyk (2003), who defines core-sense reductionism as follows:

(CR) All nomic special- and honorary-scientific facts, and all positive non-nomic special- and honorary-scientific facts, have an explanation that appeals only to (i) physical facts and (ii) necessary (i.e. non-contingent) truths. (Melnyk 2003, 83)

According to Melnyk, an 'honorary scientific fact' is a fact of an honorary science like folk psychology and folk physics. (Melnyk is not here defending philosophy of domesticated science, but trying to leave a door open for truths onto which the folk may have happened. See Melnyk 2003, 32n.) Nomic facts support counterfactuals, while non-nomic ones are to be understood as tokenings. Positive facts exclude absences and other negative facts. For the purposes of a reduction in the core sense, the physical facts can include non-nomic facts, including initial conditions. Finally the 'having' of the explanation is 'in principle' rather than in practice, although Melnyk thinks there is sufficient evidence (see Chapters 5 and 6 of Melnyk 2003) for the claim that there are indeed such explanations.

The basic idea behind CR is that science (Melnyk argues on the basis of many detailed real scientific examples) motivates realization physicalism (as defined in the previous section). Melnyk's realizationism explicitly lacks commitment to the identity of special science tokens with physical tokens. In fact Melnyk is careful to distinguish 'retentive' realizationism (where tokens of special science types are not ultimately ontologically replaced) from non-retentive forms, and argues that realization physicalism will turn out to be retentive just in case, and to the extent that, functional types realized by actual physical tokens turn out to be identical to special science types (Melnyk 2003, 32–48).

The fact that the explanations referred to in the definition of core-sense reductionism can include non-nomic facts, including initial conditions, shows that Melnyk's thesis is weaker than type reductionism. Because reduction relations need not track decomposition relations, it is also weaker than micro-reductionism (Melnyk 2003, 29–30). Furthermore, because it is a strictly empirical matter what material identities, if any, obtain between functional types (and tokens of those types) and their realizers, core-sense reductionism is metaphysically more modest even than some anti-reductionisms which include a general commitment to token identity (Melnyk 2003, 32–48).

It is still, though, not weak enough for us, and explaining why will further clarify our relationship to physicalism as promised in the previous section. We

[51] 'Physicalism as I have characterized it is a reductionist thesis. However, it is reductionist in an ontological sense, not as a thesis that all statements can be *translated* into statements about physical particles and so on' (Smart 1989, 81, his emphasis).

do not deny that core-sense reductionism may be true. It is PNC-compatibly motivated. But it involves a commitment stronger than the PPC. In cases where it isn't clear whether or not a special science is contradicting physics the metaphysician following the PPC keeps her peace, while the one persuaded of CR is committed to there being an explanation of the special science fact by physical facts and laws. In Chapter 3, we entertain a resolution to the measurement problem in quantum mechanics that involves denying that a measurement apparatus has a fundamental-physical description, from which the absence of a CR-style explanation follows. If this is right, it reflects a general fact concerning the relationship between fundamental-physical tokens and special-science tokens, namely that no special-science tokens will have fundamental-physical descriptions. If this is the case, and if the explanations expected by the advocate of CR require physical descriptions of special science tokens, then our view would turn out to be incompatible with CR. Melnyk doubts that commitment to this requirement is necessary for realization physicalism being truly a kind of *physicalism*; we're not sure about this. But at this point the question of whether, if we can be reconciled to CR, we can also be ontological physicalists of one sort after all, turns purely semantic.

We end our discussion of reductionism with a cautionary note from Nagel:

both successful and unsuccessful attempts at reduction have been occasions for comprehensive philosophical reinterpretations of the import and nature of physical science... These interpretations are in the main highly dubious because they are commonly undertaken with little appreciation for the conditions that must be fulfilled if a successful reduction is to be achieved. (Nagel 1961, 338)

1.6 FUNDAMENTAL AND OTHER LEVELS

We seek an ontological model according to which science is unifiable, and which explains the basis for such unity as it can produce. This, we claim, is the point of naturalistic metaphysics. At one level of description, the goal of unity of science will be approached through the incremental filling in of networks of Nagelian reductions—which, as we have just argued, carry no implications to the effect that we should expect micro-reductions or type reductions. On the contrary, the PNC leads us to scepticism about these kinds of putative reductions, for reasons to come in the next four chapters.

Nevertheless, enumeration of such Nagel-type inter-theoretic relations as have been justified at any given time don't exhaust what can be said about the unity of science. This is because Nagelian reductions are far from equally antecedently probable between any two scientific theories chosen randomly with respect to their domains. For example, there are interesting relationships between microeconomic generalizations and neuroscientific generalizations (Glimcher 2003, Ross 2005, ch. 8), but it is unlikely that there are interesting Nagelian relations between macroeconomic generalizations and chemical generalizations. (There may of course be lessons from one for the other about how to use mathematical tools.) These asymmetries in the relationships among the domains of the sciences look like features of the structure of the world. But we think that attempts to offer generalizations about these asymmetries in terms of micro-reduction and type reduction are forlorn, in the first case because, after a few promising starts, the attempts ceased to be supported by the course of science, and in the second case because the attempts never were supported by science. This book is an effort to justify generalizations about the wide-scope metaphysical structure of the world that does not rely on any of the intuitions underlying micro-reductionism or type reductionism.

Of special importance among these intuitions we will deny is that world comes in 'levels'. Contemporary science, we argue, gives no interesting content to this metaphor, and so a metaphysics built according to the PNC should not reflect it.[52] We are not the only philosophers who are lately denying widespread aspects of micro-reductionism. We here mention two philosophers in particular who share aspects of our anti-micro-reductionist metaphysics, partly just to orient the reader in the literature, but mainly to help explain our position by contrasting it with others that look at least superficially similar.

The first of these philosophers is Jonathan Schaffer. Schaffer (2003) argues specifically against the view he calls 'fundamentalism', that is, the thesis that science suggests the structure of the world to have a definite bottom level. Such a bottom level has been generally supposed to be physical, atomic, and provide a base for universal mereology and/or supervenience. This does duty in the formulation and development of arguments for a variety of related metaphysical theses:

[T]he *physicalist* claims that microphysical theory (or some future extension thereof) describes the fundamental level of reality on which all else supervenes; the *Humean* claims that all supervenes on the distribution of local, fundamental qualities in spacetime; the *epiphenomenalist* claims that all causal powers inhere at the fundamental level; and the *atomist* claims that there are no macroentities at all but only fundamental entities in various arrangements. (498)

Schaffer notes that 'the central connotation of the "levels" metaphor is that of (a) a *mereological structure*, ordered by the part–whole relation'. Then

The peripheral connotations of 'levels' include those of (b) a *supervenience structure*, ordered by asymmetric dependencies; (c) a *realization structure*, ordered by functional relations; and (d) a *nomological structure*, ordered by one-way bridge principles between

---

[52] Levels are sometimes supposed to be reflected in the division of science into disciplines, but note that, rather than a neat hierarchy, there is a good deal of overlap between the domains of, for example, chemistry and biology, or economics and psychology. John Heil (2003) also rejects the idea of levels of reality though not on naturalistic grounds.

families of lawfully interrelated properties. Those who speak of levels typically suppose that most if not all of these connotations comport. (ibid. 500)

Schaffer attacks the central connotation by arguing that current physics does not offer good evidence that mereological atomism is true. In Chapter 3 below, we will strengthen this PNC-compatible argument, providing evidence that according to current fundamental physics mereological atomism is false. At various points we will also give independent arguments, based on science, against (b) and (d).[53]

Schaffer then argues that even if mereological atomism is denied, physicalism (in his sense), Humeanism, epiphenomalism, and atomism all depend on the idea that there is a bottom level to nature, even if it isn't a bunch of atoms. (Perhaps, following Poland (1994), it's a set of physical attributes rather than objects.) Schaffer doubts there are sound scientific reasons for believing in such a level. In Chapter 3, we provide much more evidence for this doubt. Finally, Schaffer asks whether there might only be a lowest supervenience base, below which there are infinitely descending mereological divisions that introduce no new classifications. He adduces no scientific evidence against this picture. It, he maintains, would continue to exclude epiphenomenalism and atomism, but would readmit physicalism and Humeanism as possibilities.

We agree with Schaffer about what there isn't, and that it's science that establishes the various negative cases. On the other hand, he differs from us in accepting the levels metaphor itself. Because we deny that metaphor, when we wonder about 'fundamental' physics, we don't take ourselves to be asking about a putative physical 'bottom' to reality. Instead, as we explained in 1.4, by 'fundamental' physics we will refer to that part of physics about which measurements taken anywhere in the universe carry information.

The second partial ally we will mention here is Andreus Hüttemann, who argues against a view he calls 'microphysicalism' (2004). This is the view that properties of the components of processes and systems *determine*, *govern*, and *causally exclude* the macroproperties at the whole-process or whole-system level. Determination is a modal relation between some set of explanans and an explanandum. Microdetermination of process or system properties obtains if, given some particular set of microproperties, specified macroproperties obtain necessarily. General microdetermination follows if strong local supervenience holds in general. This does not imply that microdetermination is generally false if strong local supervenience is denied as holding in general; but denial of general strong local supervenience knocks out the main motivation for expecting general microdetermination to hold. Microgovernance holds if the laws generalizing micro-level relationships are

---

[53] We offer no direct arguments, as far as we can tell, against (c). Furthermore, Melnyk's (2003) version of ontological physicalism, discussed above, disassociates (c) from (a) and (b), but not (d).

sufficient conditions for all laws generalizing macro-relations but not vice versa. Microgovernance typically features as a premise in arguments for microcausal exclusion, the thesis referred to by Schaffer above as 'epiphenomenalism'.

Hüttemann rejects each of microdetermination, microgovernance and microcausal exclusion for specific reasons (and thus rejects the microphysicalist thesis they collectively constitute). We of course agree with his conclusion. However, an important premise in each of his three general arguments is that in cases of compound processes and systems, macroproperty relations make dispositions manifest, whereas dispositions of microcomponents are merely inferred or posited for purposes of explanation. This premise reflects a view Hüttemann shares with Nancy Cartwright (1989 and elsewhere) to the effect that what science aims to discover are neither Humean regularities nor modal structures of reality, but causal capacities (or 'Aristotelian natures', as Cartwright (1992) calls them) of types of things. This view is sharply at odds with the general one we will defend, and undergirds a thesis to the effect that both science and the world itself are strongly *dis*unified (and, in the case of science, not unifiable if one values wide scope of application). We will criticize different aspects of this thesis in later chapters. For now, we simply note Hüttemann's work as leading to an important part of our core conclusion, but by means of arguments that differ substantially from those we will give. Hüttemann, like Schaffer but unlike us, does not question the general adequacy of the levels metaphor. As a result, he now thinks (Papineau and Hüttemann 2005) that his arguments against part/whole microdetermination do not apply against microdetermination of higher levels by lower ones. His view is in this respect much more conservative than ours.

In denying 'levels physicalism' we will appear to convict ourselves, according to Papineau and Hüttemann, of endorsing what the latter (2004, 44–7) identifies as C. D. Broad's (1925) notion of emergence. One is an emergentist in Broad's sense, according to Hüttemann, just in case one holds that there is behaviour of some complex systems that is not determined by the behaviour of parts of the systems in question. Hüttemann denies that examples of chaotic systems and systems that undergo phase transitions, with which enthusiastic popular and Santa Fe Institute-inspired literature lately abounds, are instances of such emergence, since the inability of finite observers such as people to be able to predict macro-behaviour on the basis of micro-behaviour does not imply metaphysical failure of determination. We agree with him about this; our basis for denial of 'levels physicalism' in Chapter 3 will not be based on these sorts of cases. For this reason, we reject the suggestion that denial of 'levels physicalism' implies a doctrine that deserves to be called 'emergentism'. This doctrine warrants its name because it holds that 'higher' levels of organization 'emerge' indeterminably out of 'lower' level ones and then causally feed back 'downward'. Our position, denying that science suggests the world to be structured into levels at all, calls a pox on both houses in this dispute.[54] We and Hüttemann agree that quantum entanglement, a phenomenon that is important to our argument, implies failure of determination of the states of compounds from states of parts. (An analogous point can be made about General Relativity.) But here, because of his continuing allegiance to the levels metaphor, Hüttemann (2004, 52) is pleased to talk of 'emergence' whereas we never are.

What we aim to do in this book is displace the micro/macro distinction, whether conceived in terms of wholes and parts or in terms of higher and lower levels, by a new way of drawing the distinction between fundamental physics and special sciences. We carefully say 'displace' rather than 'replace'. We don't claim that our new distinction can do every philosophical job the old one did. Rather, we claim that many of the jobs to which the old distinction has been put are not worth doing, or are even destructive. We take on one traditional job—furnishing a basis for the synchronic and diachronic unity of science—as the central one, and argue that our distinction is better for it. Since we insist that, for naturalists, that is the only self-ratifying job there is for metaphysics, all other chips on the philosopher's table must then be allowed to fall as they will over the course of the inquiry.

1.7 STANCES, NORMS, AND DOCTRINES

Schaffer and Hüttemann may not go as far as we do in denying 'the hierarchical picture of the world as stratified into levels' (Schaffer 2003, 498). However, we take some heart from the fact that as we set out to try to convince readers that the familiar picture needs drastic revision in order to keep abreast of science, at least a few other philosophers have taken steps down the same path. We turn to another fellow inquirer, and disturber of philosophers' peace of mind, whose critical attitude we seek to emulate in a broader sense. We have said harsh things about much—indeed, by implication, most—current metaphysics in this chapter. We see no point in mincing words: it seems to us to be just ridiculous when philosophers look up from their desks and tell us that while sitting there and concentrating they've discovered (usually all by themselves) facts about the nature of the world that compete with the fruits of ingenious experimentation conducted under competitive pressure and organized by complex institutional processes. The individual philosophers are generally not crazy; but quirks in the history and structure of the modern academy have encouraged crazy activity and hidden its absurdity. A prominent philosopher of

---

[54] That said, let us note that the respective rhetorical standings of the two ways of expressing muddled philosophy differ, just as a result of cultural history. When a scientist refers to levels of reality she is merely declining to get involved in certain philosophical disputes. When someone pronounces for downward causation they are in opposition to science.

science who has recently said the same thing, though more graciously, is Bas van Fraassen (2002).[55] He reminds us of the vastness of the gulf between metaphysics and science that is concealed by common assertoric and citation-studded formal presentation.

For the scientist, every mistake she makes risks being exposed as clearly a mistake, either by someone's experiment or by mathematical demonstration. She knows this as she works. Knowing also that she is sure to make mistakes, she simultaneously reduces her responsibility and establishes an early-warning system against error by joining a team. (Bear in mind that she mainly joins a team for other reasons: science requires specialization of labour in experimental design, control, and analysis, and lone individuals will seldom be entrusted with the resources necessary for isolating and interrogating important parts of the world.) Very important here, as van Fraassen stresses, is that the scientist is not a monomaniac with respect to goals. She seeks truth, to some extent; but all that should be meant by this is that she must be careful never to report, as sincerely held beliefs about her domain, claims that she thinks are merely hunches; and that she must exercise responsibility in deciding when some approximation or idealization is too instrumental to be announced with a po face. (Again, her subordination of her judgement to that of her team, not individual moral heroism, does almost all the work here in normal circumstances.) In any event, it in no way diminishes the idea that science is our society's primary source of truth to remind ourselves that scientists don't treat truth as their proximate target. What they mainly aim for is that their contributions be *important*, to policy or commerce or practical know-how. Importance may be a highly inexact and multifariously realizable goal. But scientists generally know when they've achieved it one way or another and when they haven't; citation indices actually measure it quite reliably.

By contrast, van Fraassen points out, in metaphysics there is no goal but true belief itself. It is problematic, to say the least, that the metaphysician has no test for the truth of her beliefs except that other metaphysicians can't think of obviously superior alternative beliefs. (They can always think of *possibly* superior ones, in profusion.) In fact, every metaphysician not in the grip of bad faith should know that her favourite professional opinions are almost certainly not true; this is awkward when truth is all she has to work for.

The key consequence of this difference that van Fraassen emphasizes is as follows. When we discover that scientific theories are not exactly true, this should occasion no deep angst or regret; scientific theories have multiple possible sources of value which they *had to* demonstrate to have ever attracted significant investment in the first place. It is much less obvious that a false metaphysical theory is good for anything. But since all metaphysical theories are false ...

---

[55] All references to van Fraassen in the rest of this section are to his (2002).

Note that the metaphysician is the last person who can hide behind an instrumentalism of good works ('well, it's false, but it comforted us all for awhile') here. Van Fraassen makes the point vivid:

> Consider the real history of Newton's physics, compared to what might have been the history of Cartesian dualism. Newton's physics reigned dominant for two hundred years. It gave us false beliefs but many benefits. I don't think anyone will say 'It would have been better if Newton had never lived!' Imagine that Cartesian dualism had not been so conclusively rejected by the late seventeenth century but had also reigned for two hundred years. Would we say that the false beliefs that metaphysics gave us had been but a small price to pay for the ease and intuitive appeal felt in its explanation of the human condition? (16–17)

On top of all this, van Fraassen demonstrates the special discomfort that must be faced by the metaphysician who aims at naturalism.[56] A non-naturalistic metaphysician might be able to sincerely think that her intuitive insights certify their own truth (typically because, in part, there is a kind of 'truth' that she thinks transcends the grubby everyday sort we hope to get from science). But among the views that typically incline a philosopher to naturalism is the conviction that there are no secure foundations for knowledge. Van Fraassen associates this problem especially with empiricism, but even those philosophers who call themselves realists agree, these days, that even if they can know a priori that 'tiger' denotes all and only tigers in all possible worlds, the part of knowledge, even philosophical knowledge, that really matters is based on fallible processes.

In fact, no naturalistic philosopher in good faith can now deny that for any of the well-established philosophical 'isms' compatible with naturalism, the simple *truth* of that ism cannot plausibly be asserted. Van Fraassen puts it thus. 'For each philosophical position X there exists a statement X+ such that to have (or take) position X is to believe (or decide to believe) that X+.' Call X+ 'the dogma or doctrine of position X'. Then the naturalistic philosopher who generally conducts her specific philosophical exercises by reasoning in accord with the principles of position X cannot take X+ to be literally true, for each X, unless she is in bad faith (40–6).[57] Therefore, if it is possible to do philosophy in good faith, 'a philosophical position need not consist in holding a dogma or doctrine' (46).

If philosophical positions should not be doctrines, what can they be instead? Van Fraassen suggests that we conceive of them as 'stances':

> A philosophical position can consist in something other than a belief in what the world is like. We can, for example, take the empiricist's attitude toward science rather than his or her beliefs about it as the more crucial characteristic... A philosophical position can consist in a stance (attitude, commitment, approach, a cluster of such—possibly

---

[56] Van Fraassen uses 'naturalism' and 'materialism' interchangeably. See below for critical discussion of his reasons.

[57] We have put the point in somewhat stronger terms than van Fraassen does. We're confident that he'd endorse this way of putting it, however.

including some propositional attitudes such as beliefs as well). Such a stance can of course be expressed ... but cannot be simply equated with having beliefs or making assertions about what there is. (47–8)

One fully grasps the idea of a stance only after being given an example or two, but first note that to take the view that a sound philosophical position is a stance rather than a doctrine is to adopt an anti-metaphysical attitude in one sense. Let us say that someone endorses 'strong metaphysics' if she assents to the claim that it is (practically) possible for a person to set out to cultivate non-trivial doctrinal beliefs about the structure of the world that go beyond what the sciences tell us or imply, and then come to have a preponderance of true such beliefs relative to false such beliefs *because of* the activity of cultivating them. (We formulate things this way so as not to be troubled by 'broken clock' phenomena; perhaps it's probable that there are some people out there whose metaphysical beliefs are mainly true by sheer good luck, merely because there are so many people.) Then the considerations raised by van Fraassen that we have been reviewing should lead us to be sceptical about the value of strong metaphysics. (The reader will surmise that we are working up to the announcement of some less blighted programme for 'weak metaphysics'.) To agree with van Fraassen that philosophical positions ought to be stances rather than doctrines is to suppose that there is value in some philosophy, somehow conceived, but not in strong metaphysics.

We endorse van Fraassen's opinion that sound (general) philosophical positions should be stances. This at once raises the question for us, what is our stance? And then we add that our project here is to be an exercise in naturalistic metaphysics. How do we reconcile this ambition with the scepticism about strong metaphysics we have just said we share with van Fraassen?

The answer to the second question will follow swiftly from our answer to the first one. We will approach it by first reviewing the two explicit descriptions of rival stances, the 'empiricist stance' and the 'materialist stance' that van Fraassen provides. We will identify our own stance by reference to these. In this procedure we follow van Fraassen's meta-advice:

Besides the theses on which the day's materialists take their stand, and which vary with time, there is also such a thing as 'the spirit of materialism' which never dies. False consciousness can be avoided in two ways:

(1) the philosopher may lack that spirit and be genuinely concerned solely with certain definite factual questions about what there is.

or

(2) the philosopher may have the spirit and not confuse its expression with any particular view of what the world is like.

The latter, however, may never yet have been instantiated among philosophers.

Nevertheless, the second option is the really interesting one and similar to the one I would favour for any attempt to continue the empiricist tradition. The problem for materialists will then be to identify the true materialist stance and for the empiricist to identify the true empiricist stance (or the spectrum of true empiricist stances). Being or becoming an empiricist will then be similar or analogous to conversion to a cause, a religion, an ideology, to capitalism or to socialism, to a worldview ... (60–1)

It will be obvious from the above that van Fraassen does not claim to have himself fully worked out what these stances are. But it seems that the idea of a fully worked out stance is a will-o-the-wisp anyway; one begins by approximately targeting the stances and from there follows the never-to-be-completed activity of elaborating—by *discovering*—their details. So let us begin with van Fraassen's approximate targets.

Philosophers who adopt the empiricist stance, according to van Fraassen, emphasize the following attitudes. First, they hold that philosophy should emulate the norms of anti-dogmatism characteristic of science. No hypothesis should be ruled out of consideration come what may. (This, as van Fraassen argues, is why empiricists can't assert as their doctrine that no knowledge can arise from pure intuition or that all knowledge arises from observation; the doctrine refutes itself as *doctrinal*, and the empirical *stance* rules out the coherence of an empiricist dogma.) As in science, so in philosophy every proposal should be considered as seriously as the evidence and arguments for it warrant. This is the positive part of the empiricist stance. The negative part is a more specific sort of resistance to metaphysics than the generic anti-metaphysical attitude described above that favours stances over doctrines to begin with. This resistance manifests itself in tendencies for:

(a) ... rejection ... of demands for explanation at certain crucial points

And

(b) ... strong dissatisfaction with explanations (even if called for) that proceed by postulation (37).

Van Fraassen reviews the main historical philosophers who have been regarded as exemplary empiricists, or who have been regarded as moving philosophy in empiricist directions relative to the state in which they found it (for example, Aristotle after Plato), and argues that we find nothing general in common among them at the level of doctrine; all, however, can be found 'rebelling' against 'overly' metaphysical tendencies by reference to the elements of the empiricist stance.

The strong dissatisfaction with neo-scholasticism we expressed in 1.1 and 1.2 are evidently manifestations of empiricist rebellion. We admire science to the point of frank scientism. As with van Fraassen's leading empiricist examples, what most impresses us about science is not its results—marvellous though these have been—but the way in which its institutional organization selects for rationality and collective epistemic progress in the activities of a species that seems, in its more natural institutional settings, strongly disposed to superstition and fearful conservatism. Our verificationism, like all versions of that, is promoted as a bar against seeking explanation where we have good reasons to doubt that it promises anything but temporary psychological satisfaction at the expense of truth. In particular, we deny that there is value to be had in philosophers postulating explananda, without empirical constraint, on grounds that these would make various putative explanans feel less mysterious if they prevailed.

It seems, then, that by van Fraassen's lights we must be empiricists, despite the fact that in defending objective modality we apparently violate a standard empiricist doctrine.[58] After all, we evince every aspect of the empiricist stance, and that, according to him, is necessary and sufficient for being an empiricist. But before we simply announce our membership in the empiricist brigades, thanking van Fraassen for the doctrinal freedom his 'stance stance' allows us to plunder as we like from the realist warehouse, consider the rival stance he identifies.

He calls this the 'materialist stance'. Here is his account of it:

> [M]aterialism ... is not identifiable with a theory about what there is but only with an attitude or cluster of attitudes. These attitudes include strong deference to the current content of science in matters of opinion about what there is. They include also an inclination (and perhaps a commitment, at least an intention) to accept (approximative) completeness claims for science as actually constituted at any given time. (59)

Reading van Fraassen is like consulting a sophisticated caster of horoscopes, for this describes us to a T also. Yet according to him the great tension in the history of (secular) western philosophy is between the empiricist and the materialist stances. Let us reiterate words we quoted above: 'The problem for materialists will then be to identify the true materialist stance and for the empiricist to identify the true empiricist stance'; this is van Fraassen articulating his hopes for the best possible *future* for philosophy. Are we then schizophrenic, advocating both stances at once? Or do we embody the resolution of philosophy's long struggle? Perhaps we are instances of a Hegelian stance.

It is noteworthy that van Fraassen calls the second stance 'materialism' *and* equates it with 'naturalism'. Why? The philosophers he has in mind as exemplifying the materialist tradition—Holbach, Laplace, Quine, Lewis—all maintained that everything is physical. This they took as meaning the same thing as 'natural' (so that the supernatural is what some confused people imagine to exist extramaterially—so much for Hegel after all). However, van Fraassen argues, across the whole tradition of materialism 'physical' means nothing more than 'whatever physicists endorse'. In the eighteenth and nineteenth centuries this correlated with 'composed out of matter'; but as physics moved beyond this restriction, materialist philosophers could and would follow it loyally. That is no doubt what van Fraassen will suggest we are doing when we insist that naturalists should drop their commitments to atomism and, indeed (as we will argue in

---

[58] In Chapter 2 we discuss at length the relationship between constructive empiricism and the philosophy of modality. It turns out that, surprisingly, van Fraassen does not regard belief in objective modality as antithetical to his empiricism.

Chapters 2 and 3), to substantivalism altogether in the name of allegiance to physics.

So we really are, in detail, both perfect empiricists and perfect materialists according to van Fraassen's criteria. Note that this cannot be diagnosed as implying doctrinal self-contradiction, for that is a charge that is inappropriate to stances.

Well, then, let us own up to being materialist Hegelians.[59] To start to see the basis for synthesis between the empiricist and materialist stances, let us attend to the critical feature they have in common. Both demand that philosophy fashion its style of inquiry after the example of science. Empiricists emphasize the fallibilism and tolerance of novel hypotheses they find in science. Materialists emphasize scientific reluctance to take 'spooky' processes seriously. Both of these really are perennial aspects of the history of science, so scientistic philosophers of both stances come by them naturally. Thus, might the source of (secular) philosophical tension lie within science itself? The two aspects indeed look to be awkwardly related: the tendency celebrated by the empiricists wants to let a hundred flowers bloom and the tendency celebrated by materialists wants to stomp on some it regards as weeds.

This tension in science has surely been productive. Scientific institutions are hyper-conservative about taking novelty seriously, as a result of which science (unlike philosophy) stays on an equilibrium path.[60] And then this very conservatism, by ensuring that the few novelties that get through the institutional filters are extremely rigorously selected and robust, produces an engine for continuous epistemic and cultural renovation in societies that take science seriously. Perhaps, then, philosophy in aiming to emulate science should welcome incorporation of this tension in itself, if only it can do so without being foolishly incoherent.

Verificationism, we suggest, is exactly the principle that pulls this off. That it does so, we believe, has been obscured for almost a century by the fact that the logical positivists and logical empiricists associated it with a very special theory of perceptual belief and of putative differences between this and other kinds of belief.[61] This is precisely the aspect of the last century's philosophy that, we respectfully submit, van Fraassen has not shaken off; so perhaps it is unsurprising he may have a blind spot exactly here. We think it salutary to remind readers, as we did in 1.2, that verificationism did not originate with and is not special to the logical positivists and empiricists. We find it in Peirce, as we have discussed,

---

[59] We have heard that before to come to mention it. But this book will contain no dodgy economics.

[60] We mean by this what economists do. The idea is that although scientific progress is far from smooth and linear, it never simply oscillates or goes backwards. Every scientific development influences future science, and it never repeats itself.

[61] As noted in 1.2, the positivists also mistakenly took verificationism to be a theory of meaning. This is an error van Fraassen does not share with them.

and (though not of course by name) in Hume. It is revealing to ask oneself, in light of van Fraassen's labels for the stances: who is generally regarded in histories of philosophy as the prototypical strong naturalist among great philosophers? The answer is of course Hume. His Christian contemporaries denounced him as a materialist, not without grounds. Yet he is simultaneously regarded as the prototypical empiricist.

Our diagnosis of the fact that Hume can be the poster philosopher for both of van Fraassen's stances is that he was the first philosopher who made verificationism his core commitment. When Hume argues that it is never rational to take reports of miracles seriously, this isn't because God is held to transmit his agency by means other than knocking bits of matter into other bits; it is because God as a putative part of the world simultaneously transcends (by hypothesis) the domain we have procedures for systematically testing. (Remember that, in Hume's *Inquiry*, the conclusion against miracles serves as an essential premise in his subsequent argument for atheism—it doesn't go the other way around.)

It is evident from the later parts of van Fraassen (2002) that he thinks empiricism is open to accommodation with the non-secular in a way that materialism is not. This may be among the factors that he thinks recommends empiricism. By contrast, we do not think that science ought to try to accommodate the religious impulse[62]—on this point too we echo Hume. The empiricist and the materialist (or naturalist) stances synthesize when they each add a pro-attitude toward verificationist limits. The empiricist then continues to endorse tolerance, but within the limits of what is verifiable; for the materialist, verificationism is the way to keep order without resorting to the dogmatism of the *ideological conservative*.

Let us call the synthesized empiricist and materialist—and resolutely secularist—stances the *scientistic* stance. (We choose this word in full awareness that it is usually offered as a term of abuse.) Its compatibility with van Fraassen's more general 'stance stance' is not merely rhetorical. Our fundamental principles are not propositional doctrines: both the PNC and the PPC are explicitly *norms*. They are not, of course, arbitrary norms; we motivate them as descriptively manifest in scientific practice. But a person who finds science unimpressive, or demeaning (think of Coleridge or Wordsworth), or Faustian, will have no reason at all to be persuaded by this book. This does not mean we have nothing to discuss with such people; their resistance to science, which must be quite thoroughgoing if it is not to be unprincipled, will confront them with serious policy problems in the management of social affairs, and we will want to press them as hard as possible on these. But we would not try to convert them with metaphysics, for van Fraassen is right that that would require strong metaphysics, and strong metaphysics can't get off the ground.

---

[62] See Dennett (2006) for careful argument to this effect.

Within the context of our scientistic stance, we will defend various propositions that will have the grammatical form of saying what there is. These should be understood not as doctrines but as proposed provisional commitments for living out the stance. Such living requires, most importantly of all, paying close attention to the actual progress of science. It might be thought surprising that there could be people who embrace the scientistic stance in all outward respects but then don't seem to bother to follow science closely; yet neo-scholastic pseudo-naturalistic philosophers such as those we criticized fit exactly this description. Most of our argument to come will not be with them, except by example. We will instead concern ourselves mainly with genuine students of science—not least van Fraassen himself—as they try to work out how to live the scientistic stance in detail.

Our discussion will therefore be rooted much more closely in the philosophy of science than in the metaphysics literature properly so called. Some contributors to the latter might say that our book is not really about metaphysics at all, and that in this section we have supplied them with grounds for this claim: there is a kind of metaphysics, strong metaphysics, which we join van Fraassen in refusing to regard as well motivated. By 'metaphysics' we mean something more limited and carefully constrained. We refer to the articulation of a unified world-view derived from the details of scientific research. We call this (weak) metaphysics because it is not an activity that has a specialized science of its own. In case someone wants to declare our usage here eccentric or presumptuous, we remind them that we share it with Aristotle.

Does our weak metaphysic not necessarily share one feature with strong metaphysics, and one equally vulnerable to van Fraassen's criticism? Is not the only virtue we can claim for it that we purport it to be true? But then if, as we claim, it is derived from the details of current science, must we not admit it to be false, since we can expect current science to be overthrown by future science? Here we encounter a disagreement between us and van Fraassen (and many others) at the level of the philosophy of science, rather than meta-philosophy or metaphysics. We are not persuaded that the history of science is most persuasively presented as a history of revolutions in which theories are repeatedly routed by successors and found to be false. There is, we think, a defensible basis for joining the logical positivists in viewing the history of science as a history of progressive accumulation of knowledge. As we will see, presentation of this picture cannot be separated from the second-order, stance-relative metaphysics in which we are engaged: our philosophy of science and our scientistic metaphysics reciprocally support each other.

2

Scientific Realism, Constructive Empiricism, and Structuralism

James Ladyman and Don Ross

So far we have described a strong form of naturalism in metaphysical methodology, and promised to demonstrate the advantages of our metaphysics by reference to problems in the philosophy of science. In this chapter, we will begin the construction of our metaphysics by arguing that a form of structural realism is motivated by reflection on issues that arise in two different domains that have been the subject of intense scrutiny during recent decades. These are: first, problems from the history of science about the abandonment of ontological commitments as old theories are replaced by more empirically adequate ones; and secondly, questions arising from the debate between scientific realists and constructive empiricists about what really is at stake between them, and about the inadequacies of constructive empiricism. We also argue that the so-called 'semantic' or 'model-theoretic' understanding of the nature of scientific representation fits very well with the kind of structural realism we advocate.

The structure of our argument is dialectical. It should be evident from Chapter 1 why we favour this approach over strictly analytic argumentation as a way of discovering the contents of our scientistic stance. Our (weak) metaphysical view is an aspect of our understanding of science. We do not gain superior understanding of science by discerning first principles or conceptual truths and then constructing syllogisms. Rather, one metaphysical proposal constructed in accordance with the PNC is to be preferred to another to the extent that the first unifies more of current science in a more enlightening way. Thus defending a metaphysical proposal is necessarily an exercise in comparison. Degree of consilience, the property a PNC-compatible metaphysics should aim to optimize, is a comparative property of a whole theory and cannot be determined just by analysing the detailed implications of its parts. This of course does not mean that analytic arguments are not important to elements of the comparative evaluations. Theories have the good or bad properties they do in large part because of what is implied by various specific claims they motivate. The reader will thus find plenty of arguments ahead. Our point here is to explain why these will occur in the context of a kind of narrative, wherein we converge on our favoured position by presenting, comparing, criticizing, and adjusting metaphysical proposals made by others, until we find we can improve no further in light of the current state of scientific knowledge.

The dialectical narrative will pivot around van Fraassen's engagement with standard scientific realism. We take it that our remarks in the last section of the previous chapter show why this is a natural approach for us. Our metaphysic is a synthesis of constructive empiricism and scientific realism based on our broadly Peircean verificationism. We aim to show that a view defended over several years by one of us (Ladyman), ontic structural realism, just is such a synthesis. We will do this by incrementally adjusting constructive empiricism and standard scientific realism in the face of each other's objections—at least, those objections we can endorse on the basis of the PNC—until their residues form a consistent mixture that is a form of ontic structural realism.

Structural realism was suggested by John Worrall (1989) to solve the problem of theory change for scientific realism. Roughly speaking, structural realism is the view that our best scientific theories describe the structure of reality, where this is more than saving the phenomena, but less than providing a true description of the natures of the unobservable entities that cause the phenomena. Following subsequent developments and accounts of the history of structuralism about science in the twentieth century, there are now a number of variants of structural realism discussed in the literature. Ladyman (1998) argues that they can be divided into epistemological and metaphysical varieties according to how they depart from standard scientific realism, and introduces a distinction between 'ontic' and 'epistemic' structural realism.[1] French's and Ladyman's (2003a, 2003b) advocacy of ontic structural realism emphasizes the commitment to objective modal structure to which Ladyman (1998) drew attention with reference to a classification of forms of realism and empiricism due to Ron Giere (1985, 83).

If structural realism (whether epistemic or ontic) is just a response to theory change it is vulnerable to the charge that it is ad hoc. The view advocated here escapes this concern by having multiple motivations, and by appeal to the virtue of consilience in relation to a variety of problems in philosophy of science. In the next chapter, we argue that our form of structural realism is motivated by the metaphysical implications of our best current (including cutting-edge) physics to the extent that it demonstrates consensus, and is compatible with all currently serious theoretical options where there is not consensus. We add further

---

[1] For example, Esfeld (2004) is clearly addressing metaphysical issues, whereas Votsis (forthcoming) is more concerned with epistemological ones. Psillos (2001) uses 'eliminative structural realism', van Fraassen (2006) uses 'radical structuralism' for ontic structural realism.

consilience in Chapters 4 and 5 when we introduce the theory of real patterns, in combination with which the position defended here makes possible a plausible account of laws, causation, and explanation, and of the relationship between the special sciences and fundamental physics. According to our naturalism, this is the only legitimate way of arguing for a speculative scientific metaphysic.

This chapter is organized as follows. First, in 2.1, we review the definition of scientific realism, and the arguments for and against it, concluding that the history of successful novel prediction science is the most compelling evidence for some form of realism, but also (in 2.2) that the history of ontological discontinuity across theory change makes standard scientific realism indefensible. In 2.3, we introduce scientific structuralism, and then assess in detail van Fraassen's empiricism. We find much to agree with him about, including the appropriateness of the semantic approach to scientific theories (which we outline in 2.3.3), but we also argue that a commitment to objective modality is needed to make his structuralist form of empiricism defensible. In 2.3.4 we introduce some novel terminology we will make use of in the rest of the book, before turning to the exposition of structural realism in 2.4. We argue against a purely epistemic form of structural realism and conclude the chapter by introducing ontic structural realism and preparing the ground for the further defence and development of it in Chapter 3.

2.1 SCIENTIFIC REALISM

There is of course a good deal of controversy about how to define scientific realism. We note later that van Fraassen defines it as a view about the aims of science. He has dialectical reasons for characterizing it that way, but here we offer a more standard epistemological definition according to which scientific realism is the view that we ought to believe that our best current scientific theories are approximately true, and that their central theoretical terms successfully refer to the unobservable entities they posit. Hence, if the theories employ terms that purport to refer to unobservable entities such as electrons, or gravitational waves, then, realists say, we ought to believe that there really are such entities having the properties and exhibiting the behaviour attributed to them.[2] For many philosophers scientific realism is obvious and uncontroversial. Certainly, for example, in debates about physicalism, reductionism, and supervenience in the philosophy of mind, it is normally assumed by all parties that there are atoms,

---

[2] There are complications of course. Scientific realism can be considered in terms of semantic, metaphysical and epistemic components (see, for example, Psillos 1999, Ladyman 2002b, ch. 5). Some philosophers adopt a pragmatic (or otherwise epistemically constrained) conception of truth and defend belief in unobservables on that basis (see for example Ellis 1985). Van Fraassen's arguments are not directed against such views (1980, 4), but they may be vulnerable to other antirealist arguments.

molecules, ions, and so on; the question is whether that is all there is, or whether there are emergent or causally autonomous entities and properties over and above the physical stuff. Scientific explanations throughout the physical and the life sciences make essential reference to unobservable entities such as electromagnetic waves, nitrogen molecules, and gravitational fields. Hence, one might be forgiven for thinking that there is no question that scientific realism is correct.

However, there are many arguments against scientific realism. Some are perhaps merely philosophical, but some appeal to contingent facts about the history of science and our best scientific theories. We are in some ways more sceptical than van Fraassen, who argues only for the permissibility of agnosticism about theoretical entities, and in this chapter we will argue that the arguments from theory change compel us to relinquish standard scientific realism. Before discussing them it is important for the consilience case that we are building that we review the main argument in favour of scientific realism, since we will argue that the form of non-standard realism about science that we recommend is also supported by it.

2.1.1 The no-miracles argument

Ladyman (2002b, ch. 7) distinguishes between local and global appeals to inference to the best explanation (IBE) to defend scientific realism. (Psillos 1999 marks a similar distinction using the terms 'first-order' and 'second order'.) A local defence of scientific realism appeals to a particular set of experimental facts and their explanation in terms of some particular unobservable entities. Opponents of scientific realism, such as van Fraassen, deny that the local defences of realism about specific unobservables are compelling, arguing that they can in each case be reinterpreted in pragmatic terms as inferences to the empirical adequacy of the explanation in question, plus a commitment to continue theorizing with the resources of the theory (see van Fraassen 1980). Hence, the debate shifts to the global level where scientific realists argue that their philosophy of science is needed to account for science, and its history of empirical success, as a whole. This 'ultimate' argument for scientific realism was famously presented by Putnam as follows: 'The positive argument for realism is that it is the only philosophy that doesn't make the success of science a miracle' (1975a, 73). A similar form of argument is to be found in Smart: 'If the phenomenalist about theoretical entities is correct we must believe in a *cosmic coincidence*' (1963, 39).[3] This is an inference to the only explanation. Many scientific realists argue naturalistically that we are led to realism if we follow in philosophy of science the same patterns of inference that we follow in science. Scientific realism is thus seen as a scientific explanatory hypothesis about science itself: '[P]hilosophy is itself a sort of empirical science' (Boyd 1984, 65).

---

[3] Ian Hacking (1985) argues from the denial of outrageous coincidence.

The no-miracles argument is made more sophisticated by citing specific features of scientific methodology and practice, and then arguing that their successful application is in need of explanation, and furthermore that realism offers the best or only explanation because the instrumental reliability of scientific methods is explained by their theoretical reliability.[4] Richard Boyd (1985, for example) argues that in explaining the success of science, we need to explain the overall instrumental success of scientific methods across the history of science. The following are all offered as general facts that the hypothesis of scientific realism explains:

(i) patterns in data are projectable using scientific knowledge;
(ii) the degree of confirmation of a scientific theory is heavily theory-dependent;
(iii) scientific methods are instrumentally reliable.

The second point refers to the fact that scientists use accepted theories to calibrate instruments, choose relevant methods for testing new theories, describe the evidence obtained, and so on.

Another feature of scientific practice that realists have long argued cannot be explained by antirealists is the persistent and often successful search for unified theories of diverse phenomena. The well-known conjunction objection (Putnam 1975b) against antirealism is as follows: That T and T' are both empirically adequate does not imply that their conjunction T & T' is empirically adequate; however, if T and T' are both true this does imply that T & T' is true.[5] Hence, the argument goes, only realists are motivated to believe the new empirical consequences obtained by conjoining accepted theories. However, in the course of the history of science the practice of theory conjunction is widespread and a reliable part of scientific methodology. Therefore, since only realism can explain this feature of scientific practice, realism is true. Van Fraassen (1980, 83–7) responds by denying that scientists ever do conjoin theories in this manner. Rather, he argues, the process of unifying theories is more a matter of 'correction', wherein pairs or sets of theories are all adjusted so as not to contradict or be isolated from each other, than 'conjunction'. Furthermore, he argues, scientists have pragmatic grounds for investigating the conjunction of accepted theories in the search for empirical adequacy. The latter response suggests how to formulate a general argument to show that the antirealist can account for the retention of any aspect of the practice of science that realism can. Suppose some feature of scientific practice is claimed by the realist to have produced instrumental success, and realism provides an explanation or justification of that feature. The

---

[4] This ironically makes the theory-ladenness of scientific methodology and confirmation evidence for realism, where Hanson, Kuhn, and Goodman took it to tell against realism.

[5] Note, however, that if we suppose that T and T' are both approximately true; this does not imply that their conjunction T & T' is also approximately true. For example, suppose T is Kepler's laws of planetary motion, and T' is Newtonian mechanics; then T & T' is actually inconsistent and therefore cannot be approximately true.

antirealist can simply point out that the history of science provides inductive grounds for believing in the pragmatic value of that feature of scientific practice. Similarly, van Fraassen (1980) offers an account of the pragmatics of science which attempts to account for (i), (ii), and (iii) above by reference to the fact that the background theories are empirically adequate.

It certainly is true that the simplistic view of theory conjunction does not do justice to the complexity of the practice of conjoining real theories. In some cases the conjunction of two theories will not even be well-formed, as with the example of general relativity and quantum mechanics (about which more in the next chapter). However, notwithstanding recent scepticism about the unity of science (notably defended by Nancy Cartwright and John Dupré, about which more is said in Chapter 4), the unification of a range of phenomena within a single explanatory scheme is often a goal of scientific theorizing. Michael Friedman argues that theoretical explanation consists in 'the derivation of the properties of a relatively concrete and observable phenomenon by means of an embedding of that phenomenon into some larger, relatively abstract and unobservable theoretical structure' (1981, 1). His examples include the reduction of gas theory to molecular theory and of chemical theory to atomic theory. Friedman then argues that theoretical explanations of this kind confirm the theories of the unobservable world that provide the reductive bases.[6] However, he cautions against a simplistic view that regards an explanatory reduction as confirmed just because it is the best available explanation because this would lead to an infinite hierarchy of explanations. According to Friedman, explanatory reductions or embeddings must have unifying power in order to be confirmed.

Friedman's argument is as follows:

(1) Unification at the theoretical level confers a higher degree of confirmation to theories at the empirical level.
(2) Only realists have epistemic grounds for unification at the theoretical level.
(3) Therefore, we should be realists.

He further argues that there are grounds for thinking that scientists ought to conjoin theories whether they in fact do so or not. The full theoretical description T is capable of acquiring a greater degree of confirmation when conjoined with other theories than that which attaches to the claim that T is empirically adequate. The argument for this goes as follows:

A theoretical structure that plays an explanatory role in many diverse areas picks up confirmation from all these areas. The hypotheses that collectively describe the molecular

---

[6] Friedman (1974) had previously argued that unification is valuable because it increases our understanding by reducing the number of brute facts in the world. This argument was criticized by Kitcher (1976) and Salmon (1989), who argued that Friedman had not given a satisfactory account of how to individuate brute facts. Friedman's later version of the argument refers only to confirmation.

model of a gas of course receive confirmation via their explanation of the behaviour of gases, but they also receive confirmation from all the other areas in which they are applied: from chemical, thermal and electrical phenomena, and so on. (Friedman 1983, 243)

On the other hand the claim that the hypotheses are empirically adequate cannot partake of this extra confirmation from other domains. Since the phenomenological import of the hypotheses is among their consequences, if the full theoretical description becomes better confirmed, so too does the phenomenological description. Hence, claims Friedman, the realist has better-confirmed knowledge even of the observable world than the antirealist, and this is the very type of knowledge at which, according to the antirealist, science aims.

However, as pointed out by Kukla (1995), the antirealist may simply argue that realists are not epistemically entitled to extra confidence in the empirical adequacy of unified theories. Kukla also argues that even if this extra confidence in the pursuit of unification is methodologically advantageous this is not sufficient grounds for realism since the extra confidence is also warranted for his so-called 'conjunctive empiricist' (234, and see also his 1996b). This is someone who believes that not only are our best theories empirically adequate, but also that when conjoined with background theories they will produce new theories that are also empirically adequate. The conjunctive empiricist then has epistemic reasons to pursue unification because it will give her more instrumental knowledge. Now if the no-miracles argument works at all it will certainly work against Kukla's proposal, since if scientific theories are false, it can only be more miraculous to the realist that they should be extendible in this way than it was that they were instrumentally successful. Furthermore, the conjunctive empiricist's position seems entirely ad hoc. However, note that this argument for scientific realism relies on the modal power of theories that quantify over unobservables in order to explain and predict phenomena. It is because diverse phenomena are subject to the operation of the same laws and causes that unification in terms of unobservables is instrumentally reliable. If scientific laws described mere regularities in the actual phenomena in some domain, their successful extension to new domains, or the successful extension of their explanations to new domains, would be no less mysterious than for the antirealist.

2.1.2 Against the no-miracles argument

Van Fraassen has three arguments against the claim that realism is the best or only explanation of the success of science. The first is that the demand that there be an explanation for every regularity amounts to the demand that Reichenbach's principle of the common cause (PCC) be satisfied, so the realist seems to be committed to this principle as a requirement for the adequacy of scientific theories. The PCC holds that the correlations among events is either evidence of a direct causal link between them, or is the result of the action of a common cause. Van Fraassen then argues that PCC is too strong a condition on the basis of quantum mechanics (see his 1980, ch. 2, and 1991, ch. 10). Bell's theorem tells us that the correlations between measurement results on members of a pair of entangled particles cannot be screened off by conditionalizing on any common cause, and the idea of a direct causal link seems to violate relativity theory, so the realist does indeed face a problem with quantum mechanics. However, regardless of whether or how this can be solved, it is not obvious that the scientific realist is committed to PCC simply in virtue of appealing to inference to the best explanation. Perhaps there are scientific explanations that violate the principle of the common cause, or perhaps explanation in terms of common causes simply gives out eventually, but is nonetheless necessary, or at least appropriate, for the vast majority of scientific purposes.

The second argument that van Fraassen offers is to object that the realist's demand for explanation presupposes that to be a lucky accident or coincidence is to have no explanation at all, whereas (according to him) coincidences can have explanations in a certain sense (1980, 25).

It was by coincidence that I met my friend in the market—but I can explain why I was there, and he can explain why he came, so together, we can explain how this meeting happened. We call it a coincidence, not because the occurrence was inexplicable, but because we did not severally go to the market in order to meet. There cannot be a requirement upon science to provide a theoretical elimination of coincidences, or accidental correlations in general, for that does not even make sense. (1980, 25)

This seems to miss the point of the no-miracles argument. The realist's claim is just that explanation of the consistent predictive success of scientific theories in terms of coincidence or miraculous luck is an unacceptable and arbitrary explanation, especially given the availability of the realist's alternative. If one kept meeting one's friend in the market unexpectedly, and some other explanation than mere coincidence was available, then one would be inclined to adopt it.

Perhaps with this response in mind, van Fraassen also offers a Darwinian explanation for the instrumental success of science:

[T]he success of current scientific theories is no miracle. It is not even surprising to the scientific (Darwinist) mind. For any scientific theory is born into a life of fierce competition, a jungle red in tooth and claw. Only the successful theories survive—the ones which *in fact* latched on to actual regularities in nature. (1980, 40)

Realists may reply that realism is not the only but the best explanation of the success of science (see for example Boyd 1985), and then point out that van Fraassen's explanation is a phenotypic one: it gives a selection mechanism for how a particular phenotype (an empirically successful theory) has become dominant in the population of theories. However, this does not preclude a genotypic explanation of the underlying properties that make some theories successful. Realists claim that the fact that theories are approximately true is a genotypic explanation of their instrumental reliability. They argue that 'a selection mechanism may explain why all the selected objects have a certain feature without explaining why each of them does' (Lipton 1991, 170). Hence, Lipton (ibid. 170 ff.) says that the realist's explanation explains two things that van Fraassen's does not: (a) why a particular theory that was selected is one having true consequences; and (b) why theories selected on empirical grounds went on to have more predictive successes.[7]

Ultimately, van Fraassen rejects the need for explanation of the kind the scientific realist claims to offer. He identifies himself with the Kantian critique of metaphysics construed as theorizing about the nature of the world as it is in itself (while not identifying himself with Kant's transcendental turn). According to him, metaphysicians have in common with scientific realists their desire to interpret and explain what is already understood in so far as it can be. He thinks that any attempt to defend scientific realism will involve metaphysics of some kind, notably modal metaphysics. It is only on the assumption that the unobservable entities posited by realists cause the phenomena that they explain them. If unobservable entities merely happened to be around when certain phenomena were occurring then their presence would not be explanatory. Hence, he argues, scientific realism relies upon some kind of metaphysical theory of laws of nature, singular causation, or essential natures. For van Fraassen, this means it ultimately rests on explanation by posit. Here we reach an impasse with the scientific realist insisting on the need for explanations where the antirealist is content without them.

Colin Howson (2000), P. D. Magnus and Craig Callender (2004) have recently argued that the no-miracles argument is flawed because in order to evaluate the claim that it is probable that theories enjoying empirical success are approximately true we have to know what the relevant base-rate is, and there is no way we can know this. Magnus and Callender argue that 'wholesale' arguments that are intended to support realism (or antirealism) about science as a whole (rather than 'retail' arguments that are applied to a specific theory) are only taken seriously because of our propensity to engage in the base-rate fallacy. They think we ought to abandon the attempt to address the realism issue in general.

In this book we are attempting to develop a metaphysical position that allows us to make sense of science as a whole. We are thus engaging in wholesale reasoning about science. Rather than offering a principled defence of so doing, we leave it to the reader to decide whether the product of our labour vindicates our methods. In particular, we do suppose that the no-miracles argument is an important consideration in the thinking of many scientists who say they find antirealism unsatisfactory as an account of their aims. Magnus and Callender claim that the continued success of the argument in this respect, despite its fallacious nature, is to be explained in terms of human frailty (2004, 329), and has no probative force. We disagree just this far: if general and universal structures can be identified—if, to that extent, a unified account of the world

---

[7] See Stanford (2000).

can be found, based on science—then successful applications of particular instantiations of these structures in new domains, so as to generate novel but reliable predictions, will be explicable and non-mysterious. It is true that we must not insist, a priori, that the instrumental success of science can be explained. But we should surely accept an explanation, rather than preferring to believe in recurrent miracles, if one can in fact be provided without violation of the PNC.

We turn to novel predictive success below. But first we take note of another fundamental criticism of the appeal to IBE in the debate about scientific realism. This one is due to Larry Laudan (1984, 134) and Arthur Fine (1984a), both of whom point out that, since it is the use of IBE in specific instances where this involves unobservables that is in question, it is viciously circular to use IBE at the global level to infer the truth of scientific realism because the latter is a hypothesis involving unobservables. According to them, any defence of realism by appeal to the no-miracles argument is question-begging. There is a similarity here with the inductive vindication of induction. Richard Braithwaite (1953, 274-8), and Carnap (1952), argue that the inductive defence of induction is circular but not viciously so, because it is rule circular but not premise circular. In the case of IBE such a view has recently been defended by David Papineau (1993, ch. 5) and Stathis Psillos (1999). The idea is that premise circularity of an argument is vicious because the conclusion is taken as one of the premises; on the other hand, rule circularity arises when the conclusion of an argument states that a particular rule is reliable and that conclusion only follows from the premises when the very rule in question is used. Hence they argue that IBE is in just the same position as inductive reasoning: it cannot be defended by a non-circular argument. (But then we may remember Lewis Carroll's story of Achilles and the Tortoise (1895) and note that even deduction is circular.)

The no-miracles defence of realism is rule- but not premise-circular. The conclusion that the use of IBE in science is reliable is not a premise of the explanationist defence of realism, but the use of IBE is required to reach this conclusion from the premises that IBE is part of scientific methodology and that scientific methodology is instrumentally reliable. However, lack of vicious circularity notwithstanding, this style of argument will not persuade someone who totally rejects IBE. Rather, what the argument is meant to show is that someone who does make abductive inferences can show the reliability of their own methods. Hence, the realist may claim that although they cannot force the non-realist to accept IBE, they can show that its use is consistent and argue that it forms part of an overall comprehensive and adequate philosophy of science. Unfortunately for the scientific realist, the history of science gives us grounds for doubting whether scientific realism is really the best explanation of the success of science, or so we shall argue in 2.1.4. In the next section we will argue that the real miracle about the success of science is not empirical success in general, but how it is that scientific theories can tell us about phenomena we never would have expected without them.

2.1.3 Novel prediction

It has been remarked by various philosophers that the scientific realism debate has reached an impasse. (See for example Blackburn 2002.) We hope to break that impasse here in favour of a form of realism about science, but one which concedes a good deal to the constructive empiricist. It is a familiar point in philosophy that the determined sceptic cannot be rationally converted. Hence there is no valid argument for scientific realism, the premises of which will be accepted by the antirealist who is prepared to forsake all explanations of science. However, van Fraassen does offer a positive account of science that is intended to vindicate its rationality and its cumulative empirical success. In this section, we will distil from the no-miracles argument the essence of the motivation for realism about science. When we talk about the success of a scientific theory there are many things that we might have in mind, for example technological applications, provision of explanations, methodological success, contribution to unification, application to other sciences, generation of predictions, and so on. Not all of these are equally good explananda for no-miracles arguments. For example, Nancy Cartwright (1983) argues that phenomenological applications of fundamental theories are not evidence of the truth of these theories because of the idealizing and falsifying that is involved in both formulating and applying them. Similarly, technological applications of science owe a great deal to trial and error, and to auxiliary know-how. Whether or not careful work can show that these forms of success support scientific realism after all, there is one kind of success which seems to be the most compelling reason for some form of realism about scientific theories, namely novel predictive success.

Alan Musgrave (1988) says flatly that the only version of the no-miracles argument that might work is one appealing to the novel predictive success of theories. Some realists, such as Psillos (1999), have gone so far as to argue that only theories which have enjoyed novel predictive success ought to be considered as falling within the scope of arguments for scientific realism. This restricts the number of abandoned past theories that can be used as the basis for arguments against scientific realism discussed in the next section. Desiderata for an account of novelty in this context are of two kinds. First, there are more or less a priori features that are demanded of theories of confirmation such as objectivity and impersonality. Second, there is conformity to paradigmatic scientific judgements about the confirmation of specific theories in the history of science. Examples of the latter that are commonly given include the empirical success of General Relativity in predicting the deflection of light as it passes near large masses, and the empirical success of Fresnel's wave optics in accounting for the celebrated phenomenon of the white spot in the shadow of an opaque disc (see Worrall 1994). However, it is no simple matter to give a general characterization of novel predictive success.

The most straightforward idea is that of *temporal novelty*. A prediction is temporally novel when it forecasts an event that has not yet been observed. The problem with adopting this kind of novel predictive success as a criterion for realism about particular theories is that it seems to introduce an element of arbitrariness concerning which theories are to be believed. When exactly in time someone first observes some phenomenon entailed by a theory may have nothing to do with how and why the theory was developed. It is surely not relevant to whether a prediction of a theory is novel if it has in fact been confirmed by someone independently but whose observation report was not known to the theorist. (The white spot phenomenon had been observed independently prior to its prediction by Fresnel's theory.) A temporal account of novelty would make the question of whether a result was novel for a theory a matter of mere historical accident and this would undermine the epistemic import novel success is supposed to have.

It is more plausible to argue that what matters in determining whether a result is novel is whether a particular scientist knew about the result before constructing the theory that predicts it. Call this *epistemic novelty*. The problem with this account of novelty is that, in some cases, that a scientist knew about a result does not seem to undermine the novel status of the result relative to her theory, because she may not have appealed to the former in constructing the latter. For example, many physicists regard the success of general relativity in accounting for the orbit of Mercury, which was anomalous for Newtonian mechanics, as highly confirming, because the reasoning that led to the theory appealed to general principles and constraints that had nothing to do with the empirical data about the orbits of planets. Even though Einstein specifically aimed to solve the Mercury problem, the derivation of the correct orbit was not achieved by putting in the right answer by hand. Worrall suggests that realism is only appropriate in the case of 'theories, designed with one set of data in mind, that have turned out to predict, entirely unexpectedly, some further general phenomenon' (1994, 4). However, in a recent analysis, Jarett Leplin (1997) rejects this as relativizing novelty to the theorist and thereby introducing a psychological and so non-epistemic dimension to novelty.[8] Take the case of Fresnel. If we say that the fact that the white spot phenomenon was known about is irrelevant, because Fresnel was not constructing his theory to account for it although his theory still predicted it, then we seem to be saying that the intentions of a theorist in constructing a theory determine in part whether the success of the theory is to be counted as evidence for its truth. Leplin argues, and we agree, that this undermines the objective character of theory confirmation needed for realism.

This motivates the idea of *use novelty*. A result is use-novel if the scientist did not explicitly build the result into the theory or use it to set the value of some

---

[8] Our discussion of Leplin (1997) follows Ladyman (1999).

parameter crucial to its derivation. Hence, Leplin proposes two conditions for an observational result O to be novel for a theory T:

Independence Condition: There is a minimally adequate reconstruction of the reasoning leading to T that does not cite any qualitative generalization of O.

Uniqueness Condition: There is some qualitative generalization of O that T explains and predicts, and of which, at the time that T first does so, no alternative theory provides a viable reason to expect instances. (1997, 77)

Leplin explains that a reconstruction of the reasoning that led to a theory is an idealization of the thought of the theorist responsible for that theory, and is said to be 'adequate' if it motivates proposing it, and 'minimally' so if it is the smallest such chain of reasoning (1997, 68–71).

According to the two conditions above, novelty is a complex relation between a theory, a prediction or explanation, the reconstruction of the reasoning that led to it, and all the other theories around at the time, since the latter are required not to offer competing explanations of the result. Two points follow. First, if we found a dead scientist's revolutionary new theory of physics, but he left no record of what experiments he knew about or what reasoning he employed, such a theory could have no novel success. Hence, according to Leplin, no amount of successful prediction of previously unsuspected phenomena would motivate a realist construal of the theory. Second, suppose that we already knew all the phenomena in some domain. In such a case we could never have evidence for the truth rather than the empirical adequacy of any theories that we constructed in this domain, no matter how explanatory, simple, or unified they were. These consequences are surely contrary to realist thinking.

Furthermore, Leplin admits that his analysis makes novelty temporally indexed (1997, 97), but this seems to fall foul of his own claim that 'it is surely unintuitive to make one's epistemology depend on what could, after all, be simple happenstance' (1997, 43). The main problem seems to be with the uniqueness condition since it leaves too much to chance. For example, suppose a result is novel with respect to some theory, but that another theory comes along soon afterwards which also explains it. According to Leplin's view, a realist commitment to the former theory is warranted but not to the latter; yet it seems that the order of things might have been reversed, so that which theory we should believe to be true is a function of contingencies of our collective biography. Moreover, truth is taken to explain the mystery of the novel success of one theory, but the success of the other theory, which would be novel were its rival not around, is left unexplained.

Certainly scientific methodology includes far broader criteria for empirical success, such as providing explanations of previously mysterious phenomena. Indeed, Darwin's theory of evolution and Lyell's theory of uniformitarianism were accepted by the scientific community because of their systematizing and explanatory power, and in spite of their lack of novel predictive success. As we have seen, realists also often argue that the unifying power of theories, which may bring about novel predictive success but need not do so, should be taken as a reason for being realistic about them.

We suggest a modal account of novel prediction. That a theory *could* predict some unknown phenomenon is what matters, not whether it actually did so predict. The empirical success of a theory which was just a list of the phenomena that *actually* occurred would not stand in need of explanation. That theories sometimes produce predictions of qualitatively new types of phenomena, which are then subsequently observed, is what is important, not whether a particular prediction of a specific theory is novel (we return to this below).[9] Since some theories have achieved novel predictive success our overall metaphysics must explain how novel predictive success can occur, and the explanation we favour is that the world has a modal structure which our best scientific theories describe.

Of course, van Fraassen will argue that even novel predictive success does not need to be explained other than in Darwinian terms. Since science will not tell us anything one way or the other about what is epistemically 'required' (and of course van Fraassen is not here referring to a mere psychological requirement), we cannot attempt to refute him. We can only reiterate what we said above in response to the scepticism about explanation offered by Howson, Magnus, and Calander. However, when van Fraassen talks about scientific theories describing the relations among the phenomena, he comes very close to our view. The only difference between us is that we understand the relations in question to be nomological or more broadly modal, whereas he understands them to be extensional occurrent regularities. Of course, some scientific realists of a Humean inclination would argue that van Fraassen is right about this latter point. We will argue that the marriage of scientific realism and Humeanism about modality is an unhappy one. Van Fraassen is at least consistent in his disavowal of metaphysics. However, we will argue in 2.2.3.2 that even constructive empiricism incurs some metaphysical commitments. First, in the next two sections, we consider celebrated arguments against scientific realism.

2.1.4 Underdetermination

The underdetermination argument is widely discussed in the debate about scientific realism. Before considering its implications for our project we need to clarify it. We distinguish between two generic forms of underdetermination, 'weak' and 'strong'.

Weak underdetermination:

---

[9] The emphasis on new types of phenomena is crucial here since a theory that predicts another token of a well-known type of phenomenon may simply be based on generalizing from the known data.

(1) Some theory, T, is supposed to be known, and all the evidence is consistent with T.

(2) There is another theory T# which is also consistent with all the available evidence for T. T and T# are *weakly empirically equivalent* in the sense that they are both compatible with the evidence we have gathered so far.

(3) If all the available evidence for T is consistent with some other hypothesis T#, then there is no reason to believe T to be true and not T#.

Therefore, there is no reason to believe T to be true and not T#.

This kind of underdetermination problem is faced by scientists every day, where T and T# are rival theories but agree with respect to the classes of phenomena that have so far been observed. To address it scientists try to find some phenomenon about which the theories give different predictions, so that some new experimental test can be performed to choose between them. For example, T and T# might be rival versions of the standard model of particle physics which agree about the phenomena that are within the scope of current particle accelerators but disagree in their predictions of what will happen at even greater energies.[10] The weak underdetermination argument is a form of the problem of induction. Let T be any empirical law, such as that all metals expand when heated, and T# be any claim implying that everything observed *so far* is consistent with T but that the next observation will be different (Goodman 1955). If antirealism about science is to make use of underdetermination to motivate some kind of epistemic differentiation, specifically between statements about observables and statements about unobservables, then it ought to amount to more than just common-and-garden fallibilism about induction.

Let us first discuss strong underdetermination. To generate a strong underdetermination problem for scientific theories, we start with a theory H, and generate another theory G, such that H and G have the same empirical consequences, not just for what we have observed so far, but also for any possible observations we could make. If there are always such *strongly empirically equivalent* alternatives to any given theory, then this might be a serious problem for scientific realism. The relative credibility of two such theories cannot be decided by any observations even in the future and therefore, it is argued, theory choice between them would be underdetermined by all possible evidence. If all the evidence we could possibly gather would not be enough to discriminate between a multiplicity of different theories, then we could not have any rational grounds for believing in the theoretical entities and approximate truth of any particular theory. Hence, scientific realism would be undermined.

---

[10] One of us is now running an experiment to isolate divergent predictions about reward conditioning in the human brain made by different mathematical models of the process that are each equally adequate to existing data measured at the less fine-grained scales used in the advance of the planned test—but where each theory offers a different explanation of, and therefore different further predictions about successful treatment of, addiction.

The *strong form of the underdetermination argument* for scientific theories is as follows:

(i) For every theory there exist an infinite number of strongly empirically equivalent but incompatible rival theories.
(ii) If two theories are strongly empirically equivalent then they are *evidentially equivalent*.
(iii) No evidence can ever support a unique theory more than its strongly empirically equivalent rivals, and theory-choice is therefore radically underdetermined.

There are various ways of arguing that strong empirical equivalence is incoherent, or at least ill-defined:

(a) The idea of empirical equivalence requires it to be possible to circumscribe clearly the observable consequences of a theory. However, there is no non-arbitrary distinction between the observable and unobservable.
(b) The observable/unobservable distinction changes over time and so what the empirical consequences of a theory are is relative to a particular point in time.
(c) Theories only have empirical consequences relative to auxiliary assumptions and background conditions. So the idea of the empirical consequences of the theory itself is incoherent.

Furthermore, it may be argued that there is no reason to believe that there will always, or often, exist strongly empirically equivalent rivals to any given theory, either because cases of strong empirical equivalence are too rare, or because the only strongly empirically equivalent rivals available are not genuine theories.[11] Regardless of their views on (i), many realists argue that (ii) is false. They argue that two theories may predict all the same phenomena, but have different degrees of evidential support. In other words, they think that there are non-empirical features (*superempirical* virtues) of theories such as simplicity, non-ad hocness, (temporal or other) novel predictive power, elegance, and explanatory power, which give us a reason to choose one among the empirically equivalent rivals. Antirealists may agree that superempirical virtues break underdetermination at the level of theory choice, but argue, following van Fraassen, that their value is merely pragmatic, in so far as they encourage us to choose a particular theory with which to work, without giving us any reason to regard it as true. This is part of the motivation for van Fraassen's claim that our best scientific theories may just as well be empirically adequate rather than true. Contrary to how his view is usually presented by his critics, he doesn't directly appeal to strong underdetermination to argue against scientific realism and for constructive empiricism. Rather, he uses cases of strong empirical equivalence to show that theories have extra

---

[11] For discussion of these issues see Laudan and Leplin (1991), (1993), Kukla (1993), (1994), (1996a), and (1998), and Ladyman (2002b, ch. 6).

structure over and above that which describes observable events, in defence of the claim that belief in empirical adequacy is logically weaker than belief in truth simpliciter.[12]

Nonetheless, it is significant that van Fraassen concedes that he has nothing to say in the face of inductive scepticism, and we shall return to this below. Furthermore, the fact that he does not use the underdetermination argument against scientific realism does not mean that the realist cannot deploy it against constructive empiricism. The problem that antirealists face is how to overcome the weak underdetermination argument without thereby overcoming the strong one. Realists argue that the same superempirical considerations that entitle them to regard a well-tested theory as describing future observations as well as past ones also entitle them to choose a particular theory among strongly empirically equivalent ones. The particular strong underdetermination problem for scientific realists is that all the facts about observable states of affairs will underdetermine theory-choice between T_0, a full realistically construed theory, and T_1, the claim that T_0 is empirically adequate. However, all the evidence we have available now will underdetermine the choice between T_1 and T_2, the claim that T_0 is empirically adequate before the year 2010 (the problem of induction). Furthermore all the facts about all actually observed states of affairs at all times will underdetermine the choice between T_1 and T_3, the claim that T_0 describes all actually observed events. So, the realist can argue, even the judgement that T_0 is empirically adequate is underdetermined by the available evidence, and hence, the constructive empiricist ought to be an inductive sceptic in the absence of a positive solution to the underdetermination problem.

The underdetermination arguments discussed above do not seem unequivocally to support either realism or antirealism. However, there is something to the thought that genuine cases of strong empirical equivalence from science pose a particular problem for the scientific realist. Roger Jones (1991) raised the problem 'Realism about what?' by drawing attention to the existence of alternative formulations of physical theories that coexist in science. In a recent paper, Jonathan Bain (2004) analyses one of Jones's examples, namely that of Newtonian gravity, and shows that there are at least six theories that are prima facie entitled to that name. The details are technical, but the basic idea is that the theories differ with respect to their spacetime symmetries, their boundary conditions, and their dynamical symmetries. The most obvious difference is that some of them are field theories on flat spacetimes, and some are theories in which the effect of gravity is incorporated into the spacetime structure in the form of curvature as in General Relativity. In a discussion of the implications for the realism debate, Bain goes on to argue that three of these theories seem to constitute a genuine case of strong empirical equivalence, since they have differing spacetime symmetries and absolute objects. However (374), he points

[12] See M. van Dyck, (forthcoming).

out, they have the same dynamical symmetries and so a structural realist who regards dynamical structure as what is physically real should regard them as different ways of describing the same structure.

Quantum physics gives further examples of empirical equivalence. Our view is that the choice among competing ways of embedding empirically equivalent substructures in fundamental theory is a pragmatic one, but ultimately different formulations may lead naturally to the discovery of new modal relations and hence to new empirical discoveries. For example, Newton's force law suggested the mathematical form for Coulomb's law, although we know that both theories can be geometrized so that explicit forces disappear from the formalism. However, as Bain says, the standard scientific realist has the option of trying to break any cases of underdetermination by reference to the kinds of superempirical virtues mentioned above. The problem is that these sometimes seem to pull in different directions, and there is no obvious way to rank them (see Ladyman 2002b, ch. 8, section 2).

The underdetermination argument cannot be ignored by the scientific realist, but, since it also threatens any positive form of antirealism such as constructive empiricism, it does not give us compelling grounds to abandon standard scientific realism. (However, note that, as Bain says, there are grounds for thinking that underdetermination supports structural realism.) In any case, we will not discuss it further since in the next section we present what we regard as a more compelling—indeed, as the most compelling—form of argument against standard scientific realism.

2.2 THEORY CHANGE

Many realists are inclined to dismiss worries based on underdetermination as 'mere philosophic doubt'. They argue that since scientists find ways of choosing between even strongly empirically equivalent rivals, philosophers ought not to make too much of merely in-principle possibilities that are irrelevant to scientific practice. Leaving aside the question of whether all cases of strong empirical equivalence are scientifically unproblematic, and supposing (counterfactually, in our view) that the epistemological defence of scientific realism in terms of IBE is compelling, the problem of theory change at its strongest suggests that scientific realism is not supported by the facts about scientific theories because realism is not even empirically adequate. The power of the arguments against scientific realism from theory change is that, rather than being a priori and theoretical, they are empirically based and their premises are based on data obtained by examining the practice and history of science. Ontological discontinuity in theory change seems to give us grounds not for mere agnosticism but for the positive belief that many central theoretical terms of our best contemporary science will be regarded as non-referring by future science.

The notorious pessimistic meta-induction was anticipated by the ancient Greek sceptics,[13] but in its modern form, advocated notably by Laudan (1981), it has the following structure:

(i) There have been many empirically successful theories in the history of science which have subsequently been rejected and whose theoretical terms do not refer according to our best current theories. (Laudan gives a list.)

(ii) Our best current theories are no different in kind from those discarded theories and so we have no reason to think they will not ultimately be replaced as well.

So, by induction we have positive reason to expect that our best current theories will be replaced by new theories according to which some of the central theoretical terms of our best current theories do not refer, and hence, we should not believe in the approximate truth or the successful reference of the theoretical terms of our best current theories.

The most common realist response to this argument is to restrict realism to theories with some further properties (usually, maturity and novel predictive success) so as to cut down the inductive base employed in (i). However, assuming that such an account can be given there are still a couple of cases of mature theories which enjoyed novel predictive success by anyone's standards, namely the ether theory of light and the caloric theory of heat. If their central theoretical terms do not refer, the realist's claim that approximate truth explains empirical success will no longer be enough to establish realism, because we will need some other explanation for the success of the caloric and ether theories. If this will do for these theories then it ought to do for others where we happened to have retained the central theoretical terms, and then we do not need the realist's preferred explanation that such theories are true and successfully refer to unobservable entities. To be clear:

(a) Successful reference of its central theoretical terms is a necessary condition for the approximate truth of a theory.

(b) There are examples of theories that were mature and had novel predictive success but whose central theoretical terms do not refer.

(c) So there are examples of theories that were mature and had novel predictive success but which are not approximately true.

(d) Approximate truth and successful reference of central theoretical terms is not a necessary condition for the novel-predictive success of scientific theories.

So, the no-miracles argument is undermined since, if approximate truth and successful reference are not available to be part of the explanation of some theories' novel predictive success, there is no reason to think that the novel predictive success of other theories has to be explained by realism.

---

[13] See Annas and Barnes (1985).

Hence, we do not need to form an inductive argument based on Laudan's list to undermine the no-miracles argument for realism. Laudan's paper was also intended to show that the successful reference of its theoretical terms is *not* a necessary condition for the novel predictive success of a theory (1981, 45), and there are counter-examples to the no-miracles argument.[14]

2.2.1 The flight to reference

There are two basic (not necessarily exclusive) responses to this:

(I) Develop an account of reference according to which the abandoned theoretical terms are regarded as referring after all.

(II) Restrict realism to those parts of theories which play an essential role in the derivation of subsequently observed (novel) predictions, and then argue that the terms of past theories which are now regarded as non-referring were non-essential so there is no reason to deny that the essential terms in current theories will be retained.[15]

Realists have used causal theories of reference to account for continuity of reference for terms like 'atom' or 'electron', when the theories about atoms and electrons undergo significant changes. The difference with the terms 'ether' and 'caloric' is that they are no longer used in modern science. In the nineteenth century the ether was usually envisaged as some sort of material, solid or liquid, which permeated all of space. It was thought that light waves had to be waves in some sort of medium and the ether was posited to fulfil this role. Yet if there really is such a medium then we ought to be able to detect the effect of the Earth's motion through it, because light waves emitted perpendicular to the motion of a light source through the ether ought to travel a longer path than light waves emitted in the same direction as the motion of the source through the ether. Of course, various experiments, the most famous being that of Michelson and Morley, failed to find such an effect. By then Maxwell had developed his theory of the electromagnetic field, which came to be regarded as not a material substance at all. As a result the term 'ether' was eventually abandoned completely.

However, as Hardin and Rosenberg (1982) argue, the causal theory of reference may be used to defend the claim that the term 'ether' referred after all, but to the electromagnetic field rather than to a material medium. If the reference of theoretical terms is to whatever causes the phenomena responsible for the terms' introduction, then since optical phenomena are now believed to be caused by

---

[14] P. Lewis (2001), Lange (2002), and Magnus and Callender (2004) regard the pessimistic meta-induction as another instance of the base rate fallacy (see 2.1.2 above). Here we sidestep their concerns by reformulating the argument from theory change so that it is not a probabilistic or inductive argument.

[15] Stanford (2003) refers to this as 'selective confirmation'.

the oscillations in the electromagnetic field, the latter is what is referred to by the term 'ether'. Similarly, since heat is now believed to be caused by molecular motions, then the term 'caloric' can be thought to have referred all along to these rather than to a material substance. The danger with this, as Laudan (1984) pointed out, is that it threatens to make the reference of theoretical terms a trivial matter, since as long as some phenomena prompt the introduction of a term it will automatically successfully refer to whatever is the relevant cause (or causes). Furthermore, this theory radically disconnects what a theorist is talking about from what she thinks she is talking about. For example, Aristotle or Newton could be said to be referring to geodesic motion in a curved spacetime when, respectively, they talked about the natural motion of material objects, and the fall of a body under the effect of the gravitational force.

The essence of the second strategy is to argue that the parts of theories that have been abandoned were not really involved in the production of novel predictive success. Philip Kitcher says that: 'No sensible realist should ever want to assert that the idle parts of an individual practice, past or present, are justified by the success of the whole' (1993, 142). Similarly, Psillos argues that history does not undermine a cautious scientific realism that differentiates between the evidential support that accrues to different parts of theories, and only advocates belief in those parts that are essentially involved in the production of novel predictive success. This cautious, rather than all-or-nothing, realism would not have recommended belief in the parts of the theories to which Laudan draws attention, because if we separate the components of a theory that generated its success from those that did not we find that the theoretical commitments that were subsequently abandoned are the idle ones. On the other hand, argues Psillos: 'the theoretical laws and mechanisms that generated the successes of past theories have been retained in our current scientific image' (1999, 108). Such an argument needs to be accompanied by specific analyses of particular theories which both identify the essential contributors to the success of the theory in question, and show that these were retained in subsequent developments, as well as criteria for identifying the essential components of contemporary theories.

Leplin, Kitcher, and Psillos have all suggested something like this strategy, but we will only discuss Psillos's execution of it. What Leplin (1997) sets out to show is something very weak, namely that there are conditions that would justify a realist attitude to a theory, rather than that these conditions have actually been satisfied. He does allude (1997, 146) to the idea of (II) above, but doesn't develop it in any detail. Kitcher (1993) suggests a model of reference according to which some tokens of theoretical terms refer and others do not, but his theory allows that the theoretical descriptions of the theoretical kinds in question may have been almost entirely mistaken, and seems to defend successful reference only for those uses of terms that avoid ontological detail at the expense of reference to something playing a causal role in producing some observable phenomena.¹⁶ The most detailed and influential account is due to Psillos (1999), who takes up Kitcher's suggestion of (II) and develops it in some detail, and also combines it with (I). Laudan claims that if current successful theories are approximately true, then the caloric and ether theories cannot be because their central theoretical terms don't refer (by premise (ii) above). Strategy (I) accepts premise (ii) but Psillos allows that sometimes an overall approximately true theory may fail to refer. He then undercuts Laudan's argument by arguing that abandoned theoretical terms that do not refer, like 'caloric', were involved in parts of theories not supported by the evidence at the time, because the empirical success of caloric theories was independent of any hypotheses about the nature of caloric. Abandoned terms that were used in parts of theories supported by the evidence at the time do refer after all; 'ether' refers to the electromagnetic field.

Below, we identify some problems with this type of defence of realism and the claims made about 'ether' and 'caloric'. Most importantly, we argue that the notion of 'essential', deployed in (a) above, is too vague to support a principled distinction between our epistemic attitudes to different parts of theories. In our view, the problem with strategy (II) is that its applications tend to be ad hoc and dependent on hindsight. Furthermore, by disconnecting empirical success from the detailed ontological commitments in terms of which theories were described, it seems to undermine rather than support realism. As Kyle Stanford (2003) argues, this strategy seems to have 'sacrificed the substantive tenets of realism on the altar of its name' (555).¹⁷

According to Psillos: 'Theoretical constituents which make essential contributions to successes are those that have an indispensable role in their generation. They are those which "really fuel the derivation"' (1999, 110). This means that the hypothesis in question cannot be replaced by an independently motivated, non-ad hoc, potentially explanatory alternative. (Remember that the sort of success referred to here is novel predictive success.) Psillos gives as an example of an idle component of a theory Newton's hypothesis that the centre of mass of the universe is at absolute rest. However, within Newton's system this hypothesis cannot be replaced by an alternative that satisfies the above requirements. That the centre of mass of the universe should move with any particular velocity is surely more ad hoc than it being at rest. There is also a sense in which the universe being at rest is simpler than it having a particular velocity, since the latter would raise the further question of whether it had always been in motion or whether some force set it in motion. Certainly, any alternative hypotheses that might have been

---

¹⁶ For a critique of Kitcher's account of the reference of theoretical terms see McLeish (2005). She argues persuasively that there are no satisfactory grounds for making the distinction between referring and non-referring tokens. McLeish (2006) argues that abandoned theoretical terms partially refer and partially fail to refer. This strategy trivializes the claim that abandoned theoretical terms referred after all unless it is combined with strategy (II).
¹⁷ Psillos is also criticized in Lyons (2006).

entertained would not be explanatory of anything or independently motivated. It seems that this hypothesis does count as an essential contribution to the success of Newton's theory, by Psillos's criteria, and hence that he would have had us be realistic about it. However, the notions of absolute space and absolute rest have no meaning in modern physics so Psillos's criterion above has accidentally bolstered historically inspired scepticism.

This case is arguably not so serious for the realist for it does not involve a central theoretical term that was essentially deployed, yet which cannot be regarded as referring from the standpoint of later theory. Nonetheless, Psillos seeks to deal with the threat of such examples using this distinction between essential and inessential theoretical constituents, and his example reveals that his definition of the distinction does not in general capture only the theoretical hypotheses with which the realist would be happy. Another problem is the ambiguity that arises concerning the type of dependence in question when we ask if a theory's success is dependent on a particular hypothesis. We can understand this as at least either logical/mathematical dependence or causal dependence. So, when we are asked to look at the particular novel empirical success of a theory and decide which are the parts of the theory that this success depended on, then we will give different answers according to how we understand dependence. Furthermore, the realist should be careful here for it is dangerous for realism overall to disconnect the metaphysical hypotheses and background assumptions about supposed entities like caloric or the ether from what are construed as the real successes of the theories in issuing certain predictions. As we have seen, one of the central claims of contemporary realism is that we should take seriously the involvement of theoretical and metaphysical beliefs in scientific methodology, and that we cannot disconnect the success of science from the theoretically informed methods of scientists. This is meant to support realism, for, according to Boyd and others, only realism explains why these extra-empirical beliefs are so important. However, Psillos suggests that after all we need not take seriously scientists' beliefs about the constitution of the ether or caloric, because the success of the theories floats free of such assumptions. Let us now examine the examples he discusses.

2.2.2 'Ether', 'Caloric', ether and caloric

Ether theories were successful by any plausible criteria that the realist may concoct and were mature, yet it was widely believed that the ether was a material substance, and there certainly is no such thing permeating space according to Maxwell's successor theory. The hypotheses about the material nature of the ether were no presuppositional or idle posits because they motivated the use of mechanical principles, like Hooke's law, in investigations of how light waves would propagate in such a medium. This led to the fundamental departure from previous wave theories brought about by Fresnel's assumption that light propagates as a transverse wave motion. Fresnel thus got important 'heuristic mileage' out of these mechanical principles. Therefore, as Worrall (1994) has argued, although the replacement of ether theory by electromagnetic theory provides a constructive proof of the eliminability of the former, mechanics is really only eliminable from the success of Fresnel's theory in a minimal logical sense.

When we inquire into what hypotheses 'really fuel the derivation' we have no other way to address this question than by explaining how we would derive the prediction in question using our understanding of the theory. This does not show that hypotheses which we do not use in reconstructing the derivation played no role in making it possible for the scientists of the time to derive the prediction that they did. A modern scientist may not need to invoke anything about the material constitution of the ether to reconstruct Fresnel's predictions, but Fresnel did so to derive the predictions in the first place. Psillos says that 'essential constituents' of success are ones such that 'no other available hypothesis' (1999, 309) can replace them in the derivation of the novel predictions made by the theory in question. The question is: available to whom? There was no other hypothesis available to Fresnel about the nature of the ether that would have allowed him to derive the novel predictions of his theory. In general it seems true that quite often there will be no other hypothesis available at the time, but that in reconstructing derivations we may have several alternatives.

Psillos argues that there is continuity between the causal roles of attributes of the ether and those of the field. For example, the fundamental causal role of the ether is arguably to act as the repository of the energy associated with light between emission by a source and its absorption or reflection by matter. Light was known to travel at a finite velocity, so it had to be in some medium while passing through otherwise empty space. The electromagnetic field is now thought to be that medium. However, the determination of which causal role is important is done with the benefit of hindsight. Our assessment of what matters in the description of optical phenomena is very much relative to our current state of knowledge, as is any statement about the relevant causal role of some posited unobservable entity. However, we do not know what parts of current theories will be retained, in other words what the real causal roles are.

It is true that the important principles about light (that it undergoes transverse propagation, for example) are carried over into Maxwell's theory, and indeed there is a lot of continuity between the ether theories and electromagnetic field theory. However, the latter has now been replaced by quantum field theories, which may soon be replaced by a theory of superstrings or a grand unified theory of quantum gravity. It is implausible to suggest that 'ether' referred all along to a quantum field, because the latter have a completely different nature to Maxwell's electromagnetic field. For example, the latter is supposed to permeate all space and have a definite magnitude at different points, whereas the former is multi-dimensional and incorporates only probabilities for different magnitudes (at least according to the usual understanding of it).[18]

In the case of caloric, Psillos argues that 'the approximate truth of a theory is distinct from the full reference of all its terms' (1994, 161). 'Caloric', he maintains, was not a *central* theoretical term, and we should only worry about the central theoretical terms, 'central in the sense that the advocates of the theory took the successes of the theory to warrant the claim that there are natural kinds denoted by these terms' (Psillos 1999, 312). One response to Psillos on this issue is to argue that in the realism debate we ought only to be concerned with what scientists *should* believe and not with what they do in fact believe. For example, if a group of antirealists came up with a novel predictively successful theory, realists should not thereby be driven to admit that, because none of the scientists involved believes in the natural kinds denoted by the theory, the theory has no central terms. When we are concerned with philosophical disputation about science the fact that a *particular* scientist had this or that attitude to a theory is irrelevant. Indeed, while the PNC demands that metaphysics be subservient to science, it does not accord the opinions of scientists about metaphysics any special authority. It is the content of science per se that constrains metaphysics but scientists may be and often are wrong about the metaphysical implications of theories. In particular, the authors of a discovery have no privileged weight in discussions of its metaphysical interpretation. For example, we think that a PNC-compatible attitude would have required metaphysicians to take local hidden-variables interpretations of quantum mechanics seriously prior to the experiments that violated Bell's inequality in the 1970s,[19] even though most physicists ignored them, because the scientific knowledge of the time was compatible with them. What Einstein, Heisenberg, or Bohr believed about local hidden variables is epistemically irrelevant.

Thus, in response to Psillos: when a philosopher is reconstructing the history of science for interpretation of the central commitments of theories, she should take account of all the science we know now. The metaphysical opinions of historical scientists have no more authority than do the metaphysical attitudes of historical philosophers. Of course, these opinions are important for explaining

---

[18] Saatsi adopts strategy (II) with respect to the ether theory of light and argues that 'ether' was not a central theoretical term, but agrees with Psillos that there is continuity between the ether theory and Maxwell's electromagnetic field theory at the level of the causal properties of light. However, like Chakravartty (1998), he fails to consider the consequences of the fact that physics has moved on considerably since Maxwell. Our understanding of those causal properties has changed dramatically and there simply is nothing in contemporary physics that corresponds to the entities that were supposed to bear those properties. There is no consensus among those defending standard realism in the face of theory change and their manoeuvres have become increasingly ad hoc. Hence, we have token not type reference (Kitcher), partial disjunctive reference (McLeish), reference after all or terms not central after all (Psillos), and approximate truth of the causal roles but not reference (Saatsi)—better to abandon standard scientific realism given the availability of structural realism as an alternative.

[19] There were many such experiments; see Redhead (1987, 108).

why the history of science went as it did. In any case, to return to Psillos's example, he argues that in the case of caloric theory all the important predictive successes of the theory were independent of the assumption that heat is a material substance. He argues that all the well-confirmed content of the theory, the laws of experimental calorimetry, was retained in thermodynamics anyway. Here Psillos can be construed as exploiting, as scientists often do, the positive side of the underdetermination problem. When we theorize about or model the phenomena in some domain we will inevitably make a few mistakes. If there was only one (the true) theory that could describe the phenomena, then we would almost certainly never hit upon it initially; we arrive at adequate theories by modifying earlier ones in the face of new evidence. More importantly, after radical theory change we want to be able to recover the empirical success of old theories without buying into their outdated ontologies. The ever present possibility of alternative, empirically equivalent theories is therefore essential. Psillos makes a similar point with respect to ether as for caloric. 'Ether' refers, according to him, because scientists weren't committed to it being of any particular nature but were committed to its having some properties which as a matter of fact are carried over to Maxwell's theory.

We agree with Psillos that what is philosophically important about caloric theory and ether theory is that they were stations along the way to our current conceptions of heat and of the propagation of light, respectively. Directing attention to the question of whether the terms 'hooked onto objects' is a red herring (and the same red herring that ushered in application by Putnam of the Kripke/Putnam theory of reference to cases in the philosophy of science). However, Psillos's indexing the basis for his conclusions about caloric and ether to judgements of historical scientists is *also* a red herring. Suppose that scientists had been so committed to belief in the material constitution of the ether that denying its material nature would have seemed absurd to them. Imagine that this *would have* caused them to fail to achieve some derivations that were otherwise available. (What we are here imagining was no doubt true of some or many scientists.) Would we then be called upon to reverse Psillos's whiggish verdict about ether theory? In that case what is presented as the subject of a metaphysical question—Does the theory (partially) accurately describe the world?—seems to resolve into questions about the intentional characterization of some particular long-dead scientists. This is surely confused. What matters to the metaphysical question, we suggest, is the structure of the theory. Science is authoritative on the question of which current theories are appropriate objects of comparison with the structure of the old theory for purposes of metaphysically evaluating the latter. Rejecting Psillos's version of historicism does not replace appeal to history with appeal to philosophers' intuitions.

But there is another problem. Even if Psillos's defence of referential continuity for the ether theories of light were taken as correct for his reasons, this would show only that the term 'ether' refers to the classical electromagnetic field. This is not much of an advance on the position that says 'ether' doesn't refer or refers to the empty set, and for the small gain we must pay the high price of making reference too easy. Scientific realism becomes uninteresting if it is the thesis that our theoretical terms refer but to things that may be quite unlike theoretical descriptions we have of them. Rather than saying that the world is much like our best theories say it is, and that theoretical terms of such theories genuinely refer, realists would retreat to saying that some theoretical terms refer, some don't, and others approximately refer. Any hypothesis or entity will only be a legitimate object of belief if it is essential for producing not just the success, but the novel predictive success, of a theory. So the realist has to concede that a theory can be highly empirically successful despite the lack of anything in the world like the entities it postulates.[20] Hence, the danger with strategy (II) is that it may rely on hindsight in such a way as to make the pronouncements as to which terms are central and which are not depend on which have subsequently been abandoned. This would make (II) an ad hoc solution to the problem of abandoned theoretical terms.

We join Bishop (2003) in emphasizing that the debate about whether the world is really how it is described to be by scientific theories is not an issue in the philosophy of language. No matter how the realist contrives a theory of reference so as to be able to say that terms like 'ether' refer, there is no doubt that there is no elastic solid permeating all of space. The argument from theory change threatens scientific realism because if what science now says is right, then the ontologies of past scientific theories are far from accurate accounts of the furniture of the world. If that is right, even though they were predictively successful then the success of our best current theories does not mean they have got the nature of the world right either. Thus the argument from theory change can be thought of as a reductio on scientific realism. Our solution to this problem is to give up the attempt to learn about the nature of unobservable entities from science. The metaphysical import of successful scientific theories consists in their giving correct descriptions of the structure of the world. Hence, we now turn to structural realism.

From the perspective of the structural realist the problem of referential continuity across theory change is a pseudo-problem. Indeed, it has some claim to be regarded as the 'patient zero' of the neo-scholastic disease in metaphysics. Putnam's (1975b) account of the referential semantic of natural kind terms, which brought Kripke-style metaphysics into the philosophy of

---

[20] The realist does have to make this concession. Consider, for example, so-called 'rational expectations theory' in economics. This is predictively successful, and enjoyed novel predictive success, in at least some very sophisticated niche markets (for example, markets for derivatives) between phase shifts in higher-scale dynamics. Yet the main objects of predication in the theory, atomless measure spaces and infinitely lived agents, quite obviously have no empirical counterparts. (See Kydland and Prescott 1982, the most important in a series of papers that earned its authors the Nobel Prize.)

science, was a response to Feyerabend's (1962) argument that if the extensions of theoretical terms are functions of the intensional meanings entertained by scientists, then theory change implies ontological discontinuity over time, rather than progress with respect to truth. We suggest that the appropriate response to Feyerabend's problem is to give up the idea that science finds its metaphysical significance in telling us what sorts of things there are, rather than to follow Putnam in promoting a non-empiricist account of the relationship between things, terms, and theories.

2.3 STRUCTURALISM

As we have seen, in the debate about scientific realism, the no-miracles argument is in tension with the arguments from theory-change. In an attempt to break this impasse, and have 'the best of both worlds', John Worrall (1989) introduced *structural realism* (although he attributes its original formulation to Poincaré). Using the case of the transition in nineteenth-century optics from Fresnel's elastic solid ether theory to Maxwell's theory of the electromagnetic field, Worrall argues that:

There was an important element of continuity in the shift from Fresnel to Maxwell—and this was much more than a simple question of carrying over the successful empirical content into the new theory. At the same time it was rather less than a carrying over of the full theoretical content or full theoretical mechanisms (even in 'approximate' form) ... There was continuity or accumulation in the shift, but the continuity is one of form or structure, not of content. (1989, 117)

According to Worrall, we should not accept full-blown scientific realism, which asserts that the *nature* of things is correctly described by the metaphysical and physical content of our best theories. Rather we should adopt the structural realist emphasis on the mathematical or *structural* content of our theories. Since there is (says Worrall) retention of structure across theory change, structural realism both (a) avoids the force of the pessimistic meta-induction (by not committing us to belief in the theory's description of the furniture of the world) and (b) does not make the success of science (especially the novel predictions of mature physical theories) seem miraculous (by committing us to the claim that the theory's structure, over and above its empirical content, describes the world).

Hence, a form of realism that is only committed to the structure of theories would not be undermined by theory change. The point is that theories can be very different and yet share all kinds of structure. The task of providing an adequate theory of approximate truth at the level of ontology which fits the history of science has so far defeated realists, but a much more tractable problem is to display the structural commonalities between different theories.

Examples supplementing Worrall's are given by Saunders (1993a) who shows how much structure Ptolemaic and Copernican astronomy have in common, by Brown (1993), who explains the correspondence between Special Relativity and classical mechanics, and by Bain and Norton (2001), who discuss the structural continuity in descriptions of the electron.[21] Holgar Lyre (2004) extends Worrall's example by considering the relationship between Maxwellian electrodynamics and Quantum Electrodynamics (QED). Saunders (2003a and 2003b) criticizes Cao (1997) for underestimating the difficulties with a non-structuralist form of realism in the light of the history of quantum field theory.

There are in fact numerous examples of continuity in the mathematical structure of successive scientific theories, but we offer simple examples from two of the most radical cases of theory change in science, namely the transition from classical mechanics to Special Relativity, and the transition from classical mechanics to quantum mechanics.

In classical mechanics if system B is moving at constant velocity v_{AB} with respect to system A, and system C is moving at constant velocity v_{BC} with respect to system B, then system C is moving at constant velocity $v_{AC} = v_{AB} + v_{BC}$ with respect to system A. This simple addition law for relative velocities does not hold in Special Relativity, because the coordinate systems defined by the inertial frames of reference in which each of the bodies is stationary are not related by the Galilean transformations:

$$x' = (x - vt)$$
$$y' = y$$
$$z' = z$$
$$t' = t\text{[22]}$$

but rather by the (simple) Lorentz transformations:

$$x' = \gamma(v)(x - vt)$$
$$y' = y$$
$$z' = z$$
$$t' = \gamma(v)(t - vx/c^2)$$

where $\gamma(v) = \dfrac{1}{\sqrt{1-\frac{v^2}{c^2}}}$, and c is the velocity of light in a vacuum.

---

[21] Note that although Bain and Norton do not defend ontic structural realism and seem to return to 'a property predicated on individual systems', Bain thinks that 'the more appropriate reading would abstract the structure from the individuals a la French and Ladyman' (2003, 20).

[22] For simplicity, we consider the case where one body is moving along the x-axis of the other's coordinate system so that B's y and z coordinates do not change.

However, when v becomes very small compared to c, or when c tends to infinity, so that v/c tends to 0, the mathematical structure of the latter clearly increasingly approximates that of the Galilean transformations. Although the theories are quite different in their ontological import—for example, Special Relativity does not allow the definition of absolute simultaneity—there is partial continuity of mathematical structure. There are a host of such relations that obtain between classical mechanics and Special Relativity, but now consider the relationship between quantum mechanics and classical mechanics.[23]

Neils Bohr self-consciously applied the methodological principle that quantum mechanical models ought to reduce mathematically to classical models in the limit of large numbers of particles, or the limit of Planck's constant becoming arbitrarily small. This is known as the 'correspondence principle'. There are numerous cases in quantum mechanics where the Hamiltonian functions that represent the total energy of mechanical systems imitate those of classical mechanics, but with variables like those that stand for position and momentum replaced by Hermitian operators. Similarly, the equation (known as Ehrenfest's theorem), $\text{grad}V(<r>) = m(d^2 <r> /dt^2)$, where V is the potential and r is the position operator, exhibits continuity between classical and quantum mechanics. It has a similar form to the equation $F = ma$ since in classical mechanics gradV is equated with F. But the quantum equation involves the expectation values of a Hermitian operator, where the classical equation features continuous real variables for position.[24]

The most minimal form of structuralism focuses on empirical structure, and as such is best thought of as a defence of the cumulative nature of science in the face of Kuhnian worries about revolutions (following Post 1971). This is how van Fraassen sees his structural empiricism, to which we now turn.

2.3.1 From constructive empiricism to structural empiricism

Van Fraassen defines scientific realism as follows: 'Science aims to give us, in its theories, a literally true story of what the world is like; and acceptance of a scientific theory involves the belief that it is true' (1980, 8). Constructive empiricism is then contrastively defined thus: 'Science aims to give us theories which are empirically adequate; and acceptance of a theory involves as belief only that it is empirically adequate' (1980, 12). Claiming empirical adequacy for a theory is equated by van Fraassen with claiming that: 'What it says about

---

[23] See Brown (2005) and Batterman (2002). Batterman discusses numerous examples of limiting relationships between theories, notably the renormalization group approach to critical phenomena, and the relationship between wave and ray optics. We return to Batterman's work in 4.2.

[24] The Poisson bracket provides another example of structural continuity between classical and quantum mechanics. The correspondence principle and the relationship between classical and quantum mechanics is beautifully explained in Darrigol (1992).

the observable things and events in this world, is true' (1980, 12). In other words, 'the belief involved in accepting a scientific theory is only that it 'saves the phenomena', that is that it correctly describes what is observable' (1980, 4). Note that this means that it saves *all* the *actual* phenomena, past, present, and future, not just those that have been actually observed so far, so even to accept a theory as empirically adequate is to believe something more than is logically implied by the data (1980, 12, 72). Moreover, for van Fraassen, a phenomenon is simply an *observable* event and not necessarily an observed one. So a tree falling over in a forest is a phenomenon whether or not someone actually witnesses it.

Note that

(a) Scientific realism has two components: (i) theories which putatively refer to unobservable entities are to be taken literally as assertoric and truth-apt claims about the world (in particular, as including existence claims about unobservable entities); and (ii) acceptance of these theories (or at least the best of them) commits one to belief in their truth or approximate truth in the correspondence sense (in particular, to belief that tokens of the types postulated by the theories in fact exist). Van Fraassen is happy to accept (i). It is (ii) that he does not endorse. Instead, he argues that acceptance of the best theories in modern science does not require belief in the entities postulated by them, and that the nature and success of modern science relative to its aims can be understood without invoking the existence of such entities.[25]

(b) Both doctrines are defined in terms of the aims of science, so constructive empiricism is fundamentally a view about the aims of science and the nature of 'acceptance' of scientific theories, rather than a view about whether electrons and the like exist. Strictly speaking it is possible to be a constructive empiricist and a scientific agnostic, or a scientific realist and scientific agnostic.[26] That said, it is part of van Fraassen's aim to show that abstaining from belief in unobservables is perfectly rational and scientific.

(c) Empirical adequacy for scientific theories is characterized by van Fraassen in terms of the so-called 'semantic' or 'model-theoretic' conception of scientific theories, the view that theories are fundamentally extra-linguistic entities (models or structures), as opposed to the syntactic account of theories, which treats them as the deductive closure of a set of formulas in first-order logic. The semantic view treats the relationship between theories and the world in terms of isomorphism. On this view, loosely speaking, a theory is empirically

---

[25] Both the scientific realist and the constructive empiricist hold that empirical adequacy is a necessary condition for the success of science and so it is uncontroversial among them that science has been successful.

[26] See Forrest (1994).

adequate if it 'has at least one model which all the actual phenomena fit inside' (1980, 12).[27]

Initial criticism of van Fraassen's case for constructive empiricism concentrated on three issues:

(i) The line between the observable and the unobservable is vague and the two domains are continuous with one another; moreover the line between the observable and the unobservable changes with time and is an artefact of accidents of human physiology and technology. This is supposed to imply that constructive empiricism grants ontological significance to an arbitrary distinction.

(ii) Van Fraassen eschews the positivist project which attempted to give an a priori demarcation of predicates that refer to observables from those that refer to unobservables, and accepts instead that: (a) all language is theory-laden to some extent; and (b) even the observable world is described using terms that putatively refer to unobservables. Critics argue that this makes van Fraassen's position incoherent.

(iii) The underdetermination of theory by evidence is the only positive argument that van Fraassen has for adopting constructive empiricism instead of scientific realism; but all the data we presently have underdetermine which theory is empirically adequate just as they underdetermine which theory is true (this is the problem of induction), and so constructive empiricism is just as vulnerable to scepticism as scientific realism. This is taken to imply that van Fraassen's advocacy of constructive empiricism is the expression of an arbitrarily selective scepticism.

(i) is rebutted first by pointing out that vague predicates abound in natural language but clear extreme cases suffice to render their use acceptable, and there are at least some entities which if they exist are unobservable, for example, quarks, spin states of subatomic particles, and photons. Second, van Fraassen argues that epistemology ought to be indexical and anthropocentric, and that the distinction between the observable and the unobservable is not to be taken as having direct ontological significance, but rather epistemological significance. Says van Fraassen: 'even if observability has nothing to do with existence (is, indeed, too anthropocentric for that), it may still have much to do with the proper epistemic attitude to science' (1980, 19).

For van Fraassen, 'observable' is to be understood as 'observable-to-us': 'X is observable if there are circumstances which are such that, if X is present to us under those circumstances, then we observe it' (1980, 16). What we can and cannot observe is a consequence of the fact that:

---

[27] Bueno (1997) offers a more precise explication of the notion of empirical adequacy. See also Bueno, French, and Ladyman (2002). We return to the semantic approach in 2.3.3.

the human organism is, from the point of view of physics, a certain kind of measuring apparatus. As such it has certain inherent limitations—which will be described in detail in the final physics and biology. It is these limitations to which the "able' in 'observable' refers—our limitations, *qua* human beings. (1980, 17)

So we know that, for example, the moons of Jupiter are observable because our current best theories say that were astronauts to get close enough, then they *would* observe them; by contrast the best theories of particle physics certainly do not tell us that we are directly observing the particles in a cloud chamber. Analogous with the latter case is the observation of the vapour trail of a jet in the sky, which does not count as observing the jet itself, but rather as detecting it. Now if subatomic particles exist as our theories say they do, then we detect them by means of observing their tracks in cloud chambers, but, since we can never experience them directly (as we can jets[28]), there is always the possibility of an empirically equivalent but incompatible rival theory which denies that such particles exist. This fact may give the observable/unobservable distinction epistemic significance.

(ii) is rebutted by arguing that matters of language are irrelevant to the issue at hand. The observable/unobservable distinction is made with respect to entities not with respect to names or predicates.

(iii) is the most serious problem for van Fraassen. Note first that, contrary to what is often claimed, van Fraassen does not accept that inference to the best explanation is rationally compelling in the case of the observable world while denying it this status for the case of the unobservable world.[29] Furthermore, van Fraassen does not appeal to any global arguments for antirealism such as the underdetermination argument or the pessimistic meta-induction. He rejects realism not because he thinks it irrational but because he rejects the 'inflationary metaphysics' which he thinks must accompany it, that is, an account of laws, causes, kinds, and so on, and because he thinks constructive empiricism offers an alternative view that offers a better account of scientific practice without such extravagance (1980, 73). Empiricists should repudiate beliefs that go beyond what we can (possibly) confront with experience; this restraint allows them to say 'good bye to metaphysics' (1989; 1991, 480).

What then is empiricism and why should we believe it? Van Fraassen suggests that to be an empiricist is to believe that 'experience is the sole source of information about the world' (1985, 253). The problem with this doctrine is that it does not itself seem to be justifiable by experience.[30] However, as we discussed in Chapter 1, he has lately argued that empiricism cannot be reduced to

---

[28] Van Fraassen adopts a direct realism about perception for macroscopic objects: 'we can and do see the truth about many things: ourselves, others, trees and animals, clouds and rivers—in the immediacy of experience' (1989, 178).

[29] See Psillos (1996), (1997) and Ladyman et al. (1997).

[30] Jennifer Nagel (2000) discusses the problem of how an empiricist can claim to know what counts as experience.

the acceptance of such a slogan, and that empiricism is in fact a stance in Husserl's sense of an orientation or attitude towards the world.[31] Regarding empiricism as a stance is indeed, as we saw, intended to immunize it to certain sorts of criticisms that are appropriate only for doctrines. However, this move is not meant to deny the appropriateness of justification and criticism altogether. Ladyman (2000) has identified internal tensions within van Fraassen's version of empiricism which we now interpret as identifying ways in which a defender of the empiricist stance should be dialectically led, by her own lights, towards the synthesis of the empiricist and materialist stances that we promote (that is, to our 'scientistic' stance and ontic structural realism).

Constructive empiricism is supposed to offer a positive alternative to scientific realism that dispenses with the need for metaphysics. It is a positive view of science which is intended to free us from the need to articulate accounts of laws, causes, and essential properties that take seriously the apparent modal commitments of such notions. This promised liberation from metaphysics through scepticism about objective modality is fundamental to van Fraassen's empiricism: 'To be an empiricist is to withhold belief in anything that goes beyond the actual, observable phenomena, and to recognise no objective modality in nature' (1980, 202).

Ladyman (2000) argues that to apply the empiricist stance consistently one must, in spite of what van Fraassen says here, recognize objective modality in nature. This is largely because constructive empiricism recommends, on epistemological grounds, belief in the empirical adequacy rather than the truth of theories, and hence requires that there be an objective modal distinction between the observable and the unobservable. In their reply to Ladyman, Monton and van Fraassen concede that it is possible to combine constructive empiricism with a commitment to objective modality. Ladyman (2004) suggests that this is tantamount to a form of structural realism (see also Giere 1985, 83), and this is the view that we will defend here. Such modal structural realism allows us to take account of the insights of van Fraassen's advocacy of the semantic approach to scientific theories (1979) and structural empiricism (2006), by means of which he illuminates the representational character of science, and the nature of the continuity of scientific theories through time, respectively. Hence, we will argue that ontic structural realism ought to be understood as modal structural empiricism, and that this view can claim all the advantages of constructive empiricism and scientific realism without being prone to the problems that those views respectively face.

Constructive empiricism emphasizes the description science gives of relations among phenomena. Van Fraassen (2006) proposes the following '*requirement*

---

[31] See van Fraassen (1994), (1995), (1997), and (2002), and see the symposium in *Philosophical Studies* (2004) for critical comments by Peter Lipton, Paul Teller and Ladyman, and a reply by van Fraassen.

upon succession: The new theory is so related to the old that we can explain the empirical success of the old theory if we accept the new' (298). He goes on to say:

So we can reply to Worrall: *YES*, there is an accumulation of knowledge through science, but it is knowledge about the observable phenomena. You did make a good point: there is an accumulation going on throughout all those deep theoretical changes. Moreover this requirement upon theory succession should satisfy the 'No Miracle' intuition! The success of science is *not a miracle*, because in any theoretical change both the past empirical success retained and new empirical successes *were needed as credentials* for acceptance. (298–9)

So he argues that what he calls 'structural empiricism' can account for the kind of continuity of structure among scientific theories to which Worrall draws attention. Citing other examples, such as the $c \to \infty$ limit of the Special Theory of Relativity, from which Newton's mechanics and the Galilean transformations can be recovered, and the limit of quantum mechanics as Planck's constant tends towards zero, from which classical Hamiltonian mechanics can be recovered, van Fraassen argues that in each case an old theory's empirical success is explained by the successor theories, thanks to structural continuities between their representations. He also argues that structural representation and its limits are ubiquitous in science.

In his John Locke lectures (2008), van Fraassen emphasizes the history of the underdetermination problem in the debates of the early twentieth century over whether it is possible to determine empirically the correct geometry of physical space. There was then growing currency for the idea that science represents the empirical world as embeddable in certain abstract models, where the latter are structures only knowable up to isomorphism. However, Weyl argued in this context that if science describes its subject matter only up to isomorphism, then whatever it is that distinguishes colours from points in the projective plane cannot be revealed by science, and must therefore be known by intuition. Similarly, Russell famously argued with Poincaré about the geometry of physical space, contending that we could only know the real congruence relation from isomorphic ones through intuition, whereas Poincaré maintained that asking what is the right congruence relation is like asking how to spell the letter 'a' (cited in Coffa 1993, 20–1). There is an analogy here with the debate between Frege and Hilbert about the meaning of mathematical terms. The former thought that their meaning must be known through intuition in advance since the axioms could only be true or false if they already had a reference, whereas the latter thought that the terms obtained their meaning just through the axioms. Van Fraassen's solution to this problem is to appeal to pragmatics as what fixes the reference of theories over and above their structural representation of the world. Of course, he only regards them as referring to the structure of the relevant phenomena.

2.3.2 What's really wrong with constructive empiricism?

In the following sections we explain what we regard as the most serious problems for constructive empiricism and argue that supplementing van Fraassen's empiricism with a commitment to objective modality is the best way of responding to them.

2.3.2.1 Selective scepticism and common-sense realism

It is clear that Worrall intended to go further than structural empiricism since his advocacy of structural realism explicitly referred to the need for some form of realism to satisfy the no-miracles argument. Van Fraassen denies that there must be some explanation in terms of unobservables for the 'persistent similarities' in the phenomena (1985, 298). However, realists argue that the only reason for accepting the objective existence of such everyday objects as tables is to explain such persistent similarities. Therefore, it might seem that all the good arguments for the existence of tables carry over to the existence of electrons and, similarly, any argument for withholding belief in the latter likewise motivates scepticism about the former, because of the strong empirical equivalence of hypotheses like that of Descartes' evil demon. The belief in everyday objects allows us to explain many observable phenomena that would otherwise be inexplicable. Why should such explanation be ruled out for the unobservable world? So, the realist claims, the constructive empiricist is epistemically erratic, because scientific realism is nothing more than the analogue of common-sense metaphysical realism in the unobservable domain. This view is expressed by Worrall: 'Nothing in science is going to *compel* the adoption of a realist attitude towards theories ... But this leaves open the possibility that some form of scientific realism, while strictly speaking unnecessary, is nonetheless the *most reasonable* position to adopt' (1984, 67). (Note Worrall's partial anticipation here of van Fraassen's 'stance stance' to the controversy between empiricism and realism as discussed in Chapter 1.) In order to make sense of our perceptions we are not compelled to assume the existence of a real, external world; nonetheless, this seems the most reasonable position to take. So, the question at this stage in the dialectic between the scientific realist and van Fraassen is whether van Fraassen's scepticism about unobservables is like Cartesian or inductive scepticism in being radically at odds with what is reasonable according to common sense. If the realist could show that it is he would win the argument.

Since Locke's time, many have been prepared to settle for the claim that the existence of objects is the best explanation of the regularities in our experience. However, the need to infer the existence of objects only arises in the first place if direct realism is abandoned. Locke accepted that we do not directly perceive external objects, and that many of our ideas of the properties they have are mistaken. Only if we have been similarly persuaded to abandon the manifest image and hence the common-sense view of the world is there any need for us to infer the existence of everyday objects. Of course there are many techniques that can be deployed to make us doubt that our experience of the world shows us reality rather than mere appearances. Once we have been convinced by such arguments, various replacements for common sense are possible. Perhaps everything is an idea in the mind of God, or part of a computer simulation. The only reason why one would be persuaded to abandon direct realism and replace it with indirect *realism*, and not idealism or something else, is that the same scientific story that makes us think that objects don't really possess colours also seems to give us an account of what tables and the like really are.

So the explanation of our experience provided by indirect realism was always really a promissory note for the explanation that science will one day give in terms of a basic ontology of atoms. Locke only articulated causal or indirect realism because he saw his job as giving a philosophy apt for the science of his day. His commitment to the existence of mind-independent tables, trees, and stones is really a commitment to the entities of which they are composed. He concedes that many of the properties we attribute to common-sense objects are not really possessed by them unperceived. However, aggregates of corpuscles can do duty for them and they at least possess the properties of extension, position, and motion that the objects we see around us display. Hence, his indirect realism just is scientific realism. So the attempt to show that without IBE van Fraassen has no right to believe in tables is a mistake. One only needs IBE to defend indirect realism if one has already assumed that the scientific story about how collections of atoms cause ideas of colour, solidity, and so on in us is more or less true and not merely empirically adequate. It is not van Fraassen who is at odds with the common sense view of everyday objects, but the scientific realist who claims that the table is mostly empty space, and that matter, space, and time and all other physical phenomena are nothing like how they seem. Scientific realism is like a radical sceptical hypothesis, but one that we are encouraged to take seriously because of the astounding success of science.

Arguably, common sense requires that epistemologists acknowledge the veracity of many everyday first-order knowledge claims about tables, trees, and stones. As we have seen, realists argue that van Fraassen's epistemology violates this condition because they claim that without IBE we have no warrant for believing in everyday objects. When it comes to unobservable objects, van Fraassen allows that both belief and agnosticism are rational. However, it is noteworthy that he is not prepared to be so permissive when it comes to tables. Recall that he adopts a direct realism about perception for macroscopic objects: 'We can and do see the truth about many things: ourselves, others, trees and animals, clouds and rivers—in the immediacy of experience' (van Fraassen 1989, 178). The manifest image of a world of objects is not inferred. This means that many of our first-order knowledge claims about everyday objects are indeed true, and about these he is in no danger of conflicting with common sense. Indeed, he can even say that it is irrational not to believe in everyday objects.

2.3.2.2 Selective scepticism and inductive libertarianism

Van Fraassen's response to the charge that he ought to be an inductive sceptic in the face of the problem of induction, given his lack of resources for defeating underdetermination, is to argue for a variety of inductive voluntarism, according to which ampliative inferences are not irrational so long as they do not violate the constraints of consistency embodied in logic and probability theory. Hence he argues that constructive empiricism is not rationally required but permitted, and the view of science which best expresses empiricist values (see his 1989 and 2002)—that is, implements the empirical stance. There has been dispute about whether this last claim is sustainable. (See Alspector-Kelly 2001, and Ladyman forthcoming a). Consider the following range of views about a theory T which has been tested and fits the phenomena so far observed:

T is true.
The world is as if T but T is false.
T is empirically adequate.
T is empirically adequate in region of spacetime X.
T is empirically adequate up to time t.
T is adequate to all actually observed phenomena.

Strict empiricism is an unwillingness to go beyond what experience tells us. Constructive empiricism is a compromise between strict empiricism and a prior commitment to the rationality of science. Why is constructive empiricism a better expression of empiricist values than actualist empiricism which holds fast to the last of the commitments above, but which nonetheless goes beyond strict empiricism?

Imagine the following scenario. You have a light box. This is a device which is lined on the inside with light-absorbent material, and also contains a light source. There is an eye-piece to allow an observer to look inside, and finally a copper bar is set into a recess in the base of the box. If the light is on and the bar is raised the observer sees it, but if the bar is not raised then what the observer sees is the same whether or not the light is on. The room is well lit, you are a careful observer, conservative by nature, and you have devoted your life to studying the physics of light. You have learned and tested the best theories of optics, and spent twenty years working on quality control in a light-box factory. The box before you has been tested, and handmade with the finest materials, and it is switched on. (This is supposed to be a best possible case for induction so if your years spent in analytic epistemology seminars incline you to add more conditions you should feel free to do so.)

Here are some possible judgements you can consider:

(1) There is a box in front of you.
(2) Photon theory is empirically adequate with respect to the box.
(3) Photon theory is empirically accurate with respect to the box on those occasions when you actually look.
(4) Photon theory is (approximately) true and there are photons in the box.
(5) When you next raise the bar and look you will see it.

Van Fraassen argues that (1) is known because of direct realism. With such claims in such favourable conditions it seems he even thinks that knowledge will iterate so that you can know that you know there is a box in front of you. In so far as (1) is concerned van Fraassen says nothing that conflicts with common sense. He thinks the empiricist ought to believe (2), but it is not clear why they don't stop at (3) since that says as much that is in fact empirically tested as (2). Both (2) and (3) entail (5) so the empiricist will believe that. He thinks that although it is not irrational to believe (4) if you have a commitment to explanations, empiricists shouldn't endorse it. It is interesting to consider what van Fraassen can say about whether or not first-order knowledge claims involving (2), (4), and (5) can be true. Suppose we live in a world where there are photons, and scientific realists believe in them. Suppose too that they have acquired their beliefs in the same way as the constructive empiricists acquired their belief in the empirical adequacy of photon theory, but the realists just made the extra leap to belief in unobservables. (Suppose they have not read *The Scientific Image* and think IBE is compelling and that acceptance entails belief in truth.) The scientific realists are not irrational but can they be said to know (4)?

It may seem that the whole point of van Fraassen's philosophy of science is to say that, while photons might exist, and there is nothing irrational about believing in them, we can't *know* that there are photons because they are unobservable. However, he allows that in principle there is no difference between belief in (2) and belief in (4), in the sense that nothing in the nature of rationality requires anyone to believe either. Now, because of his voluntarism, van Fraassen has to allow that the person who abstains from inductive inference is perfectly rational. Hence, the person who observes the behaviour of the box, learns the theory of photons, derives predictions, confirms them, and so on, can still refuse to endorse not only (2) and (3), but also (5), and not thereby be irrational. He must also allow that even the person who refuses to form expectations according with (5) and instead predicts that the next time they raise the bar they will not see it is not thereby irrational. Whether or not we think that scientific realism is correct, the denial of (5), and the assertion of its contrary, in the face of all the evidence, without some explanation of why things will be different next time, is surely irrational. In any case, if van Fraassen denies that first-order knowledge claims about photons are true, even if photons exist, just because belief in them goes beyond the evidence, then he must similarly deny that first-order knowledge claims about (2) and (5) are also false, even if (2) and (5) are themselves true. (That he does indeed think this is suggested by this remark: 'Empirical adequacy goes far beyond what we can know at any given time' (1989, 69).)

This is an unacceptable consequence, for it means that even in ideal circumstances for induction nobody ever knows anything about unobserved cases even though their beliefs about them are true. Furthermore, it makes van Fraassen's scepticism about unobservables uninteresting because it amounts to the claim that nobody knows that electrons exist in *just the same way* that nobody knows that any scientific theory is empirically adequate, or even that the next time you raise the bar you will see it. There would then be no epistemic asymmetry between belief in unobservables and belief in empirical adequacy.

On the other hand, suppose van Fraassen allows that first-order knowledge claims like 'you know that the next time you raise the bar you will see it' or 'you know that the photon theory is empirically adequate' can be true. Since no rule compels us to arrive at these claims, and since to have inferred nothing or indeed the contrary would have been equally rational according to him, then all that could make them known is that they are in fact true beliefs. It follows that if there are electrons, and scientific realists believe in them, then *they* know they exist. Hence, van Fraassen's philosophy of science can't consist in the claim that we can't know there are unobservables. It must then be committed to the claim that we can't know that we know there are photons. However, van Fraassen concedes that we can't know that we know a theory is empirically adequate either, so again it seems as if the claim that photons exist and the claim that photon theory is empirically adequate are epistemically on a par. We can't know that we know either but if in fact the claims are true then we can know them. The mere fact that belief in photons goes further than belief in the empirical adequacy of photon theory will not be enough to show that the latter belief can be knowledge and the former cannot, because the mere fact that belief that the theory is empirically adequate goes further than the belief that the theory has been empirically adequate up to now, is not enough to show that the latter belief can be knowledge and the former cannot.

Van Fraassen's voluntarism commits him to *inductive libertarianism*: whatever constraints govern inductive inferences are not rational constraints. Inductive libertarianism cannot impugn the rationality of deviant induction and so wishful thinking is as rational as critical realism, and someone who capriciously disregards all the evidence and counter-inducts cannot be criticized for irrationality so long as their synchronic degrees of belief remain consistent. On van Fraassen's view, then, someone who forms grue-type hypotheses, finds them continually falsified, but persists in forming new ones is just as rational as the follower of standard inductive practices. The grue-type believer only becomes irrational if they fail to abandon their specific belief if it is falsified. If van Fraassen's position allows us truthfully to claim to know the results of induction, this is only because he has weakened the notion of knowledge to its breaking point.

When we test inductive inferences we can often freely choose which of the as yet unobserved cases to observe. That we are so often able to identify regularities in phenomena and then use them for prediction needs to be explained. In a world without nomic connections all our inductive knowledge would be lucky knowledge, which is to say not knowledge at all. On the other hand, if we believe that there are real necessary connections between phenomena, then we are positively justified in making inductive inferences provided we are careful in doing so. What being careful means is something we learn by induction based on sometimes bitter experience. Of course, IBE cannot be used to defend infallibilism about inductive knowledge. However, inductive inference is rationally required in a best-case scenario like that of the light box.

There is a connection between these issues concerning induction and the discussion in the previous section concerning common-sense realism. The most obvious response to phenomenalism or idealism is to appeal to the fact that you can close your eyes, open them at random, and everything is pretty much just as it was. The best reason we have for thinking the table must be there when we aren't looking is that we can predict that it will be there when we do look, and there is intersubjective agreement about this. The table is an invariant under significant transformations of both it and observers. It is supposed to be the existence of things, whether everyday or scientific, that explains all of this, but what really does the explanatory work is not the things but their stable causal powers. We only know, if we know, about unobservable objects (unlike tables) in virtue of our theories about them.[32] Since we only understand the entities as we do because of the way they are supposed to relate to our observations, belief in the unobservables entailed by our best scientific theories entails belief that those theories are at least approximately empirically adequate, in other words that there really are the relations among the phenomena the theories attribute to the world. If those relations were merely randomly correlated with the objects then the claim that the objects exist could not count as an explanation of what is observed. It is noteworthy that scientific realists often invoke thick non-Humean notions of causation in their attempts to convince us that unobservables exist. However, we need not go so far as belief in objects, observable or not. The positing of stable modal relations among the phenomena will do just as well. The history of science undermines not only materialism and classical views of space and time, but also the claim that science describes the true objects that lie beyond the phenomena. We may know little else about the nature of reality but we are warranted in supposing that it has a modal structure which is detailed to some extent by folk knowledge and by science.

---

[32] Entity realism notwithstanding.

2.3.2.3 Constructive empiricism and the metaphysics of modality[33]

Recall that van Fraassen rejects realism not because he thinks it irrational but because he rejects the 'inflationary metaphysics' which he thinks must accompany it (an account of laws, causes, kinds, and so on), and because he thinks constructive empiricism accounts for scientific practice without metaphysics. Scientific realism is associated with the view that science answers fundamental metaphysical questions concerning space and time, the nature of matter, and so on.[34] Furthermore, those realists who employ laws of nature, causes, natural kinds, and essential properties as substantial parts of their explanation of the success of both particular scientific theories and science as a whole, are thereby committed to a metaphysics of de re modality independently of their endorsement of any particular scientific theory.[35] Van Fraassen, on the other hand, in fine empiricist and positivist style, regards natural modalities of all kinds as metaphysical conceits. Following Ladyman (2000), here we argue that in order to circumscribe the observable in a principled way, as constructive empiricism requires, it is necessary to endorse some modal facts that are theory-independent.

According to constructive empiricism, the judgement that scientific theories are empirically adequate—that they 'save the phenomena'—is not supposed to concern anything beyond what is actual: 'Empirical adequacy concerns actual phenomena: what does happen, and not, what would happen under different circumstances' (van Fraassen 1980, 60). The phenomena are what is observable and van Fraassen argues that something need not actually be observed to be observable. Rather, we circumscribe the observable by considering what *would* happen were certain entities present to us: 'X is observable if there are circumstances which are such that, if X is present to us under those circumstances, then we observe it' (1980, 16).[36]

Take the cases of the moons of Jupiter and dinosaurs, and suppose that, as a matter of fact, nobody actually observes either. Van Fraassen recommends that we should believe in them anyway on the grounds that they are observable, which

---

[33] See Ladyman (2000), Monton and van Fraassen (2003), Hanna (2004), Ladyman (2004) and Muller (2005).

[34] Indeed a recent introductory text on the philosophy of science defines it as such, namely Couvalis (1997, 172).

[35] Note that natural kind terms are those that are projectible, hence the modal structure of the world is to a considerable extent reflected in the structure of everyday and scientific language and not just in causal or lawlike statements (see 4.5).

[36] Note that it seems that the second conditional must be interpreted as a subjunctive conditional because using the material conditional would have the consequence that any object that was not actually observed was observable. Strict implication clearly inherits the same problem. The if/then of conditional probability is not appropriate because whether something is observable or not has nothing to do with our degrees of belief, and if van Fraassen has another type of conditional in mind he does not say so. (As we also mention in 2.3.3, Carnap's project of eliminating theoretical terms in scientific theories by explicitly defining them in terms of actual and possible observations faced a similar problem with material implication.)

is to say that there are circumstances such that if, say, the moons of Jupiter were present to us (suppose yourself to be standing on one of them), then we *would* observe them. Hence, two questions arise about the claim that entity X is observable: (a) is it a theory-independent fact that if X was present to us we would observe it? And (b), if so how can we know such a fact?

Turning to (b) first, in one of the passages quoted above (in 2.1) van Fraassen describes the 'able' in 'observable' as referring to our limitations as measuring devices, where these are to be described by physics and biology. So, rather than the philosophical analysis of language, it is science itself that tells us what is or is not observable. Van Fraassen says: 'It is only the content of the theory, the information it contains (and not its structure), which is meant to have the proper or relevant *adequatio ad rem*' (1989, 188). But part of the content of the theory is how it circumscribes the observable, hence its content is not limited to information about the actual but also about the possible, in particular about possible observations. How can van Fraassen rely upon theoretical science, which he does not believe to be true, to determine the limits of his scepticism?[37]

If what is observable or not were a theory-dependent matter (rather than merely described by theory), then whether a particular object is observable or not would depend upon which theory we were using to describe it. If this were so then the distinction between the observable and the unobservable really would have no epistemic relevance and constructive empiricism could not be sustained. This is a point that van Fraassen concedes:

> To find the limits of what is observable in the world described by theory T we must inquire into T itself... This might produce a vicious circle if what is observable were itself not simply a fact disclosed by theory, but rather theory-relative or theory-dependent. ... I regard what is observable as a theory-independent question. It is a function of facts about us *qua* organisms in the world. (1980, 57–8)

He goes on to argue that his account of observability is not circular because he believes what is observable to be an empirical, but nonetheless a theory-independent, matter. So it seems that (a) above must be answered in the affirmative; if X is observable then it is an objective fact that if it were present to us then we would observe it.

Given our assumption about the examples above, the circumstances in which we would observe the moons of Jupiter and dinosaurs never obtain—they are counterfactual. Hence, in order to demarcate the observable in a principled way that can bear the burden placed upon it in the epistemology of constructive empiricism, and to draw this distinction independently of what has as a matter of fact been observed (to allow for belief in dinosaurs), the constructive

---

[37] Van Fraassen needs to make the observable/unobservable distinction with respect to theories, so as to demarcate their empirical substructures, and with respect to the world, so as to demarcate the unobservable objects. Otherwise the notion of empirical adequacy would be inapplicable in practice.

empiricist is committed to believing at least some counterfactuals implied by scientific theories. Unless she takes such modal facts as that expressed by the sentence 'if a dinosaur were present to me then I would observe it' to be objective and theory-independent, her epistemic attitude will depend upon a distinction that is entirely arbitrary. Yet according to van Fraassen's analysis of statements of physical necessity, a counterfactual conditional of the form 'if the moons of Jupiter were present to us (in the right kind of circumstances) then we would observe them' ought to be construed as follows: there is no world which is physically possible relative to this world in which the moons of Jupiter are present to us (in the right kind of circumstances) and we fail to observe them.

But this in turn just reduces to the logical necessity of the conditional that has the laws of nature conjoined with the relevant class of initial conditions as antecedent, and our observing the moons of Jupiter as consequent. Since the constructive empiricist doesn't believe that the laws of nature are objectively different from any other regularities, she ought no more to believe this conditional than to believe another which has different regularities in the antecedent and as consequent that the moons of Jupiter are not present to us. Yet the former conditional entails that the moons of Jupiter are observable and the latter that they are not. Which one holds follows in part from the bits of the theories in question from which the constructive empiricist withholds belief, after all: 'So far as empirical adequacy is concerned, the theory would be just as good if there existed nothing at all that was either unobservable or not actual' (1980, 197).

To summarize, although the constructive empiricist need not be committed to the full truth of theories to demarcate the observable, she is committed to belief in some of their modal implications. In other words, she is committed to belief in more than just what theories say about what is observable and actual, in order to discern what they do say about what is observable and actual. But, according to what van Fraassen says about modality, either there is simply no objective fact of the matter about what would happen under counterfactual circumstances, or, even if he were to allow that there are objective modal facts which determine the truth values of counterfactuals, it is totally incompatible with constructive empiricism to allow that we could *know* about such things since that would amount to allowing that scientific theories tell us about more than the actual phenomena. Hence, constructive empiricism is either inconsistent with van Fraassen's repudiation of objective modality, because what is observable is an objective matter after all, or it is viciously circular, because what is observable is theory-dependent. It seems that van Fraassen must abandon either constructive empiricism or his modal antirealism.

There are, moreover, other tensions between van Fraassen's views on modality and aspects of his philosophy of science. Scientists always look for theories of the observable, not the observed; in other words, theories always involve modalities.

This fact is utterly mysterious on van Fraassen's empiricist view.[38] To expand on this point: scientists almost never formulate theories that only refer to what actually happens in the world; instead theories are always modalized in the sense that they allow for a variety of different initial conditions or background assumptions rather than just the actual ones, and so describe counterfactual states of affairs. But no observable phenomena could allow us to distinguish in our epistemic attitude between theories that agree about everything that actually happens but disagree about what would happen under possible but counterfactual circumstances. Recall that for van Fraassen 'to be an empiricist is to withhold belief in anything that goes beyond the actual, observable phenomena' (1980, 202). Hence, if empiricism were correct we would expect that scientists would and should be content with theories of the actually observed (past, present, and future).

Suppose some scientists had it in their power to create conditions in a laboratory that they knew had never obtained anywhere before, and moreover, that they knew these conditions would never obtain at all unless they created them. Suppose also that their (up to now empirically adequate) theory predicted an outcome for such an experiment. If it were true that these hypothetical scientists wanted a theory adequate only to all actual observable phenomena, then they would be motivated not to carry out this experiment in case their theory was proved wrong. But of course no good scientist would refrain from doing the experiment; it seems that scientists do want theories that are adequate to all possible phenomena and not just the actual ones. Perhaps van Fraassen would explain this fact by saying that scientists want theories that are empirically strong. But the extra strength of modalized theories can only consist in their description of possible but non-actual states of affairs which, according to him, should be of no interest to an empiricist. It may be that it is modalized theories that are the ones that turn out to be empirically adequate but why this should be so is a complete mystery from the point of view of van Fraassen's empiricism.

Empirical adequacy is not achieved by a list of all the actual phenomena. If the new theory explains the success of the old theory that can only be because it reproduces the well-confirmed relations among phenomena described by the old theory. Theories must be modalized because we can choose when to perform a test of a prediction.

There are several viable interpretations of van Fraassen's views about modality,[39] but none of them involves belief in any modal statements objectively construed. According to the modal non-objectivism which van Fraassen advocates in some of his writings about modal matters (1977, 1978, 1981), what is or is not observable is a consequence of which generalizations are important to

---

[38] This argument draws upon Rosen (1984). See especially pp. 161–2.

[39] The texts he cites (1989, n. 35, 365) as presenting his views on modality are chapter 6 of his (1980) and his (1977), (1978), and (1981).

us. Clearly this provides no objective basis for demarcating the observable from the unobservable, and the distinction is theory-dependent. On the other hand, according to the alternative positions on modality that are suggested in some of his other writings, the constructive empiricist will either abstain from belief in the truth of any modal statements or believe them all to be false. So whatever the exact details of his account of modality it cannot bear the epistemic weight put upon it in constructive empiricism via the definition of observability. It seems that even the constructive empiricist must engage in some modal metaphysics in order to articulate and argue for her epistemic attitude to science. Could she simply adopt some form of epistemic realism about objective modality to save constructive empiricism? Monton and van Fraassen (2003) end their reply to Ladyman by arguing that the best reason to be a constructive empiricist is that it makes the best sense of science, and that this claim does not depend on modal nominalism. If modal nominalism is incompatible with constructive empiricism, then the latter could only make sense of science in conjunction with modal realism. However, this view, dubbed 'modal empiricism' by Ron Giere (1985, 83), is a form of structural realism because according to it the theoretical structure of scientific theories represents the modal structure of reality. Although Giere is a scientific realist and van Fraassen is not, both of them are prominent defenders of the semantic approach to scientific theories. Indeed, although van Fraassen's recent defence of his structural empiricism is clearly a response to Worrall's advocacy of structural realism, the emphasis on structure in understanding scientific knowledge and representation is notable throughout his work. We argue in the next section that the semantic approach is the appropriate framework for our form of structural realism.[40]

2.3.3 The semantic approach

The so-called 'syntactic' (or 'received') view of theories was developed by various logical empiricist philosophers, including especially Rudolf Carnap, Ernst Nagel, and Hans Reichenbach. These philosophers recognized that some elements of our theoretical knowledge seem to be independent of the empirical facts. For example, Newton's second law states that the force on a body is proportional to the rate of change of its momentum, where the constant of proportionality is called the inertial mass. This law cannot be tested in an experiment, because it is part of what gives meaning to the concepts employed to describe the phenomena. Hence, it was argued by the logical empiricists that a physical theory can be split into a part that expresses the definitions of the basic concepts and the relations among them, and a part which relates to the world. The former part also includes the purely mathematical axioms of the theory, and trivially all the logical truths

---

[40] Ladyman (1998) introduced ontic structural realism in the context of the semantic approach to scientific theories.

expressible in the language of the theory. If this part of the theory constitutes a priori knowledge, it is purely of matters of convention. The factual content of the theory is confined to the latter part, so the fundamental empiricist principle that the physical world cannot be known by pure reason is satisfied.

The logical empiricists tried to use logic to show how the theoretical language of science is related to the everyday language used to describe the observable world. They were motivated by the verification principle, according to which a (non-tautological) statement is meaningful iff it can be verified in the immediacy of experience, and by the verifiability theory of meaning, according to which the meaning of particular terms (other than logical constants) is either given in experience directly, or consists in the way in which those terms relate to what is given in experience directly.

The idea is that a physical theory will have a canonical formulation such that:

(1) L is a first-order language with identity and K is a calculus defined for L.

(2) The non-logical terms of L can be partitioned into two disjoint sets, one of which contains the *observation terms*, V_O, and the other of which contains the *theoretical terms*, V_T.

(3) There are two sublanguages of L, and corresponding restrictions of K, such that one contains no V_T terms (L_O) and the other no V_O terms (V_T). These together of course do not make up L since L also contains *mixed sentences*.

(4) The observational language is given an interpretation in the domain of concrete observable entities, processes, and events, and the observable properties thereof. An 'interpretation' of a language L (in the sense of model theory used here) is a metalinguistic attribution of a reference to each of the non-logical terms in L. If the axioms of a theory are true under some interpretation then it is a *model* for the theory.

(5) The theoretical terms of L are given a *partial interpretation* by means of two kinds of postulates:
   (a) *theoretical postulates* which define internal relations between the V_T terms and which do not feature V_O terms; and
   (b) *correspondence rules* or *bridge principles* which feature mixed sentences and relate the V_T and V_O terms. (These are also known as 'dictionaries', 'operational definitions', 'coordinative definitions', and so on depending on the author, but all these phrases refer to a set of rules which connect theoretical terms to observable states of affairs.)

The theoretical postulates are the axioms of the theory, and the purely theoretical part of the theory is the deductive closure of these axioms under the calculus K. The theory as a whole, TC, is the conjunction of T and C, where T is the conjunction of the theoretical postulates and C is the conjunction of the correspondence rules.

Initially it was required that the theoretical terms of theories be given explicit definitions. (This was originally Carnap's goal but he abandoned it after his 1936–7.)[41] An example of such a definition of a theoretical term V_T is:

$$\forall x(V_T x \leftrightarrow [Px \rightarrow Qx])$$

where P is some preparation of an apparatus (known as a test condition) and Q is some observable response of it (so P and Q are describable using only observation terms). For example, suppose it is the explicit definition of temperature: any object x has a temperature of t iff it is the case that, if x were put in contact with a thermometer then it would give a reading of t. If theoretical terms could be so defined, then this would show that they are convenient devices which are in principle eliminable and need not be regarded as referring to anything in the world. (This view is often called 'semantic instrumentalism'.)

It was soon realized that explicit definition of theoretical terms is highly problematic. Perhaps the most serious difficulty is that, according to this definition, if we interpret the conditional in the square brackets as material implication, theoretical terms are trivially applicable when the test conditions do not obtain (because if the antecedent is false the material conditional is always true). In other words, everything that is never put into contact with a thermometer has temperature t. (Interpreting the conditional as strict implication was not an option for the logical positivists since this means invoking a notion of metaphysical necessity, and in any case does not help.)

The natural way to solve this problem is to allow subjunctive assertion into the explicit definitions. That is, we define the temperature of object x in terms of what would happen if it were to be put into contact with a thermometer; temperature is understood as a dispositional property. Unfortunately this raises further problems. First, unactualized dispositions, such as the fragility of a glass that is never damaged, seem to be unobservable properties, and they give rise to statements whose truth conditions are problematic for empiricists, namely counterfactual conditionals such as 'if the glass had been dropped it would have broken' where the antecedent is asserted to be false. Dispositions are also modal, and the logical empiricists disavowed objective modality. Second, explicit definitions, dispositional or not, for terms like 'spacetime curvature', 'spin', and 'electron' have never been provided and there are no grounds for thinking that they could be.

---

[41] For an excellent account see Demopoulos (forthcoming a).

However, the advocates of the received view did not abandon the attempt to anchor theoretical terms to the observable world. Carnap's next attempt (1936–7) treats correspondence rules as 'reduction sentences' of the form:

$$Px \to [Qx \to V_x], Rx \to [Sx \to \neg V_x]$$

These do not define theoretical terms but nonetheless connect them with observational ones and so ensure the former's cognitive meaningfulness. (This also avoids the problem of trivial applicability faced by explicit definitions.) Reduction sentences, together with theoretical postulates, offer a partial interpretation for them. They are also intended to specify procedures for applying the theory to phenomena. Theoretical concepts such as those of vital forces and entelechies were criticized by the logical empiricists because their advocates failed to express them in terms of precise, testable laws.

Now according to the view developed so far, TC is only fully interpreted with respect to its V_O terms, which refer to ordinary physical objects (such as ammeters, thermometers, and the like) and their states; the V_T terms are only partially interpreted. Consider all the possible interpretations of TC such that the V_O terms have their normal meanings and under which TC is true; these are the models of TC. The problem for the advocate of this approach is now that there will be lots of these models in general so there is no unique interpretation for the theory as a whole. Hence it would seem to make no sense to talk of it being true or false of the world. Hempel (1963, 695) and Carnap (1939, 62) solved this problem by stipulating that TC is to be given an intended interpretation; theoretical terms are interpreted as (putatively) referring to the entities and so on appropriate to the normal meanings of them in scientific (and everyday) use.

This entails that the meanings of terms like 'electron' that derive from the picture of electrons as tiny billiard balls or classical point particles are important in determining the targets of theoretical reference. Once the explicit definition project is abandoned, then it must be accepted that the meanings of the V_T terms that do not have testable consequences are nonetheless important in determining their reference. As Suppe puts it:

When I give a semantic interpretation to TC, I am doing so relative to the meanings I already attach to the terms in the scientific metalanguage. In asserting TC so interpreted, I am committing myself to the meaning of 'electron' and so on, being such that electrons have those observable manifestations specified by TC. (1977, 92)

This version of the received view is committed to 'excess' or 'surplus' meaning of theoretical terms over and above the meaning given by the partial interpretation in terms of what can be observed. (Herbert Feigl (1950) recognized this explicitly and argued for the view that theoretical terms genuinely refer to unobservable entities as a consequence.)

Perhaps the most widespread criticism of the received view is that it relies upon the distinction between observational and theoretical terms. This distinction is supposed to correspond to a difference in the way language works; the former terms are more or less defined ostensibly and directly refer to observable features of the world, while the latter are defined indirectly and refer to unobservable features of the world. Examples of the former are supposed to include 'red', 'pointer', 'heavier than'; examples of the latter would include 'electron', 'charge density', 'atom'. It has been widely argued (for example by Putnam 1962) that there is no objective line to be drawn between observational and theoretical language, and that all language is theory-dependent to a degree. Moreover, eliminating theoretical terms, even if it were possible, would not eliminate talk of the unobservable, because it is possible to talk about the unobservable using V_O terms only, for example by saying there are particles that are too small to see.

Whether or not the distinction between observational and theoretical terms can be drawn in a non-arbitrary way, the received view also faces criticism concerning the correspondence rules. These were supposed to have three functions: (i) to generate (together with the theoretical postulates) a partial interpretation of theoretical terms, (ii) to ensure the cognitive significance of theoretical terms by connecting them with what can be observed, (iii) to specify how the theory is related to the phenomena. There are several problems in respect of (iii). First, if it is true that correspondence rules are part of the theory, then whenever a new experimental technique is developed in the domain of the theory, and the correspondence rules change to incorporate the new connections between theoretical terms and reality, the theory will change. We don't suppose there is any fact of the matter here; but it is surely a sound pragmatic objection that this suggests the collapse of the highly useful distinction between theories and models. Another problem, raised by Suppe (1977, 104), is that there are typically an indefinite number of ways of applying a theory, and so there ought to be an indefinite number of correspondence rules; but the formulation of the received view requires that there are only finitely many. Furthermore, a theory is applied to phenomena by using other theories to establish a causal chain between the states of affairs described by the theory and the behaviour of some measuring apparatus. For example, theories of optics are needed to link the occurrence of line spectra with changes in the energy states of an electron. The correspondence rules in this case will incorporate principles of optics and will themselves offer mechanisms and explanations for the behaviour of measuring devices. Suppe concludes that correspondence rules are not part of the theory as such, but rather they are auxiliary assumptions about how the theory is to be applied.

Many people, notably Nancy Cartwright (1983, 1989), have argued that the received view is misleading about how scientific theories are applied, because it is rarely if ever the case that it is possible to derive some concrete experimental prediction from a theory simply by the provision of auxiliary assumptions about background conditions. Rather, it is argued, the connections between abstract theory and concrete experiment are complex and non-deductive, and involve the use of many theories, models, and assumptions that are not formulated as part of the original theory.

The basis for the demise of the syntactic view and the origins of the semantic approach have been clearly set out by Fred Suppe (1977). The latter begins with Beth who was concerned to provide 'a logical analysis—in the broadest sense of this phrase—of the theories which form the actual content of the various sciences' (Beth 1949, 180), and with Patrick Suppes who pioneered the set-theoretic representation of theories later developed by Sneed, Stegmüller, and others. Two main desiderata, the need for appropriate formalizations of scientific theories ('logical analysis'), and the goal of accurate accounts of how they are regarded in practice (the 'actual content'), motivate the semantic approach. Van Fraassen elaborated and generalized Beth's approach, arguing that theories and models are essentially mathematical structures, and that experiments produce data models which are abstract structures. A theory is empirically adequate to the extent that the latter models are embeddable in the former (1980, ch. 3). (He also showed how to unite models in a common state-space in terms of which laws of coexistence and succession can be represented (see, for example, 1989).) Van Fraassen's later work incorporates Giere's understanding of a theory as consisting of a theoretical definition, which defines a certain class of systems, and a theoretical hypothesis, which asserts that certain kinds of real systems are related in certain ways to certain members of that class. For example, the theoretical definition might define the models of a simple harmonic oscillator, and the theoretical hypothesis might assert that pendula are approximately modelled by that class when variables in the model are taken to represent a pendulum's bob's mass and the length of its string (Giere 1988b). Suppe (1989) develops the notion of a relational system, which consists of a set of possible states on which various sequencing relations are imposed. Although, van Fraassen, Suppe, and Giere divide along different lines in the realism debate, each using the semantic approach for different ends, the core idea which binds them together is the claim that theories are better thought of as families of models rather than as partially interpreted axiomatic systems. That is, theories are not collections of propositions or statements, but are 'extralinguistic entities which may be described or characterised by a number of different linguistic formulations' (Suppe 1977, 221).

French and Ladyman (1997, 1998, 1999) and Bueno et al. (2002) have argued that by focusing on the relational structure of mathematical models, a unitary account of the various senses of the word 'model' as used in scientific practice can be given, as well as an illuminating account of scientific representation and idealization. In this they follow Patrick Suppes who introduced his essay 'A Comparison of the Meaning and Use of Models in Mathematics and the Empirical Sciences' with a series of quotations referring to different kinds of models, proceeding from models in the sense of model theory in mathematical logic, to ones drawn from physics and through to social science models; and claimed that

> the concept of model in the sense of Tarski may be used without distortion and as a fundamental concept in all of the disciplines from which the above quotations are drawn. In this sense I would assert that the meaning of the concept of model is the same in mathematics and the empirical sciences. The difference to be found in these disciplines is to be found in their use of the concept. (Suppes 1961, 165)

In the context of the syntactic approach, within which a theory is taken to be a set of sentences, realism amounts to the commitment to standard (correspondence) referential semantics, and to truth, for the whole theory. How should we understand questions about the relationship between theoretical objects and the world in terms of the semantic conception of theories? Giere addresses this issue in his writings on the semantic approach. Unlike van Fraassen, he seems to accept that the semantic approach transforms the terms of the scientific realism debate. In particular, Giere (1985) argues that it is unreasonable to expect all theoretical representation in science to fit the mould in which philosophers cast linguistic representation; Tarskian semantics is not appropriate for a consideration of the representative role of the tensor calculus and differential manifolds or Hilbert space. One reason for this is that Giere's 'constructive realism' stops short of asserting that real systems resemble models in all respects. Instead, he claims that models are at best similar to real systems in many or most aspects.

Giere's version of the semantic approach relies on a notion of *similarity* in specified respects and to specified degrees. French and Ladyman deploy the notion of partial isomorphism or partial homomorphism to represent the relationships between theoretical structures and models of the phenomena.[42] In his original paper on structural realism, Worrall argues that, '[m]uch clarificatory work needs to be done on this position, especially concerning the notion of one theory's structure approximating another' (1989). The approach of French and his collaborators (see also da Costa and French 2003) represents inter-theoretical relationships as partial morphisms holding between the model-theoretic structures representing the theories concerned. They offer case studies such as high-energy physics (French 1997), and the London and London theory of superconductivity (French and Ladyman 1997).[43]

---

[42] Partial isomorphism is introduced and applied to the history of superconductivity in French and Ladyman (1997), and see also French and Ladyman (1998); partial homomorphism is introduced in Bueno et al. (2002).

[43] In our view the supposed independence of models from theory is much exaggerated. For example, the recent collection edited by Morgan et al. (1999) contains numerous references to the idea that models are 'autonomous agents'. Taken literally this claim seems absurd; since models don't do anything they can't be agents. However, we take it that it is meant to suggest that models have a *sui generis* role in scientific practice, and that their development is often independent of theoretical considerations. Of course, it is quite correct to point out that in general models cannot

However the details of the semantic view are developed (and we think that lots of formal and informal approaches may be useful, perhaps, for example, using category theory rather than set theory for some purposes),[44] the important implications of the view for our position in this book are:

(a) The appropriate tool for the representation of scientific theories is mathematics.
(b) The relationships between successive theories, and theories at different scales whether spatio-temporal or energetic, are often limiting relations and similarities of mathematical structure (formally captured by structure-preserving maps or morphisms of various kinds), rather than logical relations between propositions.
(c) Theories, like Newtonian mechanics, can be literally false as fundamental physics, but still capture important modal structure and relations.

The syntactic view demands quantification over a domain of individuals, whether theoretical and observable objects in a physicalist version, or sense-data in a phenomenalist version. The semantic view encourages us to think about the relation between theories and the world in terms of mathematical and formal structures. Giere claims, and we agree, that once the semantic approach is adopted the crucial issue is whether or not theoretical models tell us about *modalities*. He argues that they do and states that (his) '[c]onstructive realism is thus a model-theoretic analogue of the view advocated by Grover Maxwell' (1985, 83). Therefore, he says, constructive realism is a species of structural realism (ibid.). Van Fraassen (1989, 1991) also emphasizes this issue, and, as we discuss in 2.3.2, argues that modality is solely a feature of models. In this respect his empiricism resembles logical empiricism, but the former's incorporation of the semantic approach is a crucial difference between them. We follow van Fraassen in this but our commitment to objective modality is a further departure from logical empiricism.

2.3.4 Real patterns, structures, and locators

It is not part of our realism that every time a scientist quantifies over something in formulation of a theory or hypothesis she is staking out an existential

---

be simply deduced from theories together with background assumptions. On the other hand, it is quite wrong to suggest that it is the norm for the development of models to be entirely independent of theoretical matters, and indeed the case of the development of the London and London model of superconductivity cited by Morrison and Suarez in their papers in the above volume is one for which French and Ladyman (1997) argue in detail that theoretical considerations played an important role.

[44] An especially useful reference is Ruttkamp (2002). Her explication of the semantic theory is both elegant and comprehensive, and develops its philosophical significance by application to the recent arguments between realists and disunity theorists like Cartwright, rather than just to the more dated realist criticisms of the logical empiricist model of scientific theories.

commitment. As we argue in Chapter 5, especially in special sciences one must often explore the way theories are practically put to work in some detail to tease out their ontological commitments; one cannot just read them off textbook formulations of theories. Indeed, we will argue that, semantic appearances notwithstanding, we should not interpret science—either fundamental physics or special sciences—as metaphysically committed to the existence of self-subsistent individuals.

To help us track the distinction between reference that is merely apparent from linguistic/propositional formulations of theories, and genuine ontological commitment in science, we will rehabilitate practice from early twentieth-century philosophy of science in distinguishing between formal and material modes of discourse. Discourse in the formal mode is always to be understood relative to a background of representational conventions (ideally, as the positivists stressed, formally axiomatized conventions, though in practice this ideal is usually only regulatory). Discourse in the material mode purports to refer directly to properties of the world. For most everyday purposes, where local misunderstanding is unlikely, scientists do not bother to draw any distinction resembling this. However, we claim that we can find differences in scientists' behaviour when they extend and modify theories which we philosophers then use the distinction to capture. Thus when we say that some bit of discourse by scientists is 'material mode' or 'formal mode' discourse we should not be interpreted as making a priori conceptual assertions; such claims are to be interpreted as (elliptical) verifiable predictions about scientific practice. (When we call some of our own discourse 'formal mode' or 'material mode' we are of course just announcing policy.)

In terms of this distinction, we will make some further terminological stipulations. We will later say that what exists are ('real') 'patterns'. 'Real patterns' should be understood in the material mode. Then 'structures' are to be understood as mathematical models—sometimes constructed by axiomatized theories, sometimes represented in set theory—that elicit thinking in the formal mode. We will argue that there is no tension between this view of structures, and the claim that 'the world has modal structure'. Objective modalities in the material mode are represented by logical and mathematical modalities in the formal mode. All legitimate metaphysical hypotheses are, according to us, claims of this kind. A metaphysical hypothesis is to be motivated in every case by empirical hypotheses that one or more particular empirical substructures are embedded in (homomorphic to) particular theoretical structures in the formal mode that represent particular intensional/modal relations among measurements of real patterns. We see no reason not to refer to physical measurement results as 'phenomena', so long as it is understood that phenomena are not sense-data in the positivists' sense. The hypothesis that some phenomena possibly instantiate a particular real pattern is motivated in each case by the representation of extensional relations among actual phenomena in a (formal-mode) data model embedded in (homomorphic to) the relevant empirical substructure. 'High-level theories' are used to

Scientific Realism, Constructive Empiricism, and Structuralism

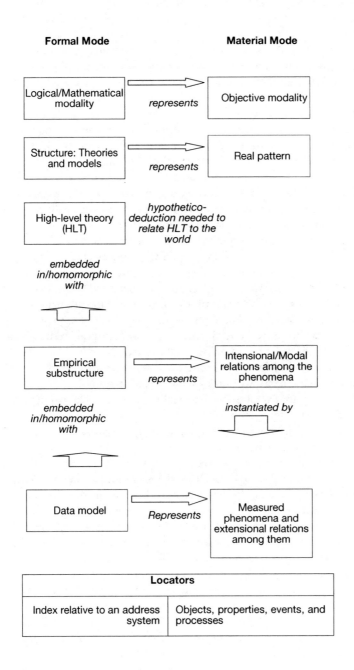

derive predictions of intensional/modal relations from structures. For clarity, we represent this network of relations in chart form on the previous page.

We have not yet explained the bottom row of the chart, which represents a terminological and conceptual innovation of ours.

Our realism consists in our claim that successful scientific practice warrants networks of mappings as identified above between the formal and the material. We will defend belief in the viability of such mappings in a way compatible with the PNC, by appeal to scientific practice.

When we go on to deny that, strictly speaking, there are 'things', we will mean to deny that in the material world as represented by the currently accepted scientific structures, individual objects have any distinctive status. Some real patterns, we will argue, behave like things, traditionally conceived, while others behave like traditional instances of events and processes. In a PNC-motivated metaphysic these distinctions lose all significance except purely practical, book-keeping, significance for human beings in certain sorts of special circumstances. From the metaphysical point of view, what exist are just real patterns. To put the point another way, to define 'real pattern'—something we will postpone until Chapter 4—is to say everything there is to be said about the criteria for existence. Science motivates no separate metaphysical theories about objects, events, and processes.

In order to be able to refer to particular real patterns before specifying the structures that represent them but without relying on the conventional distinctions between objects, events, and processes, we need a special bit of technical vocabulary. One picks out a real pattern independently of its structural description by an ostensive operation—that is, by 'pointing at it'. This is intended as evocative talk for operations of fixing, stabilizing, and maintaining salience of some data from one measurement operation to another. (So, think of 'pointing' as meaning 'directing a measurement apparatus'.) In the fully generalized sense, this means that one indicates the real pattern's *location* in some coordinate system with high enough dimensionality to permit its disambiguation from other real patterns. (What this dimensionality must be is itself an empirical question for scientific investigation, as we will see.) We will thus speak of all of the traditional kinds of objects of reference—objects, events, processes—by mentioning their *locators*.[45] A locator is to be understood as an act of 'tagging' against an established address system.[46] Such address systems can be formal representations (for example, a map or a spatio-temporal coordinate system) or could be referenced to cognitive dispositions that induce common structural reference points among a group of observers

---

[45] Floridi (2003) refers to roughly our notion of locator by co-opting 'observable' as a special term of art for the purpose. We suggest that, in light of the past century's controversies in the philosophy of science, this suggestion invites confusion.

[46] This usage alludes to the idea that locators are aspects of non-conceptual content, and as such are dynamic, pre-cognitive representational relations between subjects and spaces in which subjects orient. See the extensive literature on such content that began with Cussins (1990).

who share measurement data. (In practice, use of the former is often parasitic upon the latter, but need not be in principle—perhaps observers with very different cognitive structures, and so responsive to very different salience cues, could use mathematics to establish common index systems for locating.)

Use of a locator in a given instance involves no commitment to a type–token distinction: there are locators for each of 'Napoleon', 'French emperors', 'French people', and 'people'. This is another way of expressing our idea that everything that exists is a real pattern, and that that is all there is to say about what there is.

Nothing we have just said should be interpreted as expressing a particular epistemological thesis. That is, we are not suggesting that one begins by locating real patterns and then discovers their structural descriptions. Location is a recursive practice, and generally goes on against the background of some already developed structure. In practice, then, a locator will be a partial interpretation of a structure in the context of another, presupposed, structure. By 'partial interpretation' we mean no more, at least in the first place, than an indication of where in the universe measurements should be taken that will be relevant to assigning values to variable parameters of the structural element in question. Thus the locator '*Orycteropus afer*' (the aardvark) directs us to take measurements in certain sorts of terrestrial African habitats within a certain range of temporal coordinates, and focused on certain sorts of behavioural regularities and certain sorts of genetic processes; the locator 'market' directs us to take measurements wherever certain minimum thresholds of complexity in cognitive processing capacities, sociality, and scarcity of inputs to batteries are jointly found; and the locator 'top quark' directs us to take measurements everywhere in the universe.[47] How precisely a given locator, in the context of a given theoretical structure, tells us to focus our measurements varies with the refinement of empirical theory. One of the things that science does as it progresses, according to the account we will give, is add precision to locators. When we wish to mention a particular locator, either for exemplary purposes or because that locator is of general metaphysical significance (as the locator for the top quark probably is but the locator for aardvarks probably is not), we will write 'aardvark (L)'. So this should not be read as purporting to refer to collections of substantival, pattern-independent individual aardvarks (since we do not presume there are such things, and will indeed doubt that there are such things). It denotes, again, an element of some already partially elaborated structure plus measurement instructions.

2.4 WHAT IS STRUCTURAL REALISM?

We hope to have convinced the reader that both standard scientific realism and constructive empiricism face serious difficulties. However, a commitment

---

[47] Our usage of 'direct' here again alludes to dynamical non-conceptual content.

to objective modality will be compatible with the no-miracles argument because if science describes objective modal relations among the phenomena, as well as just what actually happens, then it is no miracle that theory-laden features of scientific methodology, such as the use of background theories in testing new ones, and the practice of theory conjunction are successful. Scientific realism without a commitment to objective modality is unable to explain the success of science, because there is no connection between unobservable entities and the phenomena we observe other than constant conjunction in the actual world, and that doesn't explain anything. If theorists are able sometimes to capture the objective modal structure of the world then it is no surprise that successful novel prediction sometimes works, and the practice of theory conjunction ought to lead to progress at the empirical level. Constructive empiricism in its structuralist form, when combined with a commitment to objective modality, does not face the problems we identified which turn on van Fraassen's eschewal of modal realism.[48] Finally, the idea that science describes the objective modal structure of the world is not undermined by theory change in the history of science, since all the well-confirmed modal relations expressed by old theories are approximately recovered in their successors.

Recent work has revealed that many different forms of structuralism and structural realism can be found in the work of some of the greatest philosophers of science. Barry Gower's (2000) historical survey of structural realism discusses how structuralism figures in the thought of Cassirer, Schlick, Carnap, and Russell. Worrall approvingly cites Poincaré as a structural realist (and Zahar 1994 and Gower seem to agree with this reading of him). However, Mary Domski (1999) has argued convincingly that Poincaré was not any kind of realist in the modern sense of the term. His structuralism was combined with neo-Kantian views about the nature of arithmetic and group theory, and with conventionalism about the geometry of space and time. Hence, she argues, Poincaré is better thought of as a structural empiricist or a structural neo-Kantian. Meanwhile, Stathis Psillos (1999) has explored the connections between structuralism and the Ramsey sentence approach to scientific theory as it figured in the development of Carnap's philosophy from logical positivism to ontologically relativist empiricism. We shall have more to say about Russell and Cassirer, and about Ramsey sentences, below.

Given the intricate connections between structuralist views about science and theories in epistemology, semantics, and metaphysics, we will not here offer a complete taxonomy, or even a partial history of them. Instead, we begin (following Ladyman 1998) by asking a fundamental question about structural realism, namely, is it metaphysics or is it epistemology? Worrall's paper is ambiguous in this respect. At times his proposal seems to be that we put an epistemic constraint on realism to the effect that we only commit ourselves to

---

[48] The problem of inductive libertarianism is not obviously removed by the adoption of objective modality but we will suggest how it can be in 4.5.

believing in the structural content of a theory, while remaining agnostic about the rest. This is suggested by his citation of Poincaré who talks of the redundant theories of the past capturing the 'true relations' between the 'real objects which Nature will hide forever from our eyes' (1905, 161). Elsewhere, he declares that 'what we call objective reality is ... what is common to many thinking beings and could be common to all; ... the harmony of mathematical laws' (1906, 14). Similarly, Zahar's structural realism is a form of Kantian transcendental idealism according to which science can never tell us more than the structure of the noumenal world; the nature of the entities and properties of which it consists are epistemically inaccessible to us.[49]

On the other hand, Worrall's position in his 1989 paper is not explicitly an epistemic one, and other comments suggest a departure from the metaphysics of standard scientific realism. For example, he says: 'On the structural realist view what Newton really discovered are the relationships between phenomena expressed in the mathematical equations of his theory' (1989, 122). If the continuity in scientific change is of 'form or structure', then perhaps we should abandon commitment to even the putative reference of theories to objects and properties, and account for the success of science in other terms. Others who have contributed to structural realism have more explicitly signalled a significant departure from traditional realist metaphysics. For example, Howard Stein:

[O]ur science comes closest to comprehending 'the real', not in its account of 'substances' and their kinds, but in its account of the 'Forms' which phenomena 'imitate' (for 'Forms' read 'theoretical structures', for 'imitate', 'are represented by'). (1989, 57)

In the next section we will argue that structural realism gains no advantage over traditional scientific realism if it is understood as merely an epistemological refinement of it. Thereafter, we will explain how structural realism can be taken as a metaphysical position. Argument for structural realism in accordance with the PNC, which requires appeal to physics, will be given in the next chapter, when we adduce support from considerations in the foundations of quantum mechanics, spacetime physics, quantum information theory, and research in quantum gravity.

2.4.1 Against epistemic structural realism (ESR)

One way of thinking about structural realism is as an epistemological modification of scientific realism to the effect that we only believe what scientific theories tell us about the relations entered into by unobservable objects, and suspend judgement as to the nature of the latter. There are various forms this might

---

[49] Thomas Ryckman (2005) calls Kantian structural realism 'transcendental structuralism'. Our view is what he calls 'transcendent structuralism', as opposed to van Fraassen's 'instrumental structuralism'.

take. For example, the unobservable objects may be thought of as unknown or as unknowable. Either way we get the claim that scientific theories tell us only about the form or structure of the unobservable world and not about its nature. For Poincaré structural knowledge in science is knowledge of first-order properties and relations. Russell (1927) and Carnap (1928) took this one step further and argued that we don't even know the relations, rather all we know is their properties and relations. On this extreme form of structuralism science only tells us about purely logical features of the world.

Ladyman (1998) discusses Russell's view and argues that in general epistemological forms of structural realism do not significantly improve the prospects of standard scientific realism, and hence that structural realism should be thought of as metaphysically rather than epistemically revisionary.[50] We will now briefly reprise this argument.

A position called structural realism, which amounts to an epistemological gloss on traditional scientific realism, was, as Worrall mentioned, detailed and advocated by Grover Maxwell (1962, 1970a, 1970b, 1972). Maxwell set out to defend a robust scientific realism, but he wanted this to be compatible with what he called 'concept empiricism' about the meaning of theoretical terms, and he also wanted to explain how we can have epistemic access to unobservable entities. The problem as Maxwell saw it was that theories talk about all sorts of entities and processes with which we are not 'acquainted'. How, he wondered, can we then know about and refer to them and their properties? The answer that Maxwell gave, following Russell's emphasis on structure and relations in his later philosophy (see, for example, Russell 1927), was that we can know about them by description, that is, we can know them via their structural properties. In fact, he argues, this is the limit of our knowledge of them, and the meanings of theoretical terms are to be understood purely structurally. Maxwell explicates this notion that the structure of the theory exhausts the cognitive content of its theoretical terms, by considering the Ramsey sentence of the theory. As is well known, Ramsey's method allows the elimination of theoretical terms from a theory by replacing them with existentially quantified predicate variables (or names in the case of Lewis 1970). If one replaces the conjunction of assertions of a first-order theory with its Ramsey sentence, the observational consequences of the theory are carried over, but direct reference to unobservables is eliminated.

If we formalize a theory in a first-order language: $\prod(O_1,\ldots,O_n;T_1,\ldots,T_m)$, then the corresponding Ramsey sentence is $\exists t_1, \ldots , \exists t_m \prod(O_1, \ldots, O_n; t_1, \ldots, t_m)$. Thus the Ramsey sentence only asserts that there are some objects, properties,

---

[50] Psillos (2001) refers to the 'upward path' to structural realism as beginning with empiricist epistemological principles and arriving at structural knowledge of the external world. The 'downward' path is to arrive at structural realism by weakening standard scientific realism as suggested by Worrall. Epistemic structural realists include Worrall, Zahar, Votsis, and Morganti (2004). The latter differs from the others by arguing that ESR should amount to agnosticism about whether there is a domain of individuals over and above relational structure.

and relations that have certain logical features, satisfying certain implicit definitions. It is a higher-order description, but ultimately connects the theoretical content of the theory with observable behaviour. However, it is a mistake to think that the Ramsey sentence allows us to eliminate theoretical entities, for it still states that these exist. It is just that they are referred to not directly, by means of theoretical terms, but by description, that is via variables, connectives, quantifiers, and predicate terms whose direct referents are (allegedly) known by acquaintance. Thus Maxwell (and Russell) claimed that knowledge of the unobservable realm is of its structural rather than intrinsic properties, or, as is sometimes said, limited to knowledge of its second-order properties. It is arguable that this is the purest structuralism possible, for the notion of structure employed refers to the higher-order properties of a theory, those that are only expressible in purely formal terms.

This is an epistemological structural realism meant to vindicate and not to revise the ontological commitments of scientific realism. On this view the objective world is composed of unobservable objects between which certain properties and relations obtain; but we can only know the properties and relations of these properties and relations, that is, the structure of the objective world. However, there are serious difficulties with this view which were originally raised by Newman (1928) and which were rediscovered by Demopoulos and Friedman (1985). The basic problem is that structure is not sufficient to pick out uniquely any relations in the world. Suppose that the world consists of a set of objects whose structure is W with respect to some relation R, about which nothing else is known. Any collection of things can be regarded as having structure W provided there is the right number of them.[51] As Demopoulos and Friedman point out, if $\prod$ is consistent, and if all its purely observational consequences are true, then the truth of the corresponding Ramsey sentence follows as a theorem of second-order logic or set theory (provided the initial domain has the right cardinality—and if it does not then consistency implies that there exists one that does). The formal structure of a relation can easily be obtained with any collection of objects provided there are enough of them, so having the formal structure cannot single out a unique referent for this relation; in order to do so we must stipulate that we are talking about the intended relation, which is to go beyond the structural description. 'Thus on this view, only cardinality questions are open to discovery!' (Demopoulos and Friedman 1985, 627); everything else will be known a priori.

This leads Demopoulos and Friedman to conclude that reducing a theory to its Ramsey sentence is equivalent to reducing it to its empirical consequences, and thus that: 'Russell's realism collapses into a version of phenomenalism or

---

[51] This is because according to the extensional characterization of relations defined on a domain of individuals, every relation is identified with some set of subsets of the domain. The power set axiom entails the existence of every such subset and hence every such relation.

strict empiricism after all: all theories with the same observational consequences will be equally true' (1985, 635). This conclusion can be reached by a different path. It has been proved, by Jane English (1973), that any two Ramsey sentences that are incompatible with one another cannot have all their observational consequences in common. In other words, if we treat a theory just as its Ramsey sentence then the notion of theoretical equivalence collapses into that of empirical equivalence. Therefore, equating the structure of a theory to what is embodied in its Ramsey sentence cannot do justice to Worrall's intention in proposing structural realism, since he is quite clear that commitment to the structure of a theory goes beyond commitment to the strictly empirical level of a theory.[52]

In any case, it is hard to see how Maxwell's epistemology and semantics could help deal with the problem of theory change that structural realism is intended to solve. After all, as Maxwell himself pointed out, his structural realism does not dispense with reference to theoretical entities, but it makes that reference a function of the (place of the theoretical terms in the) overall structure of the theory, as manifested in the Ramsey sentence. The problem of ontological discontinuity is left untouched by simply adopting Ramsification. In fact, it seems even worse if contextualism about the meaning of theoretical terms is adopted to this extent. Hence, this epistemic reading of structural realism is of no help with the problem of theory change. The Ramsey sentence of a theory may be useful to a concept empiricist because it shows how reference to unobservables may be achieved purely by description, but this is just because the Ramsey sentence refers to exactly the same entities as the original theory.[53]

---

[52] Jeff Ketland (2004) argues in detail that the Newman objection trivializes the Ramsey sentence formulation of ESR. Worrall and Zahar (2001), Cruse (2005), and Melia and Saatsi (2006) defend the Ramsey sentence approach against model-theoretic arguments by questioning the assumption that all predicates which apply to unobservables must be eliminated in favour of bound variables. Mixed predicates such as 'extended' are those that apply to both observable and unobservable objects. The Newman objection does not go through if mixed predicates are not Ramsified, because a model of the Ramsey sentence will not necessarily be one in which what is claimed regarding the mixed properties and relations holds. In response, Demopoulos (forthcoming b) points out that the Ramsey sentence of a theory with mixed predicates where the latter are not Ramsified will be true provided the original theory is satisfiable—hence the claim that the content of the Ramsey sentence is merely the observational content of the original theory plus a cardinality claim is still true when mixed predicates are considered. Melia and Saatsi (2006) also argue that intensional notions, such as naturalness and causal significance, may be applied to properties to save the Ramsey sentence formulation of ESR from triviality. (This recalls the defence of Russell's structuralism against Newman discussed in Hochberg 1994.) Demopoulos also raises two problems with this strategy: first, even non-natural relations can have significant claims made about them in a theory, and second, the cognitive significance of unramsified theories is independent of a commitment to 'real' or 'natural' relations. Hence, Demopoulos insists that the Ramsey sentence of a theory and the theory itself are importantly different.

[53] Frank Jackson (1998), Rae Langton (1998), and David Lewis (forthcoming) also advocate views similar to ESR. Jackson refers to 'Kantian physicalism' (23–4), Langton to 'Kantian Humility', and Lewis to 'Ramseyian Humility'. Jackson argues that science only reveals the causal/relational properties of physical objects, Langton argues that science only reveals the extrinsic properties of physical

With reference to the case of the transition between classical electrodynamics and quantum electrodynamics (QED), Lyre says:

> Whereas from the viewpoint of an ontology of objects, many of the entities in the transition of Maxwell's theory to QED have changed ... there is a considerable element of retention of group structure and its embedding into a larger framework which makes the scientific progress much less discontinuous than it looks on the level of objects. (2004, 664)

Worrall's approach to structural realism with its emphasis on the Ramsey sentence of a theory and the distinction between observational and theoretical terms is thoroughly embedded in the syntactic view of theories that adopts first-order quantificational logic as the appropriate form for the representation of physical theories. Since ontic structural realism is not formulated in these terms, the Newman problem does not arise for ontic structural realism. In particular, we will eschew an extensional understanding of relations without which the problem cannot be formulated. According to Zahar (1994, 14) the continuity in science is in the *intension* not the *extension* of its concepts. If we are to believe that the mathematical structure of theories is what is important, then, as Zahar suggests, we need a different semantics for theories, one that addresses the representative role of mathematics directly.

Bearing in mind that structural realism is supposed to be realist enough to take account of the no-miracles argument, and in particular the record of successful novel predictions which some theories enjoy, there is a minimal metaphysical commitment that we think structural realism must entail. This is that there are mind-independent modal relations between phenomena (both possible and actual), but these relations are not supervenient on the properties of unobservable objects and the external relations between them. Rather, this structure is ontologically basic. This is enough to make structural realism distinct from standard realism but also from constructive empiricism. From this metaphysical thesis there follow plenty of realist methodological and epistemic implications but, we hope, no unsustainable beliefs in the specific ontologies that are employed to help us grasp the structure of the world according to particular theories.

---

objects, and both then argue that their intrinsic natures, and hence the intrinsic nature of the world, are epistemically inaccessible. Jackson points out that this inference can be blocked if the natures of objects and their intrinsic properties are identified with their relational or extrinsic properties, but argues that this makes a mystery of what it is that stands in the causal relations. The metaphysical article of faith to the effect that objects and properties must have intrinsic natures prior to entering into relations is one that we reject. Our naturalism leads us to reject such metaphysical doctrines if they are not supported by science, and we argue at length in the next chapter that they are not. We return to precisely this issue in 3.4. Lewis's argument also begins from the observation that science only tells us about the causal roles occupied by fundamental objects and properties. Ney (forthcoming b) argues that on Lewis's view, unlike the Kantian view of Langton, intrinsic properties ground causal ones, and hence that the former can be known via the latter, and so Lewis's argument fails.

2.5 CONCLUSION

We have argued that a version of structural realism that incorporates a minimal commitment to objective modal structure achieves consilience arising from the following considerations:

(1) The no-miracles argument (theory conjunction and novel prediction).
(2) Theory change.
(3) Scientific representation.
(4) The problems with constructive empiricism, in particular, the modality of observability and inductive libertarianism.

The importance of modality in the understanding of scientific knowledge was well understood by Peirce:

[W]e must dismiss the idea that the occult state of things (be it a relation among atoms or something else), which constitutes the reality of a diamond's hardness can possibly consist in anything but in the truth of a general conditional proposition. For to what else does the entire teaching of chemistry relate except to the 'behaviour' of different possible kinds of material substance? And in what does that behaviour consist except that if a substance of a certain kind should be exposed to an agency of a certain kind, a certain kind of sensible result would ensue, according to our experiences hitherto. As for the pragmaticist, it is precisely his position that nothing else than this can be so much as meant by saying that an object possesses a character. He is therefore obliged to subscribe to the doctrine of a real Modality, including real Necessity and real Possibility. (Peirce, 1960–6, v. 5, 457)

In the next chapter, we argue that our form of structural realism is also motivated by the contents of our best physical theories, namely quantum theory and general relativity. In later chapters we will go on to argue that, supplemented with concepts from information theory, the position defended here makes possible a plausible account of laws, causation, and explanation, and the relationships among the special sciences, and between them and fundamental physics.

3

Ontic Structural Realism and the Philosophy of Physics

James Ladyman and Don Ross

Ontic Structural Realism (OSR) is the view that the world has an objective modal structure that is ontologically fundamental, in the sense of not supervening on the intrinsic properties of a set of individuals. According to OSR, even the identity and individuality of objects depends on the relational structure of the world. Hence, a first approximation to our metaphysics is: 'There are no things. Structure is all there is.' We of course acknowledge that special sciences are richly populated with individual objects. Thus, to accommodate their elimination from metaphysics we will owe a non-ad hoc account of the point and value of reference to and generalization over objects in sciences other than fundamental physics. We will argue that objects are pragmatic devices used by agents to orient themselves in regions of spacetime, and to construct approximate representations of the world. In keeping with the PNC, this account must not imply revision of special sciences for the sake of metaphysical comfort. We postpone this task, and then the closely related account of the metaphysical status of causation, to Chapters 4 and 5 respectively. In the present chapter our purpose is to motivate ontic structural realism from contemporary fundamental physics, as the PNC requires.

This chapter argues for a position that consists in the conjunction of eliminativism about self-subsistent individuals, the view that relational structure is ontologically fundamental, and structural realism (interpreted as the claim that science describes the objective modal structure of the world). For 'modal' read 'nomological' if you like. We do not take it to be 'causal structure', for reasons to be explained, but will argue later that causal structure is the pragmatically essential proxy for it in the special sciences (but not in fundamental physics). Part of the motivation for our denial that there are metaphysically fundamental individuals comes from our verificationism. The epistemic structural realist thinks that all we can know is structure, but it is the structure of an unknowable realm of individuals. However, we shall argue that in the light of contemporary physics the PNC entails that talk of unknowable intrinsic natures and individuals is idle and has no justified place in metaphysics. This is the sense in which our view is eliminative; there are objects in our metaphysics but they have been purged of their intrinsic natures, identity, and individuality, and they are not metaphysically fundamental.

It might be thought that verificationists ought to be agnostic about whether there are unknowable objects over and above the knowable structure of the world. We suppose that there might be other universes which have no causal, spatio-temporal, or any other kind of connection with our own. We certainly can't have any evidence that there are no such universes, by hypothesis. Does this mean we ought to be agnostic about their existence? This depends on where the burden of proof lies. Should we reject the existence of things in which we could have no reason to believe, or suspend judgement about them? Perhaps the latter is the more enlightened option, but then we ought to be agnostic about a literal infinity of matters—whatever anyone can conceive without contradicting physics. Should we be agnostic about the existence, somewhere, of two-headed gerbils that sing the blues? If the agnosticism a philosopher advises us to take up towards the unknowable noumenal objects is strictly analogous to this then we are sanguine about agreeing to it. Of course, an epistemic structural realist may insist in a Kantian spirit (recall the discussion of Kantian humility in 2.4.1, and see 3.4 below) on more than this, namely that there being such objects is a necessary condition for our empirical knowledge of the world. This claim is one about which we are not agnostic, but which we deny. The burden of proof is on us to show that ontic structural realism (OSR) as motivated by current physics is intelligible without any Kantian residue.

As we saw in the previous chapter, Worrall's motivation for introducing structural realism was the need to respond to the pessimistic meta-induction. French's and Ladyman's advocacy of OSR is motivated by two further desiderata: (a) the need for an ontology apt for contemporary physics, and a way of dissolving some of the metaphysical conundrums it presents; and (b) the need for a conception of how theories represent the world that is compatible with the role of models and idealizations in physics. Two leading philosophers of physics, Simon Saunders and Michael Redhead, share these broad motivations and arrive at interestingly comparable positions. We have already mentioned Saunders's (1993a) paper on continuity in structural representation in physics, and more recently he (2003a, b, c, d) and David Wallace (2003a) have deployed structuralism to solve the problem of how macroscopic objects with more or less determinate properties can be recovered from the Everett interpretation of quantum states (the so-called preferred basis problem). Redhead's classic paper on theories and models (1980) reflects a long-standing concern with representation in physics, and he has endorsed structural realism as a way of interpreting quantum field theory (1999).[1]

Some of the pioneers of structuralism in twentieth-century philosophy of science were also concerned with (a) above. For example, although in his discussion of Cassirer's work Gower (2000) confines himself to an epistemological reading of structural realism, French and Ladyman (2003a) have argued that Cassirer, like Weyl, was concerned to replace an ontology based on individuals with one more suited to twentieth-century science. French has recently examined the role of group theory in the development of quantum mechanics (1999, 2000), and explored the idea of the group-theoretic 'constitution' of objects as sets of invariants under symmetry transformations which can be found in the writings of Cassirer, Eddington, Schrödinger, Weyl, and others. Here we shall explain how modern physics undermines classical metaphysical intuitions, and then argue for OSR as a response.[2] As Ernan McMullin says: 'imaginability must not be made the test for ontology. The realist claim is that the scientist is discovering the structures of the world; it is not required in addition that these structures be imaginable in the categories of the macroworld' (1984, 14).[3]

The demand for an ontology based on individuals may be criticized on the grounds that it is the demand that the mind-independent world be imaginable in terms of the categories of the world of experience. OSR allows for a global relation between models and the world, which can support the predictive success of theories, but which does not supervene on the successful reference of theoretical terms to individual entities, or the truth of sentences involving them.

3.1 IDENTITY AND INDIVIDUALITY IN QUANTUM MECHANICS[4]

The debate about whether quantum particles are individuals began with the development of quantum statistics. There is a fundamental difference between the way classical statistical mechanics and quantum theory deal with the permutation

---

[1] Others who follow Ladyman (1998) in regarding structural realism as a metaphysical response to the ontological import of modern physics as well as a solution to the problem of theory change include: Bain (2003 and 2004), Esfeld (2004), Lyre (2004), Stachel (2002 and 2006).

[2] The other strand of structuralism in the philosophy of science is that associated with Russell and Carnap which we discussed in the previous chapter. Though Russell had much of interest to say about the physics of his time (see Chapter 5 below), his and Carnap's versions of structuralism were more directly motivated by epistemological and semantic problems than by ontological issues arising from physics.

[3] McMullin (1990) explains how Duhem was a realist about the relations found in laws but not about explanations in terms of an ontology. Duhem seems to have been an epistemic structural realist.

[4] The classic paper in the philosophical literature about these issues is French and Redhead (1988). See also French (1989, 1998), Teller (1989), and van Fraassen (1991).

of indistinguishable elementary particles.[5] The crucial point is illustrated by the following example. Imagine that there are two particles (1 and 2), and two boxes (A and B), where each of the particles must be in one box or the other.[6] Classically there are four possible configurations for the system:

Both 1 and 2 in A; both 1 and 2 in B; 1 in A and 2 in B; 1 in B and 2 in A.

If these are regarded as equipossible, each will be assigned a probability of $1/4$. The situation is quite different in quantum mechanics (QM), where there are only three possible states:

Both 1 and 2 in A; both 1 and 2 in B; one of 1 and 2 in A and the other in B.

Hence, if these are regarded as equipossible, each will be assigned a probability of $1/3$. In quantum statistics, then, what would be regarded as two possible states of affairs classically is treated as one possible state of affairs.[7] This is formalized by the so-called 'indistinguishability postulate':[8]

If P is the operator corresponding to the interchange of the particles 1 and 2 in a two-particle quantum state, Ψ_{12},

$$P\Psi_{12} = \Psi_{21} \text{ and } P\Psi_{21} = \Psi_{12},$$

then, for observable property, Q:

$$\langle P\Psi_{12}|Q|P\Psi_{12}\rangle = \langle \Psi_{12}|Q|\Psi_{12}\rangle, \forall Q, \forall \Psi_{12}.$$

---

[5] Both classical and quantum elementary particles of a given type are regarded as indistinguishable in the sense that they will all have the same mass, size, and shape (if any), charge, and so on; but they may in principle be distinguished by their spatio-temporal or other state-dependent properties. We may think of the former properties as 'essential', in the sense that they are characteristic of the natural kind to which a particular particle belongs, as opposed to the 'accidental' properties which are those that a particle just happens to have, such as its velocity or position at a particular time. (Note that this distinction between essential and accidental properties does not correspond to that between permanent and temporary properties; an electron might happen accidentally to have the same position throughout the history of the universe.) Some authors talk of 'identical particles' (for example, van Fraassen 1991) but they mean by this just what we mean by indistinguishable particles, that is, particles that are qualitatively identical because they are in the same state, and yet numerically (or quantitatively) distinct. Note also that if Lee Smolin (2001) is right about quantum gravity then even the properties of quantum particles that we now regard as essential will turn out to be dynamical and dependent on the contingent evolution of the universe (see 3.7.2).

[6] This example constructed for pedagogical purposes is not physically realistic as the quantum particles could be in a superposition of the state corresponding to being in box A and the state corresponding to being in box B. The point being explained here, namely that permutation of particles does not lead to physically distinct states in quantum mechanics, is nonetheless a sound one.

[7] For a detailed discussion of the relationship between metaphysics and statistics see Huggett (1997, 1999). A number of papers on the issues discussed in this section can be found in Castellani (1998).

[8] See Greenberg and Messiah (1964).

So, according to the formalism of QM, the permutation of indistinguishable particles in some state is not observable, and states which differ only with respect to the permutation of particles of the same kind are treated as the same state labelled differently.

Early in the history of QM this led some physicists to argue that quantum particles are not individuals. To understand what is at stake here we must separate the related concepts of distinguishability (or discernibility) and individuality. The former epistemic notion concerns what enables us to tell that one thing is different from another. The latter metaphysical notion concerns whatever it is in virtue of that two things are different from one another, adding the restriction that one thing is identical with itself and not with anything else.[9] There are three main candidates in the philosophical tradition for a principle of individuation for physical objects:

(1) transcendent individuality: the individuality of something is a feature of it over and above all its qualitative properties;[10]

(2) spatio-temporal location or trajectory;

(3) all or some restricted set of their properties (the bundle theory).

Ordinary everyday objects, such as leaves and snowflakes, never, it seems, possess all the same properties; they are distinguishable by both their spatio-temporal properties and their intrinsic properties.[11] The particles of classical physics of a given type were thought to share all their intrinsic properties. But classical physics assumed a principle of impenetrability, according to which no two particles could occupy the same spatio-temporal location. Hence, classical particles were thought to be distinguishable in virtue of each one having a trajectory in spacetime distinct from every other one. Thus for everyday objects and for classical particles, the principle of the Identity of Indiscernibles (PII) is true, and it is plausible to argue (with Leibniz) that individuality and distinguishability amount to the same thing. After the advent of QM the status of PII has been the subject of debate,

---

[9] The distinction is due to the scholastic philosopher Suarez.

[10] This is similar to Post's (1971) notion of transcendental individuality, also adopted by French and Redhead (1988). Different ways of cashing it out include Locke's substantial substratum; the notion of a 'bare particular'; and the notion of haecceity or thisness due to Duns Scotus and rehabilitated by Adams (1979).

[11] Intrinsic properties are normally defined to be those which an object may possess independently of everything else that exists, or independently of whether or not anything else exists (so that the property of being the only existing entity does not come out as intrinsic; see Lewis 1983, Langton and Lewis 1998, and Lewis forthcoming). Hence, the mass and charge of classical particles, and the shape and size of a person are thought to be intrinsic properties. Extrinsic properties are just those that are not intrinsic; weight, being the brother of, and so on, are extrinsic properties. Another usage common in the foundations of QM (see for example, Jauch 1968, 275) treats the intrinsic properties of a system as those that do not depend on the state of the system, and the extrinsic properties as those that do. This seems to equate intrinsic with essential. In any case we shall adopt the former definitions as the latter one would seem to beg the question of whether spatio-temporal properties are intrinsic or extrinsic since they are obviously state-dependent.

and the possibility of some principle of individuation that appeals to some feature of particles other than their qualitative properties has been taken seriously.[12]

In the formalism of QM particles are not always assigned well-defined trajectories in spacetime. Furthermore, two or more particles in an entangled state may possess exactly the same monadic and relational properties that are expressible by the formalism of the theory. Consider a pair of electrons in the orbital of a helium atom for example. They have the same energy eigenstate, and the same position state (which is not localized), but because they cannot have all the same quantum states as each other by the Pauli exclusion principle, their spin state is such that in any given direction in space they must have opposite spins as represented by the singlet state:

$$\Psi_{12} = \frac{1}{\sqrt{2}}\{|\downarrow\rangle_1|\uparrow\rangle_2 - |\uparrow\rangle_1|\downarrow\rangle_2\}$$

(Here the electrons are labelled 1 and 2, and the spin components of + and − along an arbitrary axis are represented by up and down arrows respectively.) Clearly, according to this state description there is no property of particle 1 that cannot also be predicated of particle 2.[13]

Hence, quantum particles appear sometimes to possess all the same intrinsic and extrinsic properties. If two electrons really are two distinct individuals, and it is true that they share all the same properties, then it seems that there must be some principle of individuation that transcends everything that can be expressed by the formalism in virtue of which they are individuals. If we assume for now that the quantum description is complete, then we are left with a dilemma: either PII is false, the quantum particles are individuals and there must be some principle of individuation of type (1) above; or quantum particles are not individuals and PII is moot in this context.[14]

French argues that this means that QM gives rise to a form of metaphysical underdetermination (French 1989, 1998), since there are (at least) two metaphysical packages compatible with it. According to the first of these, quantum particles have transcendent individuality. They are individuated by something

---

[12] PII is easiest to state in the contrapositive form, namely there are no two things that share all their properties. This is trivially true unless its scope is restricted to exclude identity-involving properties.

[13] In what follows we shall usually only consider fermions (such as electrons or protons), because they are regarded as the particles of matter, as opposed to bosons (such as photons) which are the quanta of the various fields representing physical forces. Bosons and fermions differ in various respects including the statistics they obey and their spin magnitudes. For our purposes it is worth noting that bosons can be in identical quantum states that are not entangled states of more than one particle. Hence, there can be indefinitely many of them on the end of a pin.

[14] Here we only consider synchronic identity. Identity over time is also highly problematic for quantum particles since QM does not attribute definite trajectories to them and so if an electron is absorbed and then later an electron is emitted by the atom, it seems that there is no fact of the matter about whether the same electron is involved in each process.

that transcends the empirical facts, and for some reason they are unable to enter into certain entangled states (the spin-statistics theorem says that bosons can only occupy symmetric states (those whose mathematical description is unchanged by the permutation of particles), and fermions can only occupy antisymmetric states—those whose mathematical description is multiplied by -1 if the particles are permuted). It is worth noting that on this view the naming or labelling of the particles is problematic. This is because a descriptivist account of reference is unworkable if PII fails, and because a rigid designation account seems to imply the wrong statistics, in particular, that we ought to count a two-particle state as distinct from the same one with the two particles interchanged. The alternative metaphysical picture abandons the idea that quantum particles are individuals, perhaps in favour of a field theoretic construal of them.

Three metaphysical and methodological reasons have been offered by various authors for preferring the non-individualistic interpretation of quantum particles: (i) PII is incompatible with the individuals package under fairly reasonable assumptions, but abandoning this framework allows that it is true;[15] (ii) positing individuals plus states that are forever inaccessible to them is ontologically profligate, amounting to the acceptance of 'surplus structure'; (iii) a principle of individuality of type (1) above must be metaphysical in the sense that it posits what van Fraassen has called 'empirically surplus factors' (1991).

None of these considerations are conclusive, for several reasons. (i) There is no empirical way to confirm PII and so concluding that quantum particles are not individuals to safeguard this principle might be merely the expression of a metaphysical or metalogical preference based on experience of macroscopic objects. (ii) There are many cases in the history of science where so-called surplus structure in the formulation of a theory has later been found to be of empirical importance (see French 1995, 1997). The surplus structure in Hilbert space allows for states that are neither symmetric nor antisymmetric and hence for particles which are neither fermions nor bosons but instead obey 'para-statistics'. (iii) Van Fraassen's point about empirically surplus factors will obviously not persuade the scientific realist. Hence, the underdetermination is not easily

---

[15] These reasonable assumptions are the Principle of Statistical Mechanics (which says that states with equal phase space volume are equiprobable), and the Completeness assumption (which states that there are no hidden variables not described by the quantum formalism). Of course, there is a version of quantum theory, namely Bohm theory, according to which QM is not complete and particles do have definite trajectories at all times. However, Harvey Brown et al. (1996) argue that the 'particles' of Bohm theory are not those of classical mechanics. The dynamics of the theory are such that the properties, like mass, charge, and so on, normally associated with particles are in fact inherent in the quantum field and not in the particles. It seems that the particles only have position. We may be happy that trajectories are enough to individuate particles in Bohm theory, but what will distinguish an 'empty' trajectory from an 'occupied' one? Since none of the physical properties ascribed to the particle will actually inhere in points of the trajectory, giving content to the claim that there is actually a 'particle' there would seem to require some notion of the raw stuff of the particle; in other words haecceities seem to be needed for the individuality of particles of Bohm theory too.

broken. Van Fraassen (1991) argues that this is all the more reason to say 'goodbye to metaphysics' (1991, 480) because it presents a challenge to standard scientific realism that cannot be met and so is a reason to embrace constructive empiricism. Ladyman (1998) and French and Ladyman (2003a, b) have argued in response that, since the locus of this metaphysical underdetermination is the notion of an individual, we should reconceive individuals in structural terms.

Saunders (2003c, 2003d, and 2006) offered an interesting new twist on these issues by pointing out that French and Ladyman assume that the identity of indiscernibles should be interpreted in terms of what Quine called 'absolute discernibility' (1960, 230, and 1976 [1981]). Two objects are absolutely discernible if there exists a formula in one variable which is true of one object and not the other. For example, ordinary physical objects are absolutely discernible because they occupy different positions in space and time. Absolutely discernible mathematical objects include i and 1, since 1 is the square of itself and i is not. Two objects are 'relatively discernible' just in case there is a formula in two free variables which applies to them in one order only. Moments in time are relatively discernible since any two always satisfy the 'earlier than' relation in one order only. An example of mathematical objects which are not absolutely discernible but are relatively discernible include the points of a one-dimensional space with an ordering relation, since, for any such pair of points x and y, if they are not the same point then either $x>y$ or $x<y$ but not both. Finally, two objects are 'weakly discernible' just in case there is two-place irreflexive relation that they satisfy. For example, Max Black's (1952) two spheres, which are intrinsically identical and a mile apart in an otherwise empty space, are obviously only weakly discernible. Clearly, fermions in entangled states like the singlet state violate both absolute and relative discernibility, but they satisfy weak discernibility since there is an irreflexive two-place relation which applies to them, namely the relation 'is of opposite spin to'. So we can regard such entities as individuals without violating PII, by adopting its weakest form: weak rather than strong or relative discernibility as the necessary condition for distinct individuality.[16]

Note that in the context of philosophy of mathematics, many philosophers have followed Russell in arguing that it is incoherent to suppose there could be individuals which don't possess any intrinsic properties, but whose individuality is conferred by their relations to other individuals:

[I]t is impossible that the ordinals should be, as Dedekind suggests, nothing but the terms of such relations as constitute a progression. If they are to be anything at all, they must be intrinsically something; they must differ from other entities as points from instants, or colours from sounds. What Dedekind intended to indicate was probably a definition by

---

[16] Elementary bosons can be in states such that two of them are not even weakly discernible, so Saunders concludes that they are not individuals. Composite bosons cannot be in exactly the same quantum states.

means of the principle of abstraction ... But a definition so made always indicates some class of entities having ... a genuine nature of their own. (Russell 1903, 249)

The argument is that without distinct individuals in the first place, there is nothing to stand in the asymmetric relations that are supposed to confer individuality on the relata. Contemporary philosophers of mathematics have been most influenced in this respect by Paul Benacerraf (1965 [1983]) who argued that objects to be properly so called must be individuals, and that therefore a structuralist construal of abstract objects like numbers must fail. According to Benacerraf, an object with only a structural character could be identified with any object in the appropriate place in any exemplary structure and could not therefore be an individual. However, we agree with Saunders that the weak notion of individuality (according to which weak discernibility is all that is necessary for individuality) he advocates is perfectly coherent. It is question-begging against the kind of structuralism we advocate to oppose it on the grounds of a mere prejudice in favour of a stronger form of discernibility. We note that while Saunders's view vindicates an ontology of individuals in the context of QM, it is a thoroughly structuralist one in so far as individuals are nothing over and above the nexus of relations in which they stand.[17]

Of course our best quantum theories are field theories and the standard scientific realist might be tempted to dismiss the problems of individuality arising for many-particle QM on this basis. However, there are several problems with this. First, as with classical mechanics, the fact that non-relativistic many-particle QM has had enormous empirical success and is a paradigm of a good scientific theory means that the standard scientific realist ought to be able to say what it would be to be a realist about it; otherwise realism will only apply to the one true theory of the world, if there is one, and, since we are clearly not there yet, would be of no relevance to actual theories. Second, we ought to be able to recover the concept of fundamental particles used widely by

---

[17] Leitgeb and Ladyman (forthcoming) consider cases from graph theory that violate even weak formulations of PII. They argue that (i) the identity or difference of places in a structure is not to be accounted for by anything other than the structure itself and that (ii) mathematical practice provides evidence for this view. Another philosopher who has applied graph theory to the metaphysics of physical reality in defence of a broadly structuralist view is Dipert (1997). He claims that the world is an asymmetric graph because he believes that facts about the numerical identity and diversity of objects must supervene on the relational facts about each node in the graph representing the world's structure, whereas in symmetric graphs there are nodes that admit of exactly the same structure descriptions. Leitgeb and Ladyman argue that the only reason to accept such a supervenience requirement is the mistaken claim that facts about numerical identity and diversity must be grounded somehow; instead, the idea of a structure that does admit of non-trivial automorphism is perfectly intelligible and even suggested by mathematical practice. Whether the empirical world has such a structure is an open question. Note that, as Leitgeb and Ladyman point out, graph theory is apt for representing many of the issues discussed in this chapter including the representation of a pair of weakly indiscernible objects with the unlabelled graph consisting of two nodes standing in an undirected relation. A further analogy is that in an unlabelled graph the nodes can be aggregated but not enumerated like fermions (see n. 18 below).

physicists and chemists from the ontology of field theory, and so questions about their nature remain meaningful. Third, quantum field theories are no easier to interpret realistically in terms of individuals than ordinary quantum theory and raise new and equally compelling interpretative problems. Quantum fields are very different from classical fields, for the quantum field consists of operators parametized with spacetime points:

> These operator values associated with the space-time points are not specific values of some physical quantity. The specific or concrete values are, as one initially expects, the states, or equivalently, the catalogue of probability amplitudes for all possible measurements. (Teller 1990, 613)

Thus, the operators represent not the values of physical quantities but those quantities themselves. Teller is clear on how the interpretative problems of non-relativistic QM (particularly the measurement problem) are inherited by quantum field theories.

This leads Redhead to argue against a 'classical-style realism of possessed values, *not* against a broader realism of physical structure' (1995, 7):

> [R]ealism about what? Is it the entities, the abstract structural relations, the fundamental laws or what? My own view is that the best candidate for what is 'true' about a physical theory is the abstract structural aspect. (1995, 2)
>
> Success requires 'explanation with reference to validating the structural framework of the theory'. (1995, 7)

Fourth, the problem of individuality is not solved by shifting to field theories; if anything it becomes more intractable. As Teller points out:

> Conventional quantum mechanics seems incompatible with a classical notion of property on which all quantities always have definite values. Quantum field theory presents an exactly analogous problem with saying that the number of 'particles' is always definite. (1990, 594)

In quantum field theories the state may be a superposition of different definite particle number states. Furthermore, for a given state of the field (how many particles there are) is dependent on the frame of reference adopted. So particles seem to lose their reality in the field theoretic approach. Teller himself advocates an interpretation in terms of 'quanta' which are excitations of the field that may be aggregated like particles (we can say there is a state with so many quanta), but cannot be enumerated (we cannot say this is the first, this the second, and so on); quanta are not individuals.[18]

---

[18] Decio Krause (1992) developed a formal framework for non-individual entities based on an extension of set theory to include sets which have a cardinality but no ordinality; sets of quanta would have this feature. A similar project has been undertaken by Chiara et al. (1998) and her co-workers. This kind of analysis is developed in breadth and detail in French and Krause (2006). Lowe also thinks that fermions like electrons are countable but not individuals since there is no fact of the matter about which is which (2003b, 78).

Ernst Cassirer rejected the Aristotelian idea of individual substances on the basis of physics, and argued that the metaphysical view of the 'material point' as an individual object cannot be sustained in the context of field theory. He offers a structuralist conception of the field:

> The field is not a 'thing', it is a system of effects (*Wirkungen*), and from this system no individual element can be isolated and retained as permanent, as being 'identical with itself' through the course of time. The individual electron no longer has any substantiality in the sense that it *per se est et per se concipitur*; it 'exists' only in its relation to the field, as a 'singular location' in it. (1936 [1956], 178)[19]

OSR agrees with Cassirer that the field is nothing but structure. We can't describe its nature without recourse to the mathematical structure of field theory. Holger Lyre also argues for structural realism in the interpretation of quantum field theory. He argues that 'the traditional picture of spatiotemporally fixed object-like entities is undermined by the ontology of gauge theories in various ways and that main problems with traditional scientific realism ... can be softened by a commitment to the structural content of gauge theories, in particular to gauge symmetry groups' (2004, 666). He goes on to note that his favoured interpretation of gauge theories (in terms of non-separable holonomies) is one according to which the fundamental objects are ontologically secondary to structure. (The notion of structure he has in mind is group-theoretic structure to which we return below.)[20]

In contemporary accounts, fields are local, in the sense that field quantities are attributed to spacetime points (or, taking into account quantum entanglement, spacetime regions). The problem of individuality now arises again as that of whether fields themselves are individuals or whether they are the properties of spacetime points; this pushes the problem back to whether the spacetime points are individuals. This latter issue is bound up with the debate about substantivalism in the foundations of General Relativity (GR) to which we now turn.[21]

---

[19] This description of the field as 'a system of effects' raises an analogue of the problem about relations without relata which we discuss below, namely, how can we have an effect without a something which is doing the effecting?

[20] See also Auyang (1995), who adopts a Kantian approach to quantum field theory. According to her, spacetime is absolute, in the sense that it is presupposed by the concept of individuals, but is not a substance. Spatio-temporal relations are only 'implicit' (138), but their structure makes 'events' numerically distinct, so events are individuated structurally. The conceptual structure of the world as a field is represented by a fibre bundle (133). Events are entities in an interacting field system (129) (which are identified by a parameter of the relevant base space in a fibre bundle formulation) and divided into kinds via groups (130–2). Auyang thinks that neither spacetime structure, nor event structure should be given ontological priority: '[t]he event structure and the spatio-temporal structure of the objective world emerge together' (135).

[21] Note that when it comes to quantum field theory in a curved spacetime, 'a useful particle interpretation of states does not, in general, exist' (Wald 1984, 47, quoted in Stachel 2006, 58). See also Malament (1996) and Clifton and Halvorson (2002), who show that there is a fundamental conflict between relativistic quantum field theory and the existence of localizable particles.

3.2 INDIVIDUALITY AND SPACETIME PHYSICS

In the context of GR the old debate between substantivalism and relationalism about space must be reconstrued in terms of spacetime.[22] To a first approximation, the former view holds that the points of the spacetime manifold exist independently of the material contents of the universe, while the latter maintains that spatio-temporal facts are about the relations between various elements of the material contents of spacetime. The main foundational novelty of GR is that spacetime itself is dynamical according to the theory. The state of a universe in GR is specified by a triple <M, g, T>, where M is a differential manifold with a topology, g is a local Minkowski metric, T is the stress-energy tensor and encodes the distribution of matter and energy in spacetime, and where g and T satisfy Einstein's field equations. The latter are preserved under all diffeomorphic coordinate transformations (general covariance). The gravitational field in GR is identified with the metrical structure of spacetime.

There has been much dispute about whether GR supports relationism or substantivalism about spacetime. The theory lacks a preferred foliation of spacetime into space and time, and its metric and geometry are dynamical, suggesting relationism (because if spacetime is not a fixed background structure there is less reason to think of it as a substance). On the other hand, the stress-energy tensor T does not uniquely determine the structure of spacetime, as there is a solution to the field equations where the whole universe is rotating, which is an absurdity from a Machian point of view and is suggestive of substantivalism.[23] Moreover, the theory provides for the transfer of energy from physical objects to spacetime itself. Another problem for the relationist is that the field equations of GR have solutions where spacetime is entirely empty of matter. Hence, the theory seems to imply that spacetime can exist and have properties and structure, independently of its material contents. Furthermore, prima facie, the theory quantifies over spacetime points and predicates properties of them. Perhaps the most compelling argument in favour of substantivalism is that spacetime in GR is physical in the sense that it acts on matter and matter acts on it (unlike Minkowski spacetime in Special Relativity).[24]

On the other hand, the main problem for substantivalism is that the general covariance of the field equations of GR means that any spacetime model and its

---

[22] There is much debate about what exactly Newton and Leibniz thought, of course, and the relationships between dynamics and geometry, and notions of absolute space and time, and ideas about substance and ontological priority, are complex. See, for example, DiSalle (1994).

[23] See Sklar (1974).

[24] Harvey Brown argues that the physical relationship between spacetime structure and matter is completely mysterious in Special Relativity. The violation of the action–reaction principle by spacetime, because it affects matter but not vice versa, has also been invoked as a reason to be suspicious about ontological commitment to Minkowski spacetime in Special Relativity. See Brown and Pooley (forthcoming), Brown (2005).

image under a diffeomorphism (an infinitely differentiable, one-one and onto mapping of the model to itself) are in all observable respects equivalent to one another; all physical properties are expressed in terms of generally covariant relationships between geometrical objects. In other words, since the points of spacetime are entirely indiscernible one from another, it makes no difference if we swap their properties around so long as the overall structure remains the same. This is made more apparent by the so-called 'hole argument' which shows that if diffeomorphic models are regarded as physically distinct then there is a breakdown of determinism:

$X = < M, g, T >$ is a solution of GR, d:M $\rightarrow$ M is a diffeomorphism.
Then $X' = < M, d * g, d * T >$ is also a solution of GR.

Let h be a hole diffeomorphism, that is h differs from the identity map inside a region R of M, but smoothly becomes the identity at the boundary of R and outside. There are arbitrarily many such hs for R. Hence, the state inside R cannot be determined by the state outside R no matter how small R is, and so determinism fails. Substantivalists cannot just bite the bullet and accept this since, as Earman and Norton (1987) argue, the question of determinism ought to be settled on empirical/physical grounds and not metaphysical ones.

There have been a variety of responses to this problem. Lewis (1986) and Brighouse (1994) suggest accepting haecceitism about spacetime points, but argue that it shouldn't worry us that haecceitistic determinism, that is determinism with respect to which points end up with which metrical properties, fails. Melia (1999) also criticizes the notion of determinism employed by Earman and Norton. Nonetheless most philosophers of physics seem to have concluded that if spacetime points do have primitive identity then the substantivalist who is committed to them must regard the failure of haecceitistic determinism as a genuine failure of determinism. Hence, others have sought to modify the substantivalism. It is fair to ask: if spacetime is a substance, what is the substance in question? Is it the manifold alone, the manifold plus the metric, or the manifold, metric, and the matter field?[25]

According to manifold substantivalism, the manifold itself represents spacetime. Earman and Norton (1987) argue that manifold substantivalism is the appropriate way to understand substantivalism in the context of GR because it embodies the distinction between the container and the contained that they regard as the definitive commitment of substantivalism. One way round the hole argument for manifold substantivalists is to adopt metric essentialism, namely the idea that the manifold M has its metrical properties essentially so that there is no other possible world where the same points have different properties (see Maudlin 1990). This means that the identity of each spacetime point

---

[25] This is often called super-substantivalism; see Sklar (1974).

depends on its metrical properties and that general covariance fails. The problem with this is that physicists seem to regard diffeomorphic models as equivalent.

We think the correct response to the hole argument is that recommended by Robert DiSalle (1994). He suggests that the *structure* of spacetime be accepted as existent despite its failure to supervene on the reality of spacetime points. A similar view has been proposed by Carl Hoefer, who argues that the problems for spacetime substantivalism turn on the 'ascription of primitive identity to space-time points' (1996, 11). Hence, it seems that the insistence on interpreting spacetime in terms of an ontology of underlying entities and their properties is what causes the problems for realism about spacetime. This is similar to the position developed by Stein (1968) in his famous exchange with Grünbaum, according to which spacetime is neither a substance, or a set of relations between substances, but a structure in its own right. It seems that the ontological problems of QM and GR in respect of identity and individuality both demand dissolution by OSR.

The analogy between the debate about substantivalism, and the debate about whether quantum particles are individuals was first explicitly made by Ladyman (1998), but others such as Stachel (2002) and Saunders (2003c) have elaborated it. However, Pooley (2006) argues that there is no such analogy, or at least not a very deep one, in part because he thinks that there is no metaphysical underdetermination in GR. According to him the standard formulations of the theory are ontologically committed to the metric field, and the latter is most naturally interpreted as representing 'spacetime structure' (89). However, there is a long-standing dispute about this between so-called 'sophisticated substantivalists' on the one hand (see for example, Brighouse 1994 and Butterfield 1989), and Earman and others on the other hand (see Earman and Norton 1987 and Belot and Earman 2000, 2001). The latter maintain that the essence of relationism is the doctrine of Leibniz equivalence according to which diffeomorphic models always represent the same physically possible situation. Hence they argue that since sophisticated substantivalists accept Leibniz equivalence, and deny manifold substantivalism, sophisticated substantivalism is a 'pallid imitation' of relationism (Belot and Earman, 2001, 249). On the other hand, Pooley argues that the only metaphysical dispute in GR concerns what he calls 'haecceitist' (manifold) and 'anti-haecceitist' (sophisticated or metric) substantivalism, and he argues that there is no underdetermination in relation to this issue since the haecceitist version is plainly wrong. He argues, contrary to Belot and Earman 2000, that the substantivalist can accept that diffeomorphic models do not describe distinct possible worlds despite differing in respect of the permutation of spacetime points, and that to do so is not ad hoc if one supposes that the numerical distinctness of the individual points is grounded in their positions in a structure (Pooley 2006, 103). On this view, in so far as there are points in GR, they are the members of equivalence classes under the diffeomorphism group.

Pooley argues that the permutation invariance that underlies the problems of individuality in QM is very different from the diffeomorphism invariance of GR, because the latter case concerns a symmetry of the theory, whereas the former case concerns symmetries of solutions of the theory as well as the theory itself. He is surely right that there are considerable formal differences between the two cases. However, for our purposes these are less important than the fact that both QM and GR give us every reason to believe that the realists must interpret the theories as describing entities whose identity and individuality are secondary to the relational structure in which they are embedded.

Mauro Dorato is also a structuralist about spacetime: 'To say that spacetime exists just means that the physical world exemplifies, or instantiates, a web of spatiotemporal relations that are described mathematically' (2000, 7). He thinks that spacetime has objective existence, but not as a substance, and that a structuralist form of realism about spacetime avoids the problems facing substantivalism and captures all the features that make relationism so attractive. However, Dorato insists that, 'to the extent that real relations, as it is plausible, presuppose the existence of *relata, then spatiotemporal relations presuppose physical systems and events*' (2000, 7; his emphasis). Dorato thinks that we may avoid the supervenience of such relations on their relata by adopting a form of Armstrong's bundle theory of individuals, but it is not clear whether such a move eliminates all non-structural elements. Indeed, he earlier remarks that '*spacetime points can only be identified by the relational structure provided by the gravitational field*' (2000, 3; his emphasis). This, of course, throws the issue back to the field, and Dorato agrees with Cao in his review of the latter's book (Dorato 1999) that the existence of spatio-temporal relations must be underpinned by the existence of the gravitational field, understood as a 'concrete' and hence, presumably, non-structural, entity.[26]

However, there is some ambiguity here. Dorato identifies Cao as an ontic structural realist, because the latter denies that the structures postulated by field theories must be 'ontologically supported by unobservable entities' (Cao 1997, 5). Dorato writes that, according to Cao,

> while structural relations are real in the sense that they are testable, the concept of unobservable entities that are involved in the structural relations always has some conventional element, and the reality of the entities is constituted by, or derived from, more and more relations in which they are involved. (2000, 3)

---

[26] Dorato gives a version of the Redhead argument here which we discuss below. He says, 'I don't know how one can attribute existence as a *set of relations* in an observable or unobservable domain without also requiring that these relations be exemplified by non-abstract *relata*, namely the field itself, to be regarded as a new type of substance, radically different from the traditional, Aristotelian *ousia*' (Dorato 1999, 3; his emphasis). That Dorato is inclined towards a form of epistemic structural realism is clear from his insistence that, although we often identify physical entities via their relations, 'epistemic strategies for identifications should not be exchanged for ontological claims' (1999, 3). Dorato argues that structural realism needs entity realism to be plausible (1999, 4).

Dorato comments that, '[i]n this respect, entities postulated by physical theories are to be regarded as a web of relations, not presupposing substance-like entities or "hangers" in which they inhere' (2000, 3). This is once again reminiscent of Cassirer who said: 'To such a point also no being in itself can be ascribed; it is constituted by a definite aggregate of relations and consists in this aggregate' (1936, 195).[27]

3.3 OBJECTIVITY AND INVARIANCE

Given that we have argued that relational structure is more ontologically fundamental than objects we need to say something about how objects ought to be regarded. In the next chapter we will offer a positive account of 'real patterns' that accounts for commitment to objects in the generalizations of the special sciences. When it comes to fundamental physics, objects are very often identified via group theoretic structure. Hence, Eddington says: 'What sort of thing is it that I know? The answer is *structure*. To be quite precise it is structure of the kind defined and investigated in the mathematical theory of groups' (1939, 147). Group theory was first developed to describe symmetry. A symmetry is a transformation of some structure or object which leaves it unchanged in some respect. A group of symmetry transformations is a mathematical object which consists of the set of transformations, including the identity transformation and the inverse of each transformation, and the operation of composing them, where the result of two composed transformations is itself in the original set. Objects can be identified in terms of which symmetry transformations leave them unchanged or invariant.[28]

For example, one of the most fundamental distinctions between kinds of particles is that between fermions and bosons. This was described group theoretically by Weyl and Wigner in terms of the group of permutations, and the former's approach to relativity theory was similarly group-theoretic.[29] In the case of QM, Weyl asserts that: 'All quantum numbers, with the exception of the so-called principal quantum number, are indices characterizing representations of groups' (1931, xxi). The central point of philosophical relevance here is that the mathematical idea of invariance is taken by Weyl to characterize the

---

[27] Esfeld and Lam (forthcoming) argue for structural realism about spacetime, as does Bain (2003): 'Conformal structure, for instance, can be realized on many different types of "individuals": manifold points, twisters or multivectors... What is real, the spacetime structuralist will claim, is the structure itself and not the manner in which alternative formalisms instantiate it' (25).

[28] The history of symmetries in science is discussed in Mainzer (1996).

[29] See Ryckman (2005) for a beautiful account of the history of relativity theory and Weyl's role in it. Ryckman argues that the work of Eddington and Weyl was profoundly influenced by the phenomenology of Husserl. The latter seems to have understood objectivity in terms of invariance.

notion of *objectivity*.[30] It is this that liberates physics from the parochial confines of a particular coordinate system. For Weyl appearances are open only to intuition (in the Kantian sense of subjective perception) and therefore agreement is obtained by giving objective status only to those relations that are invariant under particular transformations. What is particularly striking is the way that Weyl uses the insights gathered from his work on transformations and invariants in relativity theory to make his crucial contribution to the development of QM. The choice of the momentum space representation of Schrödinger amounts to the choice of a coordinate system. Weyl saw immediately that the proto-theories of Schrödinger and Heisenberg had fundamental mathematical similarities. He therefore took them to be in all important respects different versions of the same theory:

[T]he essence of the new Heisenberg-Schrödinger-Dirac quantum mechanics is to be found in the fact that there is associated with each physical system a set of quantities, constituting a non-commutative algebra in the technical mathematical sense, the elements of which are the physical quantities themselves. (Weyl 1931, viii)

Weyl here anticipates von Neumann's unification of the theories in his classic text on QM, and indeed Dirac in his book on the theory cites Weyl as the only previous author to employ the same 'symbolic method' for the presentation of the theory as he does himself. This method according to Dirac 'deals directly in an abstract way with the quantities of fundamental importance (the invariants etc., of the transformations)' and therefore it goes 'more deeply into the nature of things' (1930, viii). Dirac uses vectors, and not rays in Hilbert space like Weyl and von Neumann, in his treatment of QM, but both he and Weyl recognized that the mathematical status of the two rival theories of QM as alternative *representations* of the same mathematical structure makes preference for either eliminable once a unified framework is available.[31]

The idea then is that we have various representations of some physical structure which may be transformed or translated into one another, and then we have an invariant state under such transformations which represents the objective state of affairs. Representations are extraneous to physical states but they allow our empirical knowledge of them. Objects are picked out by the identification of

---

[30] Weyl's views have recently been revived by Sunny Auyang (1995) in an explicitly neo-Kantian project which attempts to solve the problem of objectivity in QM. Auyang seeks to extract the 'primitive conceptual structure' in physical theories and she too finds it in what she calls the 'representation-transformation-invariant structure'. This is essentially group-theoretic structure. Auyang, like Born and Weyl, thinks that such invariant structure under transformations is what separates an objective state of affairs from its various representations, or manifestations to observers under different perceptual conditions.

[31] It may not have been until the empirical success resulting from Dirac's relativistic theory of the electron, which was based on the underlying group structure, that it was widely acknowledged by physicists that the abstract structure of a group is of greater significance than particular representations of it. Indeed, in the early days of QM Weyl reports that some talked of eliminating the 'group pest' (Weyl 1931, x).

invariants with respect to the transformations relevant to the context. Thus, on this view, elementary particles are hypostatizations of sets of quantities that are invariant under the symmetry groups of particle physics. When we move to gauge quantum field theories, group theory is even more important as each theory is associated with a different symmetry group, and the unification of theories was achieved by looking for theories with the relevant combined symmetry. (For example, $U(1)$ for QED, $SU(2)$ for the theory of the weak interaction, and $SU(2) \times U(1)$ for the unified electroweak theory.) Here is Lyre again: 'a group theoretic definition of an object takes the group structure as primarily given, group representations are then constructed from this structure and have a mere derivative status' (2004, 663). He goes on to note that the objects of a theory are members of equivalence classes under symmetry transformations and no further individuation of objects is possible. Similarly, Kantorovich (2003) argues that the symmetries of the strong force discovered in the 1960s are ontologically prior to the particles that feel that force, namely the hadrons, and likewise for the symmetries of the so-called 'grand unification' of particle physics in the standard model.[32]

The founders of structuralism shared an appreciation of the importance of group theory in the ontology of physics. Ernst Cassirer held that the possibility of talking of 'objects' in a context is the possibility of individuating invariants (1944). Similarly, Max Born says: 'Invariants are the concepts of which science speaks in the same way as ordinary language speaks of "things", and which it provides with names as if they were ordinary things' (1953, 149), and: 'The feature which suggests reality is always some kind of invariance of a structure independent of the aspect, the projection' (149). He goes so far as to say: 'I think the idea of invariant is the clue to a relational concept of reality, not only in physics but in every aspect of the world' (144).[33]

Consider the following remark by Howard Stein:

if one examines carefully how phenomena are 'represented' by the quantum theory ... then ... interpretation in terms of 'entities' and 'attributes' can be seen to be highly dubious ... I think the live problems concern the relation of the Forms ... to phenomena, rather than the relation of (putative) attributes to (putative) entities ... (1989, 59)

The forms in question are given in part by the invariance structure of theories.[34]

---

[32] Although he does not cite French and Ladyman, Kantorovich does refer to van Fraassen's (2006) discussion of 'radical structuralism' (which is the latter's name for OSR) as closest to his own view.

[33] This undercuts Putnam's paradox: objects are given only up to isomorphism.

[34] The importance of group theory in the development of structuralism deserves further historical analysis. It played a crucial role in epistemological reflections on geometry in relation to Klein's Erlanger programme (Birkhoff and Bennett 1988). See French (1998, 1999, 2000) and Castellani (1998) who have explored the ontological representation of the fundamental objects of physics in terms of sets of group-theoretic invariants by Weyl, Wigner, Piron, Jauch, and others.

3.4 THE METAPHYSICS OF RELATIONS

To be an alternative to both traditional realism and constructive empiricism, structural realism must incorporate ontological commitment to more than the empirical content of a scientific theory, namely to the 'structure' of the theory. We have argued that relational structure is ontologically subsistent, and that individual objects are not. However, the idea that there could be relations which do not supervene on the properties of their relata runs counter to a deeply entrenched way of thinking. The standard conception of structure is either set-theoretic or logical. Either way it is assumed that a structure is fundamentally composed of individuals and their intrinsic properties, on which relational structure supervenes. The view that this conceptual structure reflects the structure of the world is called 'particularism' by Teller (1989) and 'exclusive monadism' by Dipert (1997).[35] It has been and is endorsed by many philosophers, including, for example, Aristotle and Leibniz.

In particular, consider the doctrine that David Lewis calls Humean supervenience:

[A]ll there is to the world is a vast mosaic of local matters of particular fact, just one little thing and then another... We have geometry: a system of external relations of spatio-temporal distance between points (of spacetime, point matter, aether or fields or both). And at these points we have local qualities: perfectly natural intrinsic properties which need nothing bigger than a point at which to be instantiated... All else supervenes on that. (1986, x)

Recall that intrinsic properties are those which may be possessed by something independently of whether or not there are other entities, such as charge, mass, and so on. Extrinsic properties are non-intrinsic properties, such as being north-east of Bristol. By 'all else' Lewis means all truths of causation, laws, and identity over time. He argues that all that exists, according to physics, is a web of intrinsic properties of objects and spatio-temporal relations, extrinsic properties are determined by that. There are no abstract entities nor any necessary connections. '[A]ll the facts there are about the world are particular facts, or combinations thereof' (ibid. 111). Lewis argues that Humean supervenience is only contingently true, and that: 'If physics itself were to teach me that it is false, I wouldn't grieve' (ibid. xi).

Indeed, it is surely natural science, and in particular mechanistic materialism, that has inspired the addition of Humean supervenience to particularism. Although Lewis considers that QM may indeed teach that Humean supervenience is false, this is a lesson he refuses to learn, on the grounds that QM is 'imbued

---

[35] Dipert argues for a structuralist metaphysics in terms of the theory of graphs. It is noteworthy that some research in quantum gravity makes use of graph theory and cognate mathematics.

with instrumentalist frivolity', 'double thinking deviant logic' and 'supernatural tales' (ibid.). Yet if we are to be scientific realists (as Lewis would claim) we should surely have our metaphysics informed by our best physics, and we can hardly object that we will only do this if the deliverances of physics coincide with our prejudices. The interpretation of quantum theory may well be fraught with difficulty but the theory has produced many novel predictions and has been well confirmed to an unprecedented degree of precision.[36] We have already explained how quantum theory challenges the assumption that the entities which physics describes are individuals. We now turn to what it has to tell us about the ontology of relations, and in particular, about whether relations are supervenient on the properties of their relata as Humean supervenience requires.

If two electrons are in a joint state that is 'entangled' (like the singlet state) then according to QM they do not have any non-relational state-dependent properties. Paul Teller proposes the existence of 'non-supervenient relations' (see, for example, 1989), that is, relations that do not supervene on the monadic properties of their relata, in the interpretation of entangled states in QM. On this view, facts about relations must be understood as irreducible to facts about the non-relational properties of individuals; hence this is opposed to particularism as defined above. As mentioned previously, these relations are part of a classical ontology of individuals in Teller's picture. However, it is worth investigating the nature of these non-supervenient relations in order to appreciate how QM challenges classical intuitions about ontology, like those which motivate Lewis's notion of Humean supervenience.[37]

Jeremy Butterfield (1992) has argued that non-supervenient relations are equivalent to what Lewis (1986) calls 'external relations'. According to Lewis, an internal relation is one which supervenes on the intrinsic properties of its relata, in the sense that there can be no difference in the relations between them without a difference in their intrinsic properties. For example, if two objects are related by 'bigger than', then this relation supervenes on the sizes of the two objects. On the other hand, an external relation is one which fails to supervene on the intrinsic properties of its relata, but does supervene on the intrinsic properties of their 'composite'. The example Lewis gives is of the spatial separation of a proton and an electron orbiting it (a hydrogen atom), where this system is understood classically as if it were like the Moon orbiting the Earth. This relation will not

[36] We could read Lewis as saying that he won't take ontological lessons from QM until there is a satisfactory resolution of the measurement problem, but Bell's theorem tells us that any such resolution must do violence to at least part of Lewis's metaphysical picture. As we pointed out in 1.2.3, Bell's theorem is not about QM, but rather reveals constraints on any empirically adequate successor to QM. Entanglement, as identified with the violation of Bell inequalities, is an empirical discovery not a theoretical posit. Lewis later engaged with QM in more depth, but his earlier work is what has been most influential. Note, while recalling our invective about domesticating metaphysics in 1.1, that his preferred solution to the measurement problem is the one that he finds to be 'a comfortable and plausible way for nature to work' (2004, 10).

[37] Michel Bitbol (forthcoming) points out that Kant called such relations 'ungrounded relations'.

supervene on the intrinsic properties of the relata (since their duplicates could be further apart). Lewis then seems simply to define the composite of the electron and nucleus as the two of them with their spatio-temporal properties, so that their spatial separation does supervene on them taken together. The composite cannot be just the set of the two of them (62).[38]

All the external relations that we can think of are such spatial or spatio-temporal relations. Consider the relation of 'being each others' mirror image'. Does this supervene on the properties of the composite of two objects? If we suppose that it picks up all the spatio-temporal relations of the objects then it would seem so. The same presumably goes for the relation of 'being inside of'. However, notice that such relations also supervene on the intrinsic and relational properties that each element of the composite has independently of the whole or the other part. So if a book is inside a bag we can imagine that this relation consists in nothing more than the positions of the two objects relative to everything else there is in the world. Similarly, the spatial separation of the electron and its proton (or the Moon and the Earth) supervenes on the relational, in particular, spatial properties each object has quite independently of the other or of the composite as a whole (the position of each relative to the Sun, say). The existence of such relations does not trouble Lewis because it does not threaten Humean supervenience, which is fundamentally the thought that there are no necessary connections between distinct existences. Spatio-temporal relations may not supervene on intrinsic properties but they do supervene on relational properties of the relata that are mutually independent.

The entangled states of QM do not supervene on the intrinsic properties of their relata, because in an entangled state with respect to some observable each particle has no state of its own with respect to that observable but rather enters into a product state. The only intrinsic properties that an entity in an entangled state has that are independent of the other entities in that state are its state-independent properties such as charge, mass, and so on, and in general, its relational properties depend on the other entities to which it is related. Hence, unlike external relations, the non-supervenient relations into which several quantum particles may enter are not even supervenient on the relational properties which their relata possess independently of each other. They are much more independent of the properties of the individual particles than spatio-temporal relations between classical objects. This would seem to refute Humean supervenience in so far as the doctrine is supposed to be inspired by science as Lewis claims.[39]

---

[38] Recall the discussion of the metaphysical notion of composition in 1.2.3.

[39] Carol Cleland (1984) introduced a distinction between a weakly non-supervenient relation, where the relation is not determined by non-relational attributes of the relata, but the latter are possessed by the relata, and a strongly non-supervenient relation, where the relata lack any relevant non-relational attributes altogether. French (1989) argues that entangled states are strongly non-supervenient. See also Esfeld (2004), who agrees that pointing to the existence of non-supervenient

To recap, standard metaphysics assumes that:

(i) There are individuals in spacetime whose existence is independent of each other.[40] Facts about the identity and diversity of these individuals are determined independently of their relations to each other.[41]

(ii) Each has some properties that are intrinsic to it.

(iii) The relations between individuals other than their spatio-temporal relations supervene on the intrinsic properties of the relata (Humean supervenience).

(iv) PII is true, so there are some properties (perhaps including spatio-temporal properties) that distinguish each thing from every other thing, and the identity and individuality of physical objects can be accounted for in purely qualitative terms.

We have argued against all these theses (except (iv) suitably modified). Both QM and relativity theory teach us that the nature of space, time, and matter raises profound challenges for a metaphysics that describes the world as composed of self-subsistent individuals. In so far as quantum particles and spacetime points are individuals, facts about their identity and diversity are not intrinsic to them but rather are determined by the relational structures into which they enter. We have argued that entanglement as described by QM teaches us that Humean supervenience is false, and that all the properties of fundamental physics seem to be extrinsic to individual objects.[42] Finally, PII is true only in its weakest form when applied to fermions, and even so, Leitgeb and Ladyman (forthcoming) note that there is nothing to rule out the possibility that the identity and diversity of individuals in a structure is a primitive feature of the structure as a whole that is not accounted for by any other facts about it.

John Stachel (2006), reviewing the discussion of the relative ontological priority of relations and things, identifies four views:

(I) There are only relations, and no relata.

(II) There are relations in which the things are primary, and their relations are secondary.

---

relations is insufficient to capture quantum entanglement because the failure of supervenience is different from the failure of spatio-temporal relations to supervene on the intrinsic properties of their relata.

[40] Einstein in a letter to Max Born (1971, 170–1; quoted in Hagar 2005, 757 and Maudlin 2002a, 48) says that the idea of independently existing objects comes from 'everyday thinking'. He also regards it as a necessary presupposition of physics. Subsequent developments seem to have proved him wrong in this second speculation.

[41] Recall from 1.2.1 Lowe's claim: 'Certainly, it seems that any satisfactory ontology will have to include self-individuating elements, the only question being which entities have this status—space-time points, bare particulars, tropes, and individual substances all being among the possible candidates' (2003b, 93)

[42] Weyl says that it has been assumed that descriptions of relations between particles refer to the intrinsic attributes of things in themselves, but that 'in quantum theory we are confronted with a fundamental limitation to this metaphysical standpoint' (1931, 76).

(III) There are relations in which the relation is primary, while the things are secondary.

(IV) There are things, such that any relation between them is only apparent.

OSR is construed as either (I) or (III). (I) seems to be incoherent, because:

(a) The Earth is bigger than the Moon.

(b) (a) asserts that there is a certain relation between the Earth and the Moon.

The best sense that can be made of the idea of a relation without relata is the idea of a universal. For example, when we refer to the relation referred to by 'larger than', it is because we have an interest in its formal properties that are independent of the contingencies of their instantiation. To say that all that there is are relations and no relata, is therefore to follow Plato and say that the world of appearances is illusory.

However, Stachel points out that (I) can be read as asserting that, while there are relata, they can be analysed into further relational structure themselves. (I) means therefore that 'it's relations all the way down'.[43]

Michael Esfeld (2004) is clear in his rejection of (I) and claims that:

(a) Relations require relata.

But he denies that:

(b) These things must have intrinsic properties over and above the relations in which they stand.[44]

Esfeld also rejects (III) and (like Pooley) seems to hold:

(V) There are things and relations but neither is primary or secondary.

However, it is not clear whether this is consistent with Esfeld's avowed determination to make his metaphysics empirically motivated, because, as discussed in 3.1, Saunders (2003a, 2003b, and 2006) shows that, because fermions satisfy PII formulated in terms of weak discernibility, it is possible to regard all facts about the identity and diversity of them as supervenient on facts about the relations into which they enter, suggesting at least that relations are primary to things.

---

[43] Cf. Saunders (2003d, 129). Stachel denies that there is any reason to think (I) holds in general. We disagree and maintain that contemporary physics gives us good reason to expect that (I) is correct. (We return to this issue in 3.7.3.) Note that Stachel is not pursuing the kind of unifying metaphysics in which we are engaged here, but rather a case by case description.

[44] This is slightly misleading since the particles in entangled states still have their intrinsic essential or state-independent properties. Esfeld's point is surely that quantum particles need not have any intrinsic properties relevant to the relation in question. What is important is that even if the state-independent properties of QM are genuinely intrinsic rather than emergent from some further structure (and that seems unlikely), the fact that particles in entangled states may have all the same properties and relations as each other means that they cannot be individuated by such intrinsic properties.

In any case, disagreements among structuralists aside, none of the philosophers we have been discussing affirm (II) or (IV), or the existence of self-subsistent individual things. Saunders clearly agrees with Esfeld that fermions are not self-subsistent because they are the individuals that they are only given the relations that obtain among them. There is nothing to ground their individuality other than the relations into which they enter. The individuality of quantum particles is ontologically on a par with (V), or secondary to (III), the relational structure of which they are parts. (In Einstein's terms, particles do not have their own 'being thus'.) Stachel (III) and says that in so far as the entities of modern physics have individuality they 'inherit it from the structure of relations in which they are enmeshed' (2006, 58). Tim Maudlin argues that this means the end of ontological reductionism, and abandoning the combinatorial conception of reality that comes from thinking of the world as made of building blocks, each of which exists independently of the others (1998, 59). As he says, and contrary to what Lewis thinks, 'The world is not just a set of separately existing localized objects, externally related only by space and time' (60).

There is thus growing convergence among philosophers of physics that physics motivates abandonment of a metaphysics that posits fundamental self-subsistent individuals. Since this is essentially a negative point, it might seem to be grist to the mill of the constructive empiricist. After all, according to van Fraassen, the opposition between constructive empiricism and standard scientific realism is really that between empiricism and metaphysics, so every retreat from a metaphysical commitment found to be unmotivated by empirical phenomena constitutes a local victory for constructive empiricism. However, we argued in the previous chapter that even the constructive empiricist cannot do without some metaphysics, in particular, without a commitment to objective modal relations. We argued in Chapter 2 that it is just such a commitment that structural realism needs in order to avoid regarding the success of scientific induction as miraculous. If science tells us about objective modal relations among the phenomena (both possible and actual), then occasional novel predictive success is not miraculous but to be expected. Furthermore, the fact that scientific theories support counterfactual conditionals is also explained. Provision of these explanations is not a matter of satisfying philosophical intuitions, but of unifying scientific practices and theories. We thus suggest that in addition to the negative thesis that physical theory should not be interpreted in terms of underlying objects and properties of which the world is made, we are motivated in accordance with the PNC to take seriously the positive thesis that the world is structure and relations. Individual things are locally focused abstractions from modal structure. By modal structure we mean the relationships among phenomena (tracked or located, for reasons and according to principles we will discuss in Chapter 4, as things, properties, events, and processes) that pertain to necessity, possibility, potentiality, and probability.[45]

Of course, all the considerations from physics to which we have appealed do not logically compel us to abandon the idea of a world of distinct ontologically subsistent individuals with intrinsic properties. As we noted, the identity and individuality of quantum particles could be grounded in each having a primitive thisness, and the same could be true of spacetime points. What we can establish is that physics tells us that certain aspects of such a world would be unknowable. Recall from Chapter 2 that Rae Langton calls accepting this limitation 'Kantian Humility' (1998), while Lewis (forthcoming) speaks of 'Ramseyan humility', and Frank Jackson accepts that 'we know next to nothing about the intrinsic nature of the world. We know only its causal cum relational nature' (1998, 24). Peter Unger (2001) also argues that our knowledge of the world is purely structural and that qualia are the unknowable non-structural components of reality. On our view, things in themselves and qualia are idle wheels in metaphysics and the PNC imposes a moratorium on such purely speculative philosophical toys. Like Esfeld (2004, 614–16), we take it that such a gap between epistemology and metaphysics is unacceptable. Given that there is no a priori way of demonstrating that the world must be composed of individuals with intrinsic natures, and given that our best physics puts severe pressure on such a view, the PNC dictates that we reject the idea altogether.

3.5 OBJECTIONS TO ONTIC STRUCTURALISM

(1) Relations are impossible without relata.

This objection has been made by various philosophers including Redhead (personal communication), Psillos (2001), Morganti (2004), and Chakravartty (1998) who says: 'one cannot intelligibly subscribe to the reality of relations unless one is also committed to the fact that some things are related' (ibid. 399). In other words, the question is, how can you have structure without (non-structural) objects, or, in particular, how can we talk about a group without talking about the elements of a group? Even many of those sympathetic to the OSR of French and Ladyman have objected that they cannot make sense of the idea of relations without relata (see, for example, Esfeld 2004, Lyre 2004, and Stachel 2006).

This objection has no force against the view propounded in this chapter. As French and Ladyman emphasize, the claim that relata are constructed as abstractions from relations doesn't imply that there are no relata; rather the

---

[45] The structure of dispositions described by Mumford (2004) and Psillos's (forthcoming) idea of nomological structure are cognate.

opposite. A core aspect of the claim that relations are logically prior to relata is that the relata of a given relation always turn out to be relational structures themselves on further analysis.[46] It may be argued that it is impossible to conceive of relational structures without making models of them on domains of individuals. For example, although GR only ever licences us to make claims about an equivalence class of models under the diffeomorphism group, in practice physicists always work within some model or other. Speculating cautiously about psychology, it is possible that dividing a domain up into objects is the only way we can think about it. Certainly, the structuralist faces a challenge in articulating her views to contemporary philosophers schooled in modern logic and set theory, which retains the classical framework of individual objects represented by variables subject to predication or membership respectively. In lieu of a more appropriate framework for structuralist metaphysics, one has to resort to treating the logical variables and constants as mere placeholders which are used for the definition and description of the relevant relations even though it is the latter that bear all the ontological weight. The same approach is followed in the interpretation of physics: we see, on a scintillation screen, for example, bright flashes of light—individual flashes. On the basis of such observable phenomena we then try to carry over our metaphysics of individuality which is appropriate for the classical domain, and quantum particles are classified via the permutation group which imposes the division into the natural kinds of fermions and bosons. However, in the light of the above discussion we maintain that the elements themselves, regarded as individuals, have only a heuristic role (see French 1999). Poincaré adopted a very similar approach in a paper of 1898.[47] He defends a group-theoretic approach to geometry and in response to the objection that in order to study the group it needs to be constructed and cannot be constructed without matter, he says 'the gross matter which is furnished us by our sensations was but a crutch for our infirmity' (1898, 41), which serves only to focus our attention upon the structure of the group.[48] We may not be able to think about structure without hypostatizing individuals as the bearers of structure, but it does not follow that the latter are ontologically fundamental.

Some of the critics cited above object that OSR is not 'worked out' as metaphysics.[49] However, it is far from clear that OSR's rivals are 'worked out' in any sense that OSR isn't. There in no general agreement among philosophers that any of the metaphysical theories of, say, universals is adequate. We ask the reader to consider whether the main metaphysical idea we propose, of existent structures that are not composed out of more basic entities, is any more obscure or bizarre than the instantiation relation in the theory of universals. We think it better

---

[46] An example from pure mathematics is that of the open set, which is a brute object in algebraic topology, but a relational structure in analysis.
[47] This example is due to Mary Domski (1999).
[48] Poincaré understands group structure in Kantian terms as a pure form of the understanding.
[49] See for example Psillos (2001).

to attempt to develop the metaphysics presented in this book than to continue to use off-the-shelf metaphysical categories inherited from the ancient Greeks that are simply not appropriate for contemporary science or mathematics.[50] We might as well just put it thus: we really mean what the PNC says.

(2) Structural realism collapses into standard realism.

Psillos (1995) claims that structural realism presupposes distinctions between the *form* and *content* of a theory, or between our ability to know the *structure* and our ability to know the *nature* of the world.[51] The (sensible) realist will not accept these distinctions, according to Psillos. He argues that one success of the scientific revolution was the banishing of mysterious forms and substances that might not be fully describable in structural terms, and the consequent concentration of the mechanical philosophers on quantitative descriptions of the properties of things. For Psillos, properties in mature science are defined by the laws in which they feature: the nature of something consists in its basic properties, and the equations expressing the laws these obey.

Similarly, Ernan McMullin, in his paper 'A Case for Scientific Realism', says:

'What are electrons? Just what the theory of electrons says they are, no more, no less, always allowing for the likelihood that the theory is open to further refinement' (1985, 15).

We have a theory of electrons that describes their behaviour in terms of the laws, interactions, and so on, to which they are subject. This description, which may be termed a structural one, gives us the last word on electrons, says the realist. This is what Psillos means when he says that 'the nature and the structure of a physical entity form a continuum' (1995). Hence, for Psillos, structural realism cannot be distinguished from traditional realism without a dubious distinction between structure and nature.[52]

So it seems the realist can claim that structural realism is no different from scientific realism in so far as it advises a structural understanding of theoretical entities, because 'restriction of belief to structural claims is in fact no restriction at all' (Papineau 1996, 12). Thus construed, it gains no advantage over traditional realism with the problem of theory change because it fails to make any distinction between parts of theories that should and shouldn't provoke ontological commitment. Realists may well think that all their knowledge of the

---

[50] Structuralism of a certain kind has become popular in metaphysics recently in the form of causal essentialism. This is the doctrine that the causal relations that properties bear to other properties exhaust their natures. See for example, Shoemaker (1980), and also Mumford (2004), who adopts a structural theory of properties, and Bird (2007) whose theory of dispositions is in some ways structuralist. Harte (2002) discusses an interesting Platonic form structuralism.

[51] Stanford (2003, 570) also argues that we cannot distinguish the structural claims of theories from their claims about content or natures.

[52] This is just the complaint of Braithwaite against Eddington's structuralism (Braithwaite 1940, 463).

unobservable is structural, whether they explicate this in the Ramsey way or not, but this will not help them with ontological discontinuity.

In reply, first consider the claim that the understanding of 'natures' in terms of forms, substances, and the likes was overthrown by the scientific revolution. How was the 'nature' of atoms understood? Atoms were understood as individuals. Boltzmann incorporated such an understanding of the nature of atoms in terms of their individuality into Axiom I of his mechanics. Second, the above discussion of individuality with respect to entities in physics, and the metaphysics of relations makes it clear that standard scientific realism has been saddled with traditional metaphysics.

(3) Structural realism might be right for physics but not for the rest of science.

Gower (2000) is one of those who makes the point that structural realism seems less natural a position when applied to theories from outside of physics. We agree that this is a feature of science that requires (metaphysical) explanation; in Chapters 4 and 5 we will provide this. In those chapters we will show how structural realism about special sciences is necessary if one is to avoid ascribing a different level of ontological seriousness to them than one ascribes to physics. Scientists don't observe any such asymmetry, so this conclusion must be avoided by the naturalist as being inconsistent with the PNC.

(4) Isn't structure also lost in theory change?

Stanford (2003, 570–2) argues that mathematical structure is often lost in theory change, as does Otavio Bueno (in private discussion). This is an important point. However, the problem is surely not analogous to the one the realist faces with respect to ontological discontinuity. The realist is claiming that we ought to believe what our best scientific theories say about the furniture of the world in the face of the fact that we have inductive grounds for believing this will be radically revised, whereas the structural realist is only claiming that theories represent the relations among, or structure of, the phenomena and in most scientific revolutions the empirical content of the old theory is recovered as a limiting case of the new theory. As Post claimed, there simply are no 'Kuhn-losses', in the sense of successor theories losing all or part of the well-confirmed empirical structures of their predecessors (1971, 229). In sum, we know that well-confirmed relations among phenomena must be retained by future theories.

(5) If there is no non-structure, there is no structure either.

In a paper delivered in Leiden in 1999, van Fraassen argues that the heart of the problem with our kind of radical structuralism is this:

It must imply: *what has looked like the structure* of something with unknown qualitative features *is actually all there is to nature*. But with this, the contrast between structure and what is not structure has disappeared. Thus, from the point of view of one who adopts this position, any difference between it and 'ordinary' scientific realism also disappears.

It seems then that, *once adopted*, it is not to be called structuralism at all! For if there is no non-structure, there is no structure either. But for those who do not adopt the view, it remains startling: from an external or prior point of view, it seems to tell us that nature needs to be entirely re-conceived. (van Fraassen 2006, 292–3)

The essence of van Fraassen's objection here is that the difference between mathematical (uninstantiated) structure and physical (instantiated) structure cannot itself be explained in purely structural terms.[53]

Van Fraassen (2006, 293–4) reiterates the point in the context of Fock space formalism used in quantum field theory: all that there is cannot merely be the structure of this space, he insists, because then there would be no difference between a cell being occupied and a cell being unoccupied. However, just because our theory talks of occupation numbers does not imply that what is occupying the cell must a non-structural object, individual or not. As Auyang points out, 'To say the field is in a state | n(k1),n(k2), … > is not to say that it is composed of n(k1) quanta in mode k1 and so on but rather n(k1) quanta show up in an appropriate measurement' (Auyang 1995, 159).

Our view is this: scientific realists take it that the appearances are caused by unseen objects and that the behaviour of these objects can be invoked to explain the appearances. But the resources of the manifest image cannot be (directly) used for satisfactory representation in physics. Hence, mathematics has an ineliminable role to play in theories. When theories are empirically adequate they tell us about the structure of the phenomena and this structure is (at least in part) modal structure. However there is still a distinction between structure and non-structure. Merely listing relations among locators does not state anything with modal force. Therefore, it doesn't specify structure in our sense and it isn't yet scientific theory as we've defended it. Physical structure exists, but what is it? If it is just a description of the properties and relations of some underlying entities this leads us back to epistemic structural realism. What makes the structure physical and not mathematical? That is a question that we refuse to answer. In our view, there is nothing more to be said about this that doesn't amount to empty words and venture beyond what the PNC allows. The 'world-structure' just is and exists independently of us and we represent it mathematico-physically via our theories.[54] (This may sound suspiciously Kantian. We can best explain why it isn't when we have introduced more theoretical resources than we have done so far. The reader is asked to be patient on this point until 6.1. But OSR as we develop it is in principle friendly to a naturalized version of Platonism, a point we will also touch on in Chapter 4.)

(6) Structural realism cannot account for causation.

---

[53] There is an analogy here with the theory of universals and the problem of exemplification.

[54] A similar complaint is made by Cao (2003). Saunders (2003d) points out that there is no reason to think that ontic structural realists are committed to the idea that the structure of the world is mathematical.

Psillos (forthcoming), along with Cao (2003), argues that OSR cannot account for causal relations. However, Saunders (2003d, 130) argues that causal structure is a species of modal structure, and as such the advocate of OSR can happily endorse the claim that the world has an objective causal structure. However, we are doubtful that fundamental physics motivates a metaphysics that requires us to acknowledge objective causal structure, for reasons we will give. This way of seeing off the criticism invites another: though physics doesn't require the metaphysician to work causation into the structural fabric, it is harder to avoid this while maintaining a realist attitude towards special sciences. We discuss these issues at length in Chapter 5.

3.6 MATHEMATICAL STRUCTURE AND PHYSICAL STRUCTURE

According to OSR, if one were asked to present the ontology of the world according to, for example, GR one would present the apparatus of differential geometry and the field equations and then go on to explain the topology and other characteristics of the particular model (or more accurately equivalence class of diffeomorphic models) of these equations that is thought to describe the actual world. There is nothing else to be said, and presenting an interpretation that allows us to visualize the whole structure in classical terms is just not an option. Mathematical structures are used for the representation of physical structure and relations, and this kind of representation is ineliminable and irreducible in science. Hence, issues in the philosophy of mathematics are of central importance for the semantic approach in general, and the explication of structural realism in particular. This is why van Fraassen says:

One of my great regrets in life is that I do not have a philosophy of mathematics; I will just assume that any adequate such philosophy will imply that what we do when we use mathematics is all right. (1994, 269)

The problem for him is that he is a nominalist. But it is worth briefly considering how some recent proposals in the philosophy of mathematics relate to structural realism.

Some philosophers of mathematics have recently argued that mathematical entities such as sets and other structures are part of the physical world and not therefore mysterious abstract objects (see, in particular, Maddy 1990 and Resnik 1990). This may suggest a kind of Pythagoreanism, asserting the identity of structures in mathematics and physics, and abandoning the distinction between the *abstract* structures employed in models and the *concrete* structures that are the objects of physics.[55] The canonical concrete substance in physics is

---

[55] We briefly reconsider this idea in 4.4.

matter, but matter has become increasingly ephemeral in modern physics, losing its connection with the impenetrable stuff that populates the everyday world. Resnik (1990), Hale (1990), and Tymoczko (1991) have all made the point that the ontology of modern physics seems to be increasingly abstract and mathematical: 'If you think mathematical objects are weird then take a look at a physics textbook' seems to be a simple common form of argument. As van Fraassen points out, it is often not at all obvious whether a theoretical term refers to a concrete entity or a mathematical entity (1994, 11). Consider, for example, the term 'wave function'. This may all merely be a matter of failing to distinguish epistemology from metaphysics: the fact that we only *know* the entities of physics in mathematical terms need not mean that they are actually mathematical entities. But there are other grounds for dispensing with the abstract/concrete distinction.

The distinction is usually made in terms of either causal power or spatio-temporality. Hence, it is said that concrete objects have causal powers while abstract ones do not, or, on the other hand, that concrete objects exist in space and time while abstract ones do not. These categories seem crude and inappropriate for modern physics. Causation is problematic in the microscopic domain where, for example, the singlet state in the Bohm-EPR experiment fails to screen off the correlations between the results in the two wings of the apparatus, and thus fails to satisfy the principle of the common cause. Nobody has so far proposed an acceptable account of causation for this situation. Yet there is nothing special about the set up that prevents the problem generalizing to all entangled states. Furthermore in quantum field theory there is no absolute direction of causation for gauge interactions and causality seems to be even less well understood. If we consider the spatio-temporal criterion of concreteness then we are faced with the problem that the structure of spacetime itself is an object of physical investigation. Yet it can hardly be *in* spacetime. (Again note that we return to the status of causation in fundamental physics in Chapter 5, sections 1–4.)

Moreover, the dependence of physics on ideal entities (such as point masses and frictionless planes) and models also offers another argument against attaching any significance to the abstract/concrete distinction. It is worth noting also that the statistics obeyed by electrons are analogous to those obeyed by pounds sterling in a British bank account rather than by actual physical pounds. For example, if two people have two pounds between them in their bank accounts but no pounds are divided, then there are three possible distributions: one of them has both, the other has both, or they have one each. This is like the distribution of electrons in boxes described above. (Actual physical pound coins of course obey classical statistics.) These considerations are not compelling but do suggest the possibility of a rapprochement between the objects of physics and of mathematics.

Mathematical structuralists have long argued that the applicability of mathematics is due to the world instantiating mathematical structure.[56] Mathematical objects are usually regarded as having no essential properties or natures. Rather they merely appear as arguments in formulae. This is regarded by mathematical structuralists as evidence for their view that mathematics is the study of structures, where these are understood to be fully characterized when the relations obtaining between the objects making up the structure are characterized, without any need to say anything whatsoever about the nature of objects themselves.[57]

3.7 FURTHER REFLECTIONS ON PHYSICS

So far we have motivated OSR by means of a consilience argument that appeals to considerations from the realism debate in general philosophy of science, and also to the philosophy of quantum physics and GR. However, although the latter are stable components of advanced physics, it might be objected that, given the PNC and PPC, we ought to be able to show that OSR is motivated by cutting-edge physics. There is a quick response to this, namely, to claim that since there is no agreed-upon theory of quantum gravity, and no unified theory of all the fundamental forces recognized by physics, there are as yet no lessons to be drawn for metaphysics from these parts of physics. This would be a fair response up to a point but we nonetheless concede that we ought to have something more to say. Hence, we will briefly explain some of the issues in the search for a theory of quantum gravity, and attempt to draw some lessons for our proposed metaphysics from the developments to date. In particular, there are two related and fundamental metaphysical questions that we must address, because it is alleged by some that contemporary physics has decided them for us, namely, whether or not all times are real, and whether or not all physically possible occurrences are real.

---

[56] For an excellent survey see Reck and Price (2000). The most well-known advocates of structuralism in the philosophy of mathematics are Parsons (1990), Resnik (1997), and Shapiro (1997). Recent surveys include Hellman (2005) and MacBride (2005). Ladyman (2005) deploys Saunders's (2003a) version of PII mentioned above to defend mathematical structuralism against the identity problem raised by Keränen (2001) and MacBride (2005). See also Leitgeb and Ladyman (forthcoming).

[57] These observations lead to two versions of mathematical structuralism: a realist view according to which mathematical structures exist in their own right independently of being instantiated by a concrete structure; and an eliminativist position according to which statements about mathematical structures are disguised generalizations about sets of objects that exemplify them (see Shapiro 1997, 149–50). The relationship between ontic structural realism and ante rem structuralism has been explored by Psillos (forthcoming), Busch (2003), and Pooley (2006). As Pooley points out, Shapiro does not deny that mathematical objects exist, but he does deny that they have a nature over and above their relationship to the other objects in the relevant structure.

The former question is often claimed by philosophers to have been answered in the affirmative by Special Relativity (SR), and the latter is claimed by some influential philosophers of physics to have been answered in the affirmative by QM. Furthermore, some philosophers of physics argue that these issues are directly analogous, and that the block universe of Minkowski spacetime, in which all times are real, must be married with the Everettian multiverse, in which all physical possibilities are real. However, we argue that these are both scientifically open questions. If this is right then our naturalism requires that the metaphysics espoused in this book must be neutral with respect to them. We will go further and suggest grounds for questioning whether they could in principle be answered by physics at all, and tentatively propose quietism about them. Finally, we reflect on the implications of quantum information theory and recent proposals for the interpretation of QM both for OSR, and to support the use we make of the notion of information in the next chapter.

3.7.1 Still or sparkling?

There are several distinct though often conflated issues in the metaphysics of time:

(i) Are all events, past, present and future, real?
(ii) Is there temporal passage or objective becoming?
(iii) Does tensed language have tenseless truth conditions?

Call the view that all events are real 'eternalism', the view that only the present is real 'presentism', and the view that all past and present events are real 'cumulative presentism'. (The latter is defended by Michael Tooley (1997) although not under this name.) Those who believe in the passage of time or objective becoming often also believe that the process of becoming is that of events coming into existence and going out of existence, but this need not be so; to suppose there is becoming, one need only believe that there is some objective feature of the universe associated with the passage of time. Objective becoming could be like a light shining on events as they are briefly 'present', and is therefore compatible with eternalism.[58] On the other hand, both presentism and cumulative presentism entail a positive answer to question (ii), since if events do come into existence, whether or not they then stay existent or pass out of existence, this is enough to constitute objective becoming. Presentism and becoming have also been associated with the idea that tensed language does not have tenseless truth conditions. However, this is not a necessary connection as Michael Tooley (1997) argues. So even though the standard opposition is between those who answer 'no' to (i), 'yes' to (ii), and 'no' to (iii) on the one

---

[58] Tim Maudlin (2002b) argues that the passage of time may be an objective feature of the universe even if eternalism is true.

hand (the defenders of McTaggart's 'A-series'), and those who answer 'yes' to (i), 'no' to (ii), and 'yes' to (iii) (the defenders of McTaggart's 'B-series'), a variety of more nuanced positions are possible. We of course do not believe that philosophy of language can have any bearing on the metaphysical issues addressed by (i) and (ii), and hence we will have nothing to say about (iii).

There is a further celebrated question about time:

(iv) Does time have a privileged direction?

Clearly if (i) or (ii) are answered positively then that is enough to privilege a particular direction in time. However, eternalism and the denial of objective becoming are compatible with time having a privileged direction, since there could be some feature of the block universe that has a gradient that always points in some particular temporal direction. For example, the entropy of isolated subsystems of the universe, or the universe itself, might always increase in some direction of time. Another well-known possible source of temporal direction was proposed by Reichenbach (1956) who argued that temporal asymmetry is grounded in causal asymmetry: in general, correlations between the joint effects of a common cause are screened off by the latter but the joint causes of a common effect are uncorrelated. Although some have claimed that Reichenbach's Principle of the Common Cause is violated, not least by the behaviour of entangled states in QM (see for example van Fraassen 1991), we take it that such considerations are sufficient to show that conceptually the question of the direction of time must be separated from questions about eternalism. However, it may be that no physical meaning can be attached to the idea of the direction of time in the whole universe, because no global time coordinate for the whole universe can be defined, or because many can and there is no principled way to choose between them. The latter case is implied by SR, to which we turn.

The status of time in SR differs from its status in Newtonian mechanics in that there is no objective global distinction between the dimensions of space and that of time. Spacetime can be split into space and time, but any such foliation is only valid relative to a particular inertial frame, which is associated with the Euclidean space and absolute time of the coordinate system of an observer. This seems to imply eternalism, since if there is no privileged foliation of spacetime, then there is no global present, and so the claim that future events are not real does not refer to a unique set of events.[59] Furthermore, many have argued that, since SR implies the relativity of simultaneity, whether or not two events are simultaneous is a frame-dependent fact, and therefore there is no such thing as becoming.[60] However, this is too quick. It is possible to advocate a form of becoming that is

---

[59] Of course, one could argue that the very notion of reality must be relativized to observers, but this is to give up on the kind of metaphysics to which we are committed.

[60] The literature on these topics is voluminous but among the most influential papers are Gödel (1949), Putnam (1967), and Stein (1968) and (1989).

relative to observers or events, so strictly speaking it is only absolute becoming that is ruled out by the lack of absolute time in SR. Since the light cone structure of Minkowski spacetime is Lorentz-invariant, it can be regarded as absolute. It is easy then to define a notion of the open future of an event E, since any event E' in the forward light cone of E will have events in its backwards light cone that are not in the backwards light cone of E, meaning that there is a sense in which it can be claimed that E' is not determinate at E. This notion of becoming is objective in the sense that all observers will agree about which events are in the open future of a given observer at a particular point in his or her history, because all observers agree about the light cone structure of spacetime.[61]

In any case there is a fundamental problem with drawing metaphysical conclusions about the nature of time from SR, namely that it is a partial physical theory that cannot describe non-inertial frames of reference, or gravity.[62] For that we must turn, in the first instance, to GR, and the implications of that theory for time are not clear. This is because GR gives us field equations that are compatible with a variety of models having different global topological features, and different topological structures may have very different implications for the metaphysics of time.[63] For example, if the topology of the universe is globally hyperbolic then it is possible to define a single global foliation of spacetime for it; otherwise it may not be. Clearly we must then turn to cosmological models of the actual universe, of which there are many compatible with the observational data. As yet there is no agreement about which of these is the true one. Highly controversial issues about the cosmological constant and so-called dark energy, dark matter, and the nature of singularities, as well as the various approaches to the search for a theory of quantum gravity, all bear on the question of whether spacetime will turn out to admit of a global foliation, and hence on whether absolute time is physically definable. Even if it does turn out to be definable, there remains the question of whether such a definition ought to be attributed any metaphysical importance. For example, it is possible to define something called 'cosmic time' which is based on the average properties of the universe's global matter distribution under the expansion of the universe. Some have argued that we can regard Cosmic Time as giving us a privileged foliation (Lucas and Hodgson 1990). However, others argue that the fact that such a foliation can be

---

[61] Another possibility is to argue that while Special Relativity is empirically adequate, the empirical evidence is nonetheless compatible with the existence of a privileged foliation. This would be to advocate what Sklar (1974) calls a compensatory theory along the lines of that originally proposed by Lorentz and Fitzgerald. This is the strategy adopted by Tooley (1997) for example. The PNC forbids the revision of scientific theories on purely philosophical grounds, so the proposal of a privileged foliation contra to SR requires a scientific motivation. One possible scientific motivation is the adoption of a solution to the measurement problem that posits a preferred frame of reference (see below). Another is the identification of foundational problems with the account of relativistic phenomena in the Minkowski spacetime framework (see Bell 1987 and Brown 2005).

[62] Tooley (1997) and Sider (2001) both confine discussion to SR.

[63] Belot (2005a) argues that which category our universe falls into is an open question.

defined gives us no reason to regard it as having an objective significance (Berry 1989), not least because it is based on averages that have nothing to do with the phenomenological experience of everyday simultaneity for objects whose states of motion are not the same as the state of motion of the galaxies in our region of the universe with respect to which cosmic time is defined (see Bourne 2004).

Non-relativistic many-particle QM does not directly bear on the philosophy of time since the status of time in the formalism is not novel in the same way as in relativity. However, it has often been argued that quantum physics is relevant to questions about the openness of the future, becoming, and the direction of time, because of the alleged process of collapse of the wave function. Since Heisenberg (1962) it has been popular to claim that the modulus squared of the quantum mechanical amplitudes that are attached to different eigen states in a superposition represent the probabilities of genuinely chancy outcomes, and that when a measurement is made there is an irreversible transition from potentiality to actuality in which the information about the weights of the unactualized possible outcomes is lost forever. Hence, measurement can be seen as constituting irreversible processes of becoming that induce temporal asymmetry. However, that quantum measurements need not be so understood is shown by the time-symmetric treatment of quantum measurements in the formalism of Aharonov et al. (1964).[64] Similarly, if there is no collapse, as in the Everett interpretation, then again there is no temporal asymmetry in QM.

The upshot seems to be that the status of the arrow of time in QM is open. The tension between SR and QM is made into a definite contradiction if collapse of the wave function is regarded as an objective physical process, as in the dynamical collapse theories along the lines developed by Ghiradi et al. (1986), or if non-local hidden variables are introduced as in Bohm theory, since both imply action at a distance and pick out a preferred foliation of spacetime (Timpson and Brown forthcoming, Maudlin 1994).[65] The real questions concern what happens to time if quantum theory is married with GR, and we return to that issue below. (Since relativistic quantum field theory is based on the background of Minkowski spacetime the status of time in the former is the same as in SR.)

There is also a vast literature about whether or not the second law of thermodynamics represents a deep temporal asymmetry in nature. The entropy of an isolated system always increases in time, and so this seems to be an example of the arrow of time being introduced into physics. If the whole universe is regarded as an isolated object, and if it obeys the second law, then it would seem that there is an objective arrow of time in cosmology. However, it is not clear what the status of the second law is with respect to *fundamental* physics.

---

[64] See also Leggett (1995) and Stamp (1995).
[65] Myrvold (2002) shows that physical collapse models can be consistent with relativity theory if collapse is regarded as foliation relative feature of the world. It also seems that it may be possible to construct a version of GRW which is Lorentz-invariant after all (see Maudlin 2007).

One possibility is that the second law holds only locally, and that there are other regions of spacetime where entropy is almost always at or very near its maximum. (Boltzman himself thought this was the case in his later years.) Even if thermodynamics seems to support the arrow of time, it is deeply puzzling how this can be compatible with an underlying physics that is time-symmetric. Conservative solutions to this problem ground the asymmetry of the second law in boundary conditions rather than in any revision of the fundamental dynamics. The most popular response is to claim that the law does indeed hold globally but that its so doing is a consequence of underlying time-reversal invariant laws acting on an initial state of the universe that has very low entropy.[66] This is called the 'Past Hypothesis' by Albert (2000). A much more radical possibility (see for example, Prigogine 1980) is that the second law is a consequence of the fact there is a fundamental asymmetry in time built into the dynamical laws of fundamental physics. Given the outstanding measurement problem in QM those who propose radical answers to problems in thermodynamics and cosmology often speculate about links between them and the right way of understanding collapse of the wave function. Roger Penrose (2001), for example, suggests that gravity plays a role.

It has also been suggested that the local thermodynamic arrow of time is a consequence of the expansion of the universe (Gold 1962). Certainly, if cosmological explanations of the second law are sought, its origin must have something to do with gravity. This is because gravity is universally attractive, and dominates the effects of the other forces at large scales. It is the force that seems to lead to the creation of ordinary matter. Penrose (2004, 706 and 728) argues that what he calls 'gravitational entropy' must be taken into account when considering the entropy of the universe as a whole, and maintains that the gravitational entropy of a system is higher when it is less uniform and lower when it is more uniform, which is just the opposite of how entropy varies with thermal uniformity. According to his calculations, the initial low entropy state of the universe is so special that the phase space volume associated with it is a maximum of 1 part in 10 to the 10 to the 123 of the total phase space volume. Given this and the fact that the fundamental constants are fine-tuned to the extent that minute changes in their values would result in a universe that could not support our material being, anthropic reasoning often rears its ugly head in this arena. On the other hand, Lee Smolin (2000) speculates that universes are ten a penny, being born in the black holes of their parents and inheriting their laws and constants with minor mutations, and so being subject to natural selection since different universes will have different degrees of fertility. Clearly, both the

---

[66] It is necessary to posit this because standard arguments in statistical mechanics which show that it is overwhelmingly likely that a typical state of an isolated system will evolve into a higher entropy state in the future also show that it is overwhelmingly likely that the state in question evolved from a past state that had higher entropy too. See, for example, Albert (2000, ch. 4).

question of eternalism, and the arrow of time, lead inexorably to cosmology and thence to the realm of quantum gravity that we discuss below.

3.7.2 Quantum gravity[67]

A theory of quantum gravity must do all of the following: say what happens in nature at the Planck length (10^{-33} cm); recover GR as a low-energy limit; and provide a background spacetime, at least phenomenologically, for conventional quantum theories. What else it should do is a matter of great contention. Some, such as advocates of string theory and M-theory like Brian Greene (2004), think it must also unify the four fundamental forces of nature; others, such as Smolin (2006), argue that, in the first instance at least, it need only amount to a quantized version of GR. A further question is whether quantum gravity must also be a cosmological theory of one (unique and actual) universe, rather than allowing for models representing a variety of universes. Partly because the most obvious phenomenological domain of quantum gravity is at such high energies as to be experimentally inaccessible, and partly because the theoretical search has led to so many bizarre and conceptually diverse alternatives, the debate about quantum gravity even among physical cosmologists often becomes explicitly philosophical. This also happens because quantum gravity must reconcile a number of profound tensions between GR and quantum theories.

Most obviously, quantum physics is the physics that best describes the phenomena when we look at very short length scales, and GR is physics that was specifically designed with a distinction between local and global properties of spacetime in mind, and sought to describe deviations from the topological, geometrical, and metrical properties of Minkowski spacetime that only show up in large-scale structure—this is the scale tension. Second, GR depends on the identification of inertial and gravitational mass, and the equivalence between accelerating and gravitational frames, whereas quantum theory was originally developed to account for the interaction between electro-magnetic radiation and matter. Initially, it was only the energy states of matter that were quantized, but subsequently it has proved possible, with differing degrees of success, to quantize all the fundamental forces, with the exception of gravity—this is the force tension. Third, relativistic theories obey the condition that there are invariant and hence objective causal pasts for events, whereas in QM there are non-local correlations that some regard as evidence of action at a distance—this is the causal tension. Finally, there is the radically different status of time in quantum theory versus GR. In the former, time is a parameter external to all physical systems; in the latter it is a coordinate with no particular physical significance. More specifically there is something called 'The Problem of Time' the upshot of

---

[67] Introductions to the philosophy of quantum gravity include Callender and Huggett (2001) especially the introduction, and Rovelli (2004, ch. 1). Wallace (2000) is also very helpful.

which is that theories of quantum gravity are in danger of saying there can be no change in the universe over time (we return to this below).

The approach of string theorists is to treat GR as the low-energy limit of a quantum field theory of strings (which are two dimensional time-like world sheets within a background ten or more dimensional space), and hence to assume that quantum theory is basically correct and that GR must give the ground. Depending on who one listens to, string theory has either already led us a considerable distance down the road to a complete theory of quantum gravity, or it has achieved absolutely nothing that counts as physics rather than mathematics. String theory and the related perturbative quantum GR are both sometimes referred to as covariant quantum gravity, and both involve a particle interpretation of GR in terms of massless spin-2 bosons called 'gravitons'. The latter approach is perturbative since it begins by treating GR as if it were a linear theory and then adding non-linear parts as corrections. The linear theory in question is obtained by supposing that the fields are very weak so that spacetime is nearly flat. Hence the background structure in question is that of Minkowski spacetime. The biggest problem for this approach is that the infinite quantities thrown up in calculations are not renormalizable. String theory's achievement was to solve the renormalization problem by treating vibrating strings as fundamental entities and recover GR on the basis of the fact that one of the string vibration modes corresponds to massless spin-2 bosons. String theory then became superstring theory as super-symmetry (that treats bosons and fermions as equivalent) was added. String theorists have followed the methodology that was used in the construction of quantum field theories, namely the search for fundamental symmetries. If the string theory vision is correct then the ultimate fundamental physics will describe the universal symmetries of the universe. (Another important commonality between string theory and classical and quantum physics is that they posit a continuous space and time, which is departed from by some rival programmes—see below.)

Lee Smolin is highly critical of string theory and argues that it is not falsifiable in the sense that it makes no 'falsifiable predictions for doable experiments' (2006, 197). He claims there are no fundamental global symmetries, on grounds that those theories that posit them are not fully empirically adequate. His view is that the two big ideas that drive string theory, namely unification and symmetry, have run their course, and that there have been no substantially new results in particle physics since 1975. He points out that the standard model has so many adjustable parameters that any likely experimental data from particle accelerators can be accommodated by it.[68] He also emphasizes that there are at least 10^{100} possible string theories, and argues that all the ones that have been studied disagree with the data. Super-symmetric string theory has

---

[68] There are at least nineteen free parameters in the standard model plus almost as many from cosmology.

105 free parameters. Hence, Smolin claims, partisans will be able to maintain that whatever comes out of the next generation of particle accelerators confirms super-symmetry.[69]

He also criticizes string theory for being 'background dependent' in the sense that it relies on a background spacetime structure. Smolin predicts that the correct theory of quantum gravity will be relational in the sense that it won't posit any background structure which does not change with time but which is necessary for the definition of kinematical quantities and dynamical laws (ibid. 199). Newtonian mechanics, SR, quantum theories including quantum field theories, and string theory are all background-dependent and rely on various structures that are outside the scope of the dynamics of the theory. For example, in ordinary QM the spacetime and the algebraic structure of Hilbert space are part of the background structure. On the other hand, GR, understood as a cosmological theory of the whole universe, is a relational theory in the sense that the physically important structural features of the theory are dynamical. The equivalence class of diffeomorphic models that describes the world will not fix the values of fields at points (as noted in 3.2), but it will fix the dimension and topology of the spacetime, as well as the light cone structure (and hence the causal order of events) and a measure of the spacetime volume of sets that the light cone structure defines. Indeed, it turns out that one cannot say what the observables are without solving the dynamics, because all observables describe relations between degrees of freedom and the question of the physical identity of spacetime points is inextricable from how the values of observables are associated with them. The only background structure in GR consists of the dimensionality, the differential structure, and the topology.

String theorists now seem to have accepted that background independence is a desideratum. Brian Greene speculates about a background free version of string theory, and the search for so-called M-theory is partly motivated by the need for a way of thinking about strings that does not treat them as vibrations in a background spacetime. However, no such theory yet exists.[70] Meanwhile, Smolin has inspired a significant minority of researchers to seek background independence in other approaches to quantum gravity, and he shows how this notion plays out in the context of a variety of these theories. He suggests that the history of physics testifies to the success of the pursuit of background independence. It is true that progress has sometimes been made in physics by eliminating background structure. SR eliminated the background structure of

---

[69] Curiel (2001) bemoans the paucity of contact with observation in quantum gravity research.
[70] There are five versions of string theory but it was known that there were pairs that were dual, that is, inter-transformable. Edward Witten convinced string theorists that these five versions were mutually inter-translatable versions of a single theory after all and that theory has been dubbed M-theory (see Greene 2004). M-theory ups the number of spatial dimensions to ten and hence the total dimensionality of spacetime to eleven. There are tentative suggestions of links between M-theory and loop quantum gravity (see below).

absolute space and time, and then GR eliminated the background structure of Minkowski spacetime. On the other hand, there have been many background-dependent theories that have been highly successful, including quantum theories. Smolin himself concedes that GR is background-dependent in certain respects. Furthermore, consider the success of the pursuit of symmetry and unification of forces that motivates string theory, in generating the standard model and the unified field theories based on the knitting together of the symmetry groups previously discovered to be governing the separate forces. Smolin and Greene's dispute can be construed as concerning which of the following two desiderata for fundamental physics holds trumps: symmetry or background independence. The empirical evidence is equivocal, to say the least.

Among background-independent approaches the most well-known is canonical quantum gravity. This approach seeks a quantum theory of gravity, but not necessarily a unification of all the fundamental forces. The idea is to quantize GR after it has been formulated as a phase-space theory subject to a Hamiltonian constraint so that sets of temporally successive three-spaces are models of the field equations of GR. This gives rise to the famous Wheeler–DeWitt equation, and the infamous Problem of Time. Essentially the latter arises because to treat GR as a phase-space theory means first splitting spacetime into slices of space and time. We can do this in many ways, each corresponding to a different definition of time, but the diffeomorphism invariance of GR means that each slicing is related to the others by a gauge transformation. The further complication is that successive slices can themselves be related by gauge transformations. The invariance of the state of the universe under gauge transformations thus means that the physical states of the universe must be time-independent, and so nothing changes, assuming that only gauge invariant quantities are physically real.

The latest version of the canonical programme is loop quantum gravity. The pioneers of this approach include Abhay Ashtekar, Carlo Rovelli, John Baez, and Lee Smolin. It is based on the reformulation of GR as a theory of connections on a manifold, rather than as a theory of the metrics on a manifold. This is usually advocated as a fully relational approach, although Rickles (2005) argues that the hole problem arises in loop quantum gravity too and that it is not necessarily fully relational. He argues that theories like loop quantum gravity admit of multiple interpretations and that there are grounds for pessimism about whether quantum gravity will settle the substantivalism/relationism debate as Belot and Earman (2001) suggest. Canonical quantum gravity in general, and loop quantum gravity in particular, is only partly relational like GR in so far as it similarly takes for granted the topology, dimensionality, and differential structure of a manifold. Other approaches include:

- Causal set theory:[71] this is a background-independent approach motivated by the assumption that at the Planck scale spacetime geometry will be discrete, and

---

[71] See Dowker (2003), Sorkin (1995), and the discussion in Stachel (forthcoming).

by the fact that a discrete causal structure of events is almost sufficient to define a classical General Relativistic spacetime (see Malament 2006). The formalism models spacetime as a partially ordered set of primitive elements with a stochastic causal structure representing the probabilities for 'future' elements to be added to a given element. The 'volume' of spacetime is then recovered from the number of elements. The 'dynamical structure' is compatible with eternalism because the whole of spacetime can be considered as a single mathematical structure, and temporal relations regarded as just the order of elements. The probabilistic structure is required to be local. It can be shown that a classical spacetime can always be approximated by a causal set. However, the converse does not hold and this is a major problem for this approach (see Smolin 2006). Note that a causal set is a partially ordered set where the intersection of the past and future of any pair of events is finite: '[T]he fundamental events have no properties except their mutual causal relations' (210).

- Causal triangulation models: these models use a combinatorial structure of a large number of 4-simplexes (the 4-d version of a tetrahedron), from which a classical spacetime will emerge as a low energy limit, and from which quantum theory can be recovered if background assumptions are made. Interestingly, the latest simulations of spacetime emerging dynamically from these models generate it as four-dimensional on large scales, but two-dimensional at short distances (and it is known that a quantum theory of gravity is renormalizable in two dimensions).[72]
- Topological quantum field theories, twister theory, and non-commutative geometry: these approaches are highly abstract and speculative at present.[73]

These are all known as 'covariant' approaches because they do not invoke a preferred foliation of spacetime. Another covariant approach worth mentioning has the unique feature that the temporal dimension is abandoned altogether:

- Barbour's relationism:[74] time is supervenient on change, but change is just differences between distinct instantaneous three dimensional spaces.

All of these research programmes use new and highly abstract mathematical structures to describe the universe, and theorists hope to get the familiar behaviour of spacetime and quantum particles to emerge as limiting behaviour. It seems clear that we cannot yet say what the metaphysical implications of quantum gravity are, but the possibilities range from eleven dimensions to two, from a continuous fundamental structure to a discrete one, and from a world with universal symmetries to one with none.

---

[72] See http://arxiv.org/abs/hep-th/0505113, and also Ambjorn et al. (2004).
[73] See Baez (2001), Penrose (2004, ch. 3), and Connes (1994) respectively.
[74] This may be a misnomer since Barbour's ontology seems to include substantival space, albeit without time. A comprehensive discussion of Barbour's theory is Butterfield (2002).

For the project of this book, the main lesson of quantum gravity is that one way or another the world is not going to be describable at the fundamental level by means of the familiar categories from classical physics that derive from the common-sense world of macroscopic objects. It seems that in all the non-perturbative approaches to quantum gravity familiar macroscopic four-dimensional spacetime is dynamically emergent rather than fundamental, and in M-theory there are many more dimensions that we usually think. Thus it appears overwhelmingly likely that some kind of mathematical structure that resists domestication is going to be ineliminable in the representation of the world in fundamental physics. With respect to the metaphysics of time, it seems that it is an open question whether there is an objective global asymmetry in time, and whether such dynamical structure as there is in the universe reflects a fundamentally tensed reality or whether eternalism is true. (Mathematically, perhaps the real issue is between three—or more—plus one dimensions and four.) If M-theory is the correct theory of quantum gravity then there will be universal symmetries that are not time-dependent, and they may define a background independent structure. If there is no background-independent structure and asymmetry in time is part of fundamental physics, then it may be that there is 'dynamics all the way down' and reality is fundamentally tensed.

One possible motivation for the dynamical view is a principle that van Fraassen sometimes seems to endorse, namely that there is nothing that's both perfectly general about all of reality and also true. This coheres with the idea that there is dynamics all the way down in the universe, since any fundamental properties that hold generally would necessarily be time-independent and hence amount to background structure. Consider Smolin: 'The universe is made of processes, not things' (2001, 49). Smolin insists that a lesson of both relativity theory and quantum theory is that processes are prior to states. Classical physics seemed to imply the opposite because spacetime could be uniquely broken up into slices of space at a time (states). Relativity theory disrupts this account of spacetime and in QM nothing is ever really still it seems, since particles are always subject to a minimum amount of spreading in space and everything is flux in quantum field theory, within which even the vacuum is the scene of constant fluctuations.

Smolin and Rovelli seem to have accepted that we will not be able to describe the whole universe at once. (We mention Rovelli's radical relationism again below.) On the other hand, we claim it is a metaphysical residue of obsolete physics to suppose that the universe is 'made of' anything, whether objects or processes, and that such homely metaphors should be relinquished. Either way, the block view cannot be ruled out a priori, and has generated some empirical successes in the history of physics. Given that we seem for now to have reached an impasse it is worth noting that each view must accommodate the success of the other: the block view, if it is right, must eventually explain the second law of thermodynamics, and the emergence of the phenomena that are described by the dynamical laws of physics; and the process view must explain how something like the block universe of SR and GR emerges from a structure that is ultimately dynamical. One possibility is that neither the dynamical or block extremes are the whole story. Another is raised by attempts to recover the emergence of a foliation of spacetime and the appearance of dynamics from an account of observers as information-processing systems (see Saunders 1993b and 1995 and Hartle 2004). (We return to the Everett interpretation in the next section.)

However, one of the lessons of quantum gravity is that some philosophers—especially standard scientific realists—have jumped to overly strong metaphysical conclusions on the basis of not taking account of all the possibilities still held open by physics. There are two leading examples where what some philosophers treated as decisive rulings from physics are now questioned. One of these is the case we have just been discussing, namely, whether or not we live in a block universe. The other is the alleged discovery by quantum theory that the world is not deterministic. In Bohm theory and the Everett approach, the world comes out deterministic after all. Clearly, theories that seem to wear their metaphysical implications on their sleeves often turn out to admit of *physical* reconstruction in different terms. Many physicists have attempted to resolve tensions between QM and GR by seeking what can be regarded as the key metaphysical truth that lies behind each theory's empirical success. For example, Barbour and Smolin think that relationism is the basis for the success of relativity theory. Often it is argued that the truth of quantization that lies behind the empirical success of QM means that theorists should pursue discrete structures of space and time as in the programmes of causal set theory. If each instant is ontologically discrete then why should the timeline be continuous? On the other hand, Hardy (2005) in his discussion of quantum versus classical probabilities, and Deutsch (2004) in his discussion of quantum versus classical computation, argue that there is a sense in which QM is more in keeping with continuity than classical physics. The key insight of QM might also be regarded as the superposition principle, and the consequent problem of entangled states, but there are conceptual and empirical problems with the idea of the structure of spacetime being subject to entanglement (see Penrose 2001).

So it is not clear which aspects of the metaphysical foundations of contemporary physics—for example, continuous space and time, or four-dimensionalism—will be preserved in quantum gravity.[75] However, there are certain modal relations that will be preserved, just as those of classical mechanics were preserved in QM and relativity theory. For example, the approximate validity of the Galilean transformations at low relative velocities of the classical law of gravitation in the latter case, and the recovery of the probabilities for the position of a classical harmonic oscillator from n quantum harmonic oscillators as n becomes large, or the form of various Hamiltonian functions, in accordance with Bohr's

---

[75] Monton (forthcoming) points out that presentism is an open question in so far as a fixed foliation theory of quantum gravity has not been ruled out.

correspondence principle, in the former. Note that in these cases, although relations are preserved, the relata and even their logical type may not be. For example, recall from 2.3 that $\text{grad}V(<r>) = md^2 <r>/dt^2$, exhibits continuity between classical and quantum mechanics having a similar form to the Newtonian equation $F = ma$. But the quantum equation has as its arguments the expectation values of Hermitian operators, whereas the classical equation features continuous real variables. Further examples abound in the case of quantum fields where more and more interactions are taken into account by considering higher-order perturbation theories. Everyone working on the next stage knows they must recover the last stage as a low energy limit. Indeed one approach to quantum field theory known as the effective field theory approach seeks only low energy limits rather than a fundamental description.[76] This is true of quantum gravity also, where the recovery of classical GR and/or QM as a low energy limit is the prime methodological principle. Note that what is being recovered in all these cases is a theoretical structure that encodes the modal structure of the phenomena at a high level of abstraction and not merely the empirical structure or actual data.

We cannot say whether the correct theory of quantum gravity will be closely related to one or more of the existing research programmes, or whether some completely new approach will be necessary. Furthermore, there is no guarantee that the problem of quantum gravity will be solved in the near future, or indeed at all. Notwithstanding the conviction of the string theorists on the one hand, and the arguments of Cartwright and others on the other, it seems to us to be an open question what the outcome of the search for a theory of quantum gravity will be, and profoundly contrary to the spirit of scientific inquiry to declare the mission impossible. Given the diversity of philosophical and foundational presuppositions and implications currently abroad—for example, ranging from a return to absolute time to nihilism about time, and from ten-dimensional continuous spaces to discrete graphs—there is little positive by way of implications for metaphysics that we can adduce from cutting-edge physics. However, there is often an emphasis on modal or causal structure in quantum gravity; for example, Bell-locality is imposed *ab initio* in causal set theory.

However, when it comes to negative lessons, there is more to be said. None of the existing contenders for a theory of quantum gravity is consistent with the idea of the world as a spatio-temporal manifold with classical particles interacting locally. This is not surprising to anyone who has thought about the

---

[76] Effective field theories are only valid at certain energy scales and the move to different scales may require the introduction of completely new physical processes (cf. Hartmann 2001 and Castellani 2002). Cao and Schweber (1993) defend the claim that there will be a never-ending tower of effective quantum field theories and no fundamental theory. Note that the level structure here is given by energy scales not spatial scales although these are not unrelated. The standard model is widely regarded as an effective field theory and not a fundamental one. Wallace (2006) argues that the information that we get from effective field theories is structural information.

implications of Bell's theorem. As we said in 1.2.3 and 3.4, it is important that this is not a theorem about QM, but rather tells us something about any possible empirically adequate successor to QM, namely that it cannot be both local and posit possessed values for all measurable observables. We are justified in treating as unmotivated the idea that any theory of quantum gravity will be a local realist theory, and we should restrict consideration in metaphysics to theories that are compatible with the violation of Humean supervenience implied by entanglement. The structuralism that we have defended in this chapter fits naturally with cutting-edge physics. Here is Stachel: 'Whatever the ultimate nature(s) (quiddity) of the fundamental entities of a quantum gravity theory turn out to be, it is hard to believe that they will possess an inherent individuality (haecceity) already absent at the levels of both general relativity and quantum theory' (2006, 58). As the editors of a recent collection on the foundations of quantum gravity say:

There is a common core to the views expressed in these papers, which can be characterized as the stance that relational structures are of equal or more fundamental ontological status than objects. (Rickles et al. 2006, p. v)[77]

3.7.3 Everett–Saunders–Wallace quantum mechanics and Ontic Structural Realism

There is a close analogy between the debate about time's passage, and the debate about the collapse of the wave function that has been remarked upon by a number of authors (including Rovelli 1997 and Wallace 2002), but perhaps the clearest and most systematic investigation is due to Saunders (1995). According to his account, the notions of the present and of passage in Minkowski spacetime are directly analogous to the notions of actuality and of collapse in Everettian QM. Saunders coined the expression 'the quantum block universe' (1993b) to refer to the picture of reality that we get by fusing SR with QM without collapse. The idea is that just as in SR our notions of the present and passage must be understood as entirely contextual and perspective-relative, so in Everettian QM our notions of the unique outcome of a measurement and the collapse of the wave function must be understood as thoroughly relativized to a branch of the universal wave function. Just as, contrary to appearances all times are real, so all possible outcomes of a measurement are real. What we refer to as the actual is the branch in which we happen to find ourselves.

It is well known that the Everett interpretation faces two major problems. The first is the probability problem that concerns how to recover the Born rule that is used to interpret quantum states in terms of the probabilities of

---

[77] Dawid (forthcoming) argues that string theory supports a position akin to ontic structural realism.

measurement outcomes in a setting where it must be supposed that all the outcomes associated with non-zero amplitudes occur. The second is the problem of the preferred basis, namely what basis should be used for the representation of a quantum state given that infinitely many are possible, and only some correspond to macroscopic objects like pointers in measurement devices having definite positions.

David Deutsch and David Wallace have pioneered an interesting approach to the probability problem using decision theory but we will have nothing to say about this here since our concern is with metaphysics.[78] One solution to the problem of the preferred basis is to regard the position basis as privileged. However, Saunders and Wallace reject this approach because there is no Lorentz-invariant way to define the worlds that would result from doing so exactly. Instead they defend the idea that decoherence generates an approximate basis, and that this is sufficient to recover the appearances of a definite macroworld of localized pointer positions. We briefly explain these issues below with a view to establishing that (a) if the Everett interpretation of QM is correct then OSR is the right account of the ontology of scientific theories, and (b) it is an open question whether a no-collapse interpretation of QM commits us to the existence of more than one macroworld.

Wallace (2002) argues that the idea of a foliation of Minkowski spacetime defined by the Lorentz frame appropriate to our current state of motion is only sufficient to generate a three-dimensional space that approximates to our everyday conception of space. Analogously, the distinct 'branches' or 'worlds' within which macroscopic objects are always in definite states are only approximate according to Wallace. Wallace (2003a) argues for a view of our macroscopic everyday ontology that is closely related to the theory being developed in this book.[79] In particular, he argues that there is a false dichotomy between macroscopic objects being exactly recoverable from the formalism of QM, and their being illusory. The positive account he gives of macroscopic objects is based on the use of decoherence theory to explain the appearance of an approximate basis in which such objects have the relatively determinate states that they are observed to have. Wallace is keen to avoid both versions of the Everett interpretation which require a preferred basis in the physical world to be written directly into the formalism, and many minds versions of the Everett interpretation, which require a basis with respect to which macroscopic objects

---

[78] See Deutsch (1999), Wallace (2003b and 2006), and Greaves (2004). Of course, metaphysics and epistemology are not entirely independent here since many will reject Everettian metaphysics if they believe that it can't be used to recover the rationality of ordinary reliance on the Born rule in epistemic practice. However, evaluation of the proposed resolution of the Everettian probability problem would take us too far from our primary concerns. We are happy to concede for the sake of argument that the probability problem can be solved by the Everettian.

[79] He cites Worrall (1989)and Ladyman (1998) on structural realism, and also Dennett (1991a) whose account of 'real patterns' we take up in the next chapter.

have determinate properties to be given by the quantum mechanical properties of the brains of conscious observers.[80]

The point about decoherence theory is that it shows in quantitative terms how interference between terms that represent distinct states of macroscopic objects very quickly becomes negligible. Macroscopic objects are systems with a very large number of correlated degrees of freedom and the idea is that when a microscopic system interacts with a measuring device decoherence is responsible for the rapid appearance of a definite outcome. Some people have advocated decoherence theory as a solution to the measurement problem in itself, but the difficulty with this is that nothing in decoherence theory suffices to explain why there should be a single outcome of a measurement of a particle that exists in a superposition of two quantum states with respect to the observable being measured.[81] The other complaint about decoherence is that although it shows that the entangled system will very quickly evolve into a state that is for all practical purposes indistinguishable from a proper mixture rather than a superposition, nonetheless the interference terms are still there, albeit very small. This means that in principle it is possible to conduct an interference experiment on the composite system and detect their effects. Neither of these problems are faced by an advocate of the Everett interpretation like Saunders or Wallace who will concede both that there is no single outcome for a measurement where a system starts off in a superposition with respect to the observable in question, and that in principle interference never goes away.

The issue that Wallace addresses is whether an approximate basis in which macroscopic systems take on the appearance of definite states is sufficient to recover both the appearances, and the objectivity of discourse and theorizing about our everyday ontology. His claim is that it is a fallacy to demand an exact basis for the recovery of everyday objects and the distinct branches or worlds of the universal wave function that is presumed to describe the whole universe. Furthermore, Wallace argues that strict criteria of identity over time for worlds and indeed people are also neither available nor necessary. His account of what it is for higher-order objects to exist is in terms of the emergence of patterns or structures (2003a, 91). He also points out that some entities in physics, for example, quasi-particles, are not posits of fundamental physics but rather they are real patterns whose existence consists in their explanatory and predictive utility. Wallace also explicitly adopts functionalism in the philosophy of mind, so mental states, such as the conscious experience of observing a certain outcome of a quantum measurement, are themselves patterns or structures. Hence, indeterminacy about the identity over time of

---

[80] The many minds theory has two variants due to Michael Lockwood (1989) and Albert and Loewer (1988). For discussion of all aspects of the Everett interpretation see Barrett (1999) and the symposium in *The British Journal for the Philosophy of Science* (1996), 47.

[81] See Adler (2003).

macroscopic branches carries over to indeterminacy about the identity over time of people.[82]

One question that Wallace doesn't answer is whether he is proposing a two-tier ontology in terms of fundamental physics on the one hand, and emergent approximate structures on the other, or whether he is happy with the idea that it is real patterns all the way down. On the face of it he seems to be offering a traditional realist view of the wave function, and then to be offering the real patterns account for higher-order ontology. For example, he says: 'A tiger ... is to be understood as a pattern or structure in the physical state' (2003a, 92). On the other hand, he points out that quantum particles are themselves emergent patterns or structures of the quantum field, and he is clearly aware of the possibility that there is no underlying stuff.

This raises the issue addressed by Jonathan Schaffer (2003), namely whether or not there is a 'fundamental level' to reality (recall the discussion of 1.6). Schaffer argues that there are no good empirical grounds for believing that there is a fundamental level, and good philosophical grounds for denying it. Arguably we have inductive grounds for denying that there is a fundamental level since every time one has been posited, it has turned out not to be fundamental after all. Call 'fundamentalism' the view that there is a fundamental level to reality. Ned Markosian (2005) defends 'ontological fundamentalism', the view that the entities at the most fundamental level are the only real ones (or at least are more real than any others).[83] This is a familiar idea which can be captured by Reichenbach's (1957) distinction between illata and abstracta. The former are the things that exist at the fundamental level, the latter are those things that only exist because we conceptualize mereological sums of the illata as if they were genuine objects for pragmatic purposes. We will argue in the next chapter that the illata/abstracta distinction ought to be denied. The tentative metaphysical hypothesis of this book, which is open to empirical falsification, is that there is no fundamental level, that the real patterns criterion of reality is the last word in ontology, and there is nothing more to the existence of a structure than what it takes for it to be a real pattern. Hence, particles or spacetime points are just patterns that behave like particles or spacetime points respectively, just as 'A tiger

---

[82] Saunders (1993b) showed that observers understood as information-processing systems will pick out a particular consistent history space because they must have memories, and memories encoded in states that are not diagonal in the decoherence basis do not persist. Recently, Jim Hartle (2004) has argued that the emergence of a local now for observers is due to their information-processing properties.

[83] Recall from 1.2.3, that Markosian himself believes that there is a fundamental level consisting of simple objects, and that all other objects are mereologically composed of simples. He denies that this gives us grounds for denying the existence of higher-order objects. Note also that Schaffer seems to assume that mereological atomism and fundamentalism amount to the same thing, but this is not so since there could be a fundamental level which consists of one indivisible entity.

is any pattern which behaves as a tiger' (Wallace 2003a, 93).[84] As Saunders puts it: 'I see no reason to suppose that there are ultimate constituents of the world, which are not themselves to be understood in structural terms. So far as I am concerned it's turtles all the way down' (2003d, 129).

One reason why we believe this is that we reject any grounds other than explanatory and predictive utility for admitting something into our ontology. However, we also have a more basic problem with the idea of a fundamental level, namely, its presupposition that reality is structured into levels in the first place. The standard way in which these levels are distinguished is according to size. So, for example, the domains of different special sciences are identified with different scales, the atomic for physics, the molecular for chemistry, the cellular for biology, and so on. A moment's reflection makes the limitations of this obvious since economics can be applied to an ant colony or the world economy, and evolutionary theory can be applied to entities of any size (even, according to Smolin (2000), to the whole universe). Furthermore, in accordance with physics, we regard the structure of space and the metric used to measure length as themselves emergent structures. Hence we can hardly treat them as a fundamental framework within which to describe the levels against which everything else exists.

Wallace's citation of Ladyman (1998) recognizes the accord between his views and OSR, while structuralism has been at the heart of Saunders's work since at least his (1993a and 1993b). Two important questions remain however:

(i) How does the idea of modal structure figure in the context of the Everett interpretation?
(ii) Is the metaphysical view of this book committed to the Everett interpretation?

On the Everett interpretation, what we would normally regard as a non-realized and non-actual possibility, such as an electron being found to be up in the x-direction when we found it to be down in the x-direction, is realized (and hence indexically actual just as in a Lewisian possible world) in another branch of the universal wave function. Hence, there is a sense in which some of what are from the everyday perspective possible but non-actual events are realized after all. For the Everettian, more of the modal structure of the world than we ordinarily think is realized. Note too, however, that many non-actual possibilities will not be compatible with the wave function; for example the electron being in a definite state of up in the z-direction given that it is up in the x-direction. Yet there is a sense in which the electron could have been up in the z-direction; for example,

---

[84] Recall that the most plausible interpretation of GR has it that spacetime points cannot be identified as real objects independently of the physical properties associated with them, and the most plausible view of quantum particles is that their individuality is conferred on them by the relations into which they enter.

if we had chosen to measure z-spin instead. In general it seems that lots of the counterfactual possibilities that we quantify over in the special sciences will not have counterparts in other branches, because they will not be compatible with the universal wave function. So although on the Everett interpretation, there is more to reality than the actuality that meets the eye, it is still the case that the actual (in the non-indexical sense) universal wave function rules out some possibilities, and hence that not all the modal structure of the world is realized. Furthermore, it is also the case that the higher-order modal structure tracked by the special sciences, for example, as embodied in the second law of thermodynamics or Newton's law of gravitation, is an objective feature of the world, and hence we take it that an Everettian must still admit the idea of the modal structure of reality in order to understand laws and (the appearance of) dynamics, even if they ultimately think in terms of a timeless block universe without collapse.

The question that remains for us is (ii) and this brings us to the hardest challenge for our project, just because it is the hardest challenge for the philosophy of physics. As Saunders (2003a and 2003b) says, one can only pronounce on the ontology of quantum theories up to a point before one must have something to say about the measurement problem.

3.7.4 Remarks on the measurement problem

The measurement problem is neatly posed as a trilemma:

(1) All measurements have unique outcomes.
(2) The quantum mechanical description of reality is complete.
(3) The only time evolution for quantum systems is in accordance with the Schrödinger equation.

The problem is that QM often attributes to quantum objects superpositions with respect to properties that we can measure. In such a case, say a particle in a superposition of spin in the x-direction, QM does not attribute a definite state of the property in question to the quantum system. But Schrödinger time evolution is linear implying that if:

(a) system in state 'up' coupled to apparatus in ready state will evolve to system in state 'up' and apparatus in state 'reads up'

and

(b) system in state 'down' coupled to apparatus in ready state will evolve to system in state 'down' and apparatus in state 'reads down'

then,

(c) system in a superposition of state 'up' and 'down' and apparatus in ready state will evolve to a superposition of system in state 'up' and apparatus in state 'reads up', and system in state 'down' and apparatus in state 'reads down'.

We don't seem to observe superpositions of macroscopic objects like measurement devices contradicting (1), and so we have a problem if we continue to assume that the particle and the apparatus really don't have definite states in accordance with (2), and that the time evolution is always in accordance with (3). Everettians deny (1), Bohmians and advocates of modal interpretations deny (2), and dynamical collapse theorists deny (3).

It is often claimed by philosophers that Bohm theory offers the best prospects for realism about the quantum world. Bohm theory is a relatively straightforward modification of QM such that all particles have well-defined trajectories at all times (and indeed definite possessed values for all observables at all times), and such that the evolution of all quantum systems is entirely deterministic and causal, albeit subject to a non-local potential. However, philosophers and physicists have often rejected Bohm theory because of its alleged ad hocness, lack of simplicity relative to the standard formalism, and also because of its manifest non-locality and hence incompatibility with relativity theory.[85] On the other hand, modal interpretations work by postulating that some property of quantum systems is always definite and finding one for any given case so that macroscopic measurements end up being definite. These interpretations seem to be largely ignored by physicists. In general, the denial of (2) is scientifically unmotivated and ad hoc, and only motivated by philosophical rather than physical considerations. It also seems to result in theories that are empirically indistinguishable from regular QM and which are extremely difficult to turn into adequate field theories. This response to the measurement problem seems to be the most problematic to us, falling foul of the PNC.

With respect to (3), we have already mentioned that it may be incompatible with relativity theory. However, given that physical models that may give rise to new empirical predictions are proposed for dynamical collapse, it does not seem to be as much of a purely philosophical response as (2). We think it is an open question whether collapse is a genuine physical process and it is one that we as philosophers of science are not qualified to pronounce upon. If physics comes to recognize collapse as real then we would not be prepared to gainsay that decision despite our suspicion that collapse models are also ad hoc and physically unmotivated. It seems that some form of Everettian QM is the most natural for us to adopt, and indeed we regard it as a genuinely open question whether (1) is in fact false. However, we do have some grounds for not embracing the Everett interpretation.

First, and most importantly from the point of view of our naturalism, no quantum theory is yet globally empirically adequate theory and so we have no reason to regard any interpretation of it as the last word in metaphysics. It might be objected that lack of a globally empirically adequate theory has

---

[85] Everettians also often argue that Bohm theory features all the ontology of the Everett interpretation while adding particles into some branches. See Brown and Wallace (forthcoming).

not prevented us from drawing metaphysical conclusions about the status of individuals in fundamental physics.[86] However, there is an important difference since in that case we found a consilience between quantum theories and GR in favour of rejecting self-subsistent individuals in favour of relational structure. That consilience is extended by Wallace and Saunders since they advocate a structuralist Everett metaphysics; however, we are not yet convinced that an Everettian plurality is the most consilient way of looking at contemporary physics. We agree that it might be, but so far the main evidence for the Everettian position comes only from quantum theories. The Everett interpretation has important cognate offspring in cosmology and quantum gravity research, but as yet there is no way of knowing whether its radical metaphysical plurality of macroworlds will be posited by future physics. Second, given that we don't know the wave function of the actual universe, nor precisely how to recover the states of macroscopic objects, it seems to be an open question whether the true complete wave function will give rise to more than one complete macroworld. Multiple branches might be an artefact of incomplete descriptions and of the use of QM to represent the states of macroscopic objects.

Note that the way we set up the measurement problem relies on the idea that the state of an apparatus for measuring, say, spin in the x-direction, is a quantum state that can be represented in the usual way by a ket vector | reads 'up'>. The usual rationale for treating this as a quantum state is that the apparatus is supposed to be made of a very large number of quantum particles, but nonetheless is still essentially the same kind of thing as the electron it is measuring. However, on the view of higher-order ontology sketched above (and explained in detail in the next chapter), there is no reason to regard the measuring device as something that exists at all from a microscopic perspective. We have also made clear our hostility to the idea that macroscopic objects are fundamentally made of microscopic ones. Hence, the application of the quantum formalism to macroscopic objects is not necessarily justified, especially if those objects are importantly different from microscopic objects, as indeed they are, in not being carefully isolated from the environment. From the point of view of the PNC, the representation of macroscopic objects using quantum states can only be justified on the basis of its explanatory and predictive power and it has neither. In fact, QM is explanatory and predictively inaccurate at this scale since it entails that there ought to be superpositions that are not in fact observed. The predictive success of QM in this context consists in the successful application of the Born rule, and that is bought at the cost of a pragmatic splitting of the world into system and apparatus.

In sum, then, we deny that measurement devices are the mereological sums of quantum particles. Rather, they are real patterns and their states are legitimate posits of science in so far as they enable us to keep track of the phenomena. They do not enable us to do this if we regard them as quantum states, and therefore

---

[86] We are grateful to David Wallace for pressing us on this point.

so regarding them is not warranted. It may turn out that the correct theory of quantum gravity is one that requires the Everett interpretation, or it may be even more relational as is suggested by Rovelli, whose view we shall mention below. (This is also the prediction of Smolin 2001.) However, at the moment it must be regarded as an open question whether there is more than one macroworld.

3.7.5 Quantum information theory

Extravagant claims have been made about information because the concept is now apparently indispensable in so many fields of inquiry. For example, statistical mechanics and thermodynamic entropy is often explained in information theoretic terms (Jaynes 1957a, b, and Brillouin 1956), genes are often characterized as entities that code for particular proteins, and mathematical logic can be understood in terms of information-processing (Chaitin 1987). Some have claimed that most if not all of the important laws of physics can be derived by information theoretic reasoning (Frieden 1998). In the next chapter we explain why we are sceptical about the 'grand synthesis' of science in information theory. However, recently the emergence of quantum information theory has led to information theoretic characterizations and interpretations of QM, and in this section we explore the relationship between OSR and some of these developments. We offer a far from comprehensive discussion of a rich and fascinating new area of physics that is ripe for systematic philosophical investigation, and seek only to draw attention to points of consilience with OSR and the themes of this book.[87]

In classical information theory the Shannon entropy, $H(x)$, is a measure of the amount of information in a signal. $H(x)$ quantifies the minimum number of sequences that need to be distinguished for the transfer of information from a certain source, x. The content of both sources and channels can be measured.[88] Quantum information theory (QIT) is largely concerned with features of many-particle QM that arise because of entanglement.[89] The origins of QIT go back to discussions about the production of entropy in measurement and von Neumann's use of an information theoretic quantity for the thermodynamic entropy of a quantum system. One of the most important aspects of quantum information is that there is a kind of conservation principle applicable to it that states that the degree of entanglement in a quantum system cannot be increased by local operations and classical communication (see Nielsen and Chuang 2000, section 12.5). QIT has been operationalized to the extent that there are measures of quantum information that are based on the amount of noise applied locally

---

[87] The essentials of quantum information theory are explained in Nielsen and Chuang (2000). A superb investigation of many of the interesting philosophical and foundational issues raised by quantum information theory is Timpson (2004).
[88] See Shannon and Weaver (1949). A good introduction is Pierce (1980).
[89] Note that it is possible to have entanglement without non-locality and non-locality without entanglement.

to a subsystem to destroy correlation with the other subsystem(s). The main aim of QIT is to determine exactly how entanglement can be used as a resource for information processing, coding, and communication. Some authors argue that the notion of information is ineliminable in explanations of otherwise puzzling quantum phenomena. For example, fullerene (a molecule of sixty carbon atoms) in the two-slit experiment will produce an interference pattern. However, as the temperature of the molecules is increased, the interference pattern breaks down. This may be explained by pointing out that at higher temperatures there is an enhanced probability for the emission of thermal photons that carry 'which-path' information, and this makes the latter lose its wavelike properties in accordance with the uncertainty relation (Hackermüller et al. 2004).

There are various views about the status of QIT:

(1) Quantum information is something completely new in physics (Jozsa 1998). (Bub 2004, 262 also claims that information is a new physical entity.)
(2) Quantum information is just classical information stored in quantum systems (Duwell 2003).[90]
(3) Quantum information is just subjective degrees of belief (Fuchs and Peres 2000).

According to (3) the quantum state of a system is just a probability distribution in disguise. This is an instrumentalist approach that is metaphysically unhelpful. (For a critique see Hagar 2005.) The conflict between inflationary (1) and deflationary (2) approaches to the ontology of information is what concerns us here.

The Clifton–Bub–Halvorson theorem (CBH) (2003) (see also Halvorson 2004) purports to show that information-theoretic constraints are sufficient to recover the essential algebraic structure of QM (usually identified with the algebra of bounded operators on a Hilbert space).[91] The constraints are as follows:

(i) Superluminal information transfer via measurement is impossible;
(ii) perfectly broadcasting the information contained in an unknown physical state is impossible;
(iii) no bit commitment.[92]

Developing the ideas in Clifton et al. (2003), Bub (2004) argues that in the light of this we should think of QM as a theory about the possibilities and impossibilities of information transfer rather than a theory about the mechanics

---

[90] Duwell no longer argues that quantum information does not exist.
[91] The general framework that they used is that of C* algebra, the technicalities of which we ignore. Suffice it to say that this is a general framework within which all existing theories of classical and quantum physics can be represented. That notwithstanding it is possible to question whether the C* algebraic framework is overly restrictive (see Timpson 2004, section 6.2.2, and Halvorson 2004, 292).
[92] This is a cryptographic protocol that we do not explain here.

of waves or particles. Interestingly from our point of view, Bub argues that the information-theoretic interpretation of QM that he defends demands that state preparation and measurement devices be understood as black boxes from the point of view of QM. He argues that no account of measurement interactions in terms of QM can be acceptable because it would allow the information-theoretic constraints above to be violated (section 3.4). QM on this view is a theory about the statistical relationships between preparation devices and measurement devices.

Timpson (2004, 218) criticizes this approach for failing to provide an interpretation of QM in the sense of an account of the 'nature' of the world that explains why the statistics come out as they do. (We return to the idea of brute statistical regularities below.) Yet from the point of view of the theory developed in this book, attempts to inquire into the 'deep nature' of the world over and above the attempt to characterize its modal structure are not strictly speaking meaningless, but they are nonetheless scientifically worthless. We therefore note with approval that Bub rejects the search for such natures and argues that quantum theories are not about the mechanics of some imagined microworld at all, and that he defends the claim that the measurement problem is a pseudo-problem (262) precisely because he rejects the demand that QM be applicable to state preparation and measurement devices (just as we argue in 3.7.4).

All this notwithstanding, there is arguably a lot more to quantum physics than the bits of the formalism recovered in axiomatizations such as that of CBH. Furthermore, the standard response to claims that everything is information is an insistence that information must be encoded in the physical properties of the world and that therefore there must be physical systems and properties for there to be any information. Furthermore, Duwell (2003) argues that there is an important difference between classical and QIT, namely that a quantum source produces a physical sequence of quantum systems, whereas a Shannon source doesn't produce a physical system. He argues that the criterion of success for the transfer of quantum information is property transfer, and hence that quantum information is the intrinsic and relational properties of a quantum system understood as dispositions for measurement behaviour. Similarly Timpson (2006, 588) argues that 'Information is an abstract (mass) noun and does not refer to a spatio-temporal particular, to an entity or a substance.'[93] Timpson attempts to diffuse debates about 'where the information travels' in protocols like quantum teleportation, and claims that it is not matter that is transported but the quantum state.[94] He argues that talk of information moving is based on hypostatization. He points out that other physical entities, such as

---

[93] Timpson (2004 and 2006) considers the implications for 'information transfer' of a variety of interpretations of QM. Valentini (2002) shows that if Bohm theory is correct superluminal signalling is possible in principle with a collection of particles that is not distributed in accordance with the modulus squared of the wave function.

[94] However, he concedes that if what grounds individuality is form, then it is fair to say that objects are transported (see his n. 5, and Vaidman 1994).

energy or heat, that might be argued to be similarly abstract, are subject to local conservation laws.

However, from our perspective worries about the insubstantial nature of information are moot because we reject the dichotomy between the abstract and the concrete, and between the substantival and the structural, in the first place. Some uses of information theory are indeed purely formal. However, as Collier (1990, 1996) has argued, the sciences of dynamical systems make continuous and irreducible reference to what he calls 'bound' information, that is, information that is realized directly in what everyone regards as physical systems. Here we go a bit further still in our arguments: there are no entities in the material mode according to us, merely the world and its real patterns. In the next chapter we will use informational concepts to define these real patterns. In doing this, we do not intend to be casting a bet that John Wheeler (1990) is right to think that information theory will become the over-arching conceptual framework for fundamental physics; this is partly because, for reasons we will present in 4.3, when Wheeler suggests this he is clearly doing philosophy rather than physics, and we suspect the philosophy in question is naïve. However, we think there is at present no rival to information theory as a conceptual apparatus for generalizing a family of notions, including especially causation and inductive projection, in ways that unify these notions with concepts in fundamental physics. This is fully explained in the next chapter. For now, we note only that using information theory to express metaphysical hypotheses is PNC-compatible because, as we will see, Bub echoes many serious physicists in suggesting that information will come to be regarded as a primitive in physics. Yet more physicists we cite in Chapter 4 who do not go this far nevertheless regard informational properties as genuine physical properties, applicable not merely in thermodynamics (which might be a special science; see 4.3), but in quantum physics. Hence, information may in some sense be all there is at the scale where we describe most generally the modal structure of the world.

Modality plays an important role in the foundations of QM. For example, it is possible to prove a Bell inequality from the assumptions of locality and counterfactual definiteness, which states that the result that would have been obtained in a given measurement had it been performed is determinate (Redhead 1987). Henry Stapp (1987) also proved that the simultaneous or conjunctive satisfaction of locality conditions across different possible worlds, defined with respect to alternative and incompatible measurements that could be performed on a pair of systems in the singlet state, is enough to derive a contrary empirical prediction to that of QM. Again, Bell's theorem is remarkable because it is not really a theorem about QM at all, but rather it places constraints on any viable successor of QM. Even van Fraassen (1991) concedes that it restricts the 'way the world could be', given that QM is empirically adequate in its domain. He argues that it shows that no causal account of the correlations between entangled systems is possible. Note that this is a modal claim, albeit a negative one. Hence, even van Fraassen concedes that Bell's theorem reveals something about the modal structure of the world. He concludes that we must accept that some correlations between physical systems are brute and not explicable in terms of either a direct causal link or a common cause. These claims are challenged by Cartwright (1989), who argues that we should relax the requirements we place on causation rather than abandoning causal explanation. Nonetheless, others have similarly argued that the correlations among entangled systems are brute, including Arthur Fine (1987) who like van Fraassen thinks of this as a deflationary and antirealist move. On the contrary we claim that the acceptance of such brute correlations, properly understood, entails a metaphysical commitment to the objective modal structure of the world, because the correlations in question are probabilistic and hence modal and not merely occurrent.[95] Arguably, what is really novel about the conceptual structure of QM is that it seems to make use of probabilities irreducibly and not (merely) epistemically.[96] Information theory makes irreducible use of probabilities too. Codes must have strong probabilistic features.[97]

If we ask what information is, one common answer is that it is correlation. To say that A carries information about B, is to say that there is a correlation between the state of A and the state of B such that the probability of A being in a certain state is not equal to the probability of it being so conditional on the state of B. This leads some to replace talk of information in the interpretation of QM with talk of correlation. Consider David Mermin's 'Ithaca' interpretation of QM, according to which: 'Correlations have physical reality; that which they correlate does not' (1998, 2). Mermin argues that the physical reality of a system consists in:

(a) the (internal) correlations among its subsystems;
(b) its (external) correlations with other systems, viewed together with itself as subsystems of a larger system

Non-separable or entangled states have the following feature: only the joint state of the whole system completely determines the probabilities of state-dependent properties and correlations among them. It is possible to write down a probabilistic state of a subsystem (the reduced density matrix), but it will fail to encode information about the subsystem's correlations with other subsystems. The state does not factorize, as it is usually put. Pure states are those states that describe systems with no external correlations, whereas mixed states describe states with external correlations. The singlet state is a pure state and expresses the

---

[95] There is one proviso needed of course. If, contrary to the argument of 3.7.4, there are multiple macroworlds, then the probabilistic structure is occurrent after all.

[96] Lucien Hardy (2003) derives much of the algebraic structure of QM by adding one new axiom to a classical probability theory of measurement. Note that van Fraassen (1980) accepts that probability is an ineliminable modality in science.

[97] Dan Hausman (1998, 64–5, and 251–2) also thinks that the connections between outcomes of measurements on entangled states are modal but not causal. He refers to them as nomic (N) connections. Hausman's N-connections are modal relations.

information about subsystem correlations that is not stored in an arbitrarily small spacetime region. In this sense, quantum information is not instantiated locally. It is a remarkable fact that the subsystem correlations (for any one resolution of a quantum system into subsystems) are sufficient to completely determine the quantum state of the whole system.[98] Hence, Mermin argues there is nothing more to quantum states than correlations.[99]

Mermin cites the example of the electromagnetic field that was widely but falsely believed to be a substance (the ether), and suggests that similarly we should abandon the search for a substratum for quantum information in physics. However, he thinks there is a reality of qualia not described by physics. Thus his is a Kantian or epistemic form of structural realism, because there is more to reality than physical reality and an unknowable nature to correlata.[100] While we reject this aspect of his account, we note the consilience between Mermin's idea of correlations 'all the way down', and OSR as defended by us and Saunders. That a practising physicist entertains the idea of modal structure ungrounded in substances or natures of fundamental entities is further PNC-compatible evidence that OSR should be taken seriously despite its incompatibility with the intuitions fostered in philosophers by the combination of the parochial demands placed on our cognition during our evolution, and an education in the classical texts of the metaphysical tradition.

Without committing ourselves to any particular interpretation of QM, we claim that it describes the modal structure of the world, and not merely its occurrent structure.[101] Information is apt for characterizing QM because it is a modal concept. Tim Maudlin shows that information is modal by arguing that information can be transmitted without any energy or matter being transmitted (2002a). The point is easily established by the following example: Bob tells Alice that if he gets spin up he will phone her at 6pm, and if he gets spin down then he won't. If Bob doesn't call then Alice gets information about the measurement without any energy or matter being transmitted. As Maudlin says, in such a case 'information is transmitted only because of the *modal* structure of the situation'

---

[98] Note that whether and to what extent a system is entangled depends not only on the division of the system into subsystems, but also on the frame of reference with respect to which it is considered.

[99] Carlo Rovelli (2004, section 5.6) also defends the idea that QM is about information but with a radical twist. He claims that the information in question is that which systems have about each other, but that different systems may attribute incompatible states to the same target system and there is no fact of the matter about which one is correct. This is a more radical relative state theory than Everett's because even the universe itself lacks a definite state. This is related to Smolin's suggestion that it is impossible to describe the whole universe at once, not because of epistemological difficulties but because there is no fact of the matter about its state. Hence, this approach seems to require intuitionistic logic. We do not endorse this view or reject it, but we note that it is compatible with OSR while going further.

[100] Cf. Wheeler (1990) which is discussed in 4.3, and the discussion of Lewis et al. in 3.4. In 4.4 we explain why we reject qualia.

[101] Note that Esfeld (2004) and Mermin (1998) explicitly commit themselves to objective probabilities.

(2002a, 163, his emphasis). Alice only knows the result of the measurement because of what would have happened had the result been otherwise.[102]

3.8 CONCLUSION

Fundamental physics suggests that the world is not composed of an aggregate of self-subsistent individual objects. The PNC and the PPC suggest that lessons can be drawn for the special sciences from this, and we will argue in the next chapter that the special sciences should not be founded upon an individualist metaphysics either. As we saw in Chapter 1, some philosophers argue that there are no self-subsistent individuals described by the special sciences because such individuals as there may be are made of atoms. Of course, we deny this and think it is only defended by those oblivious to the implications of our best physics. There is no single account of what individuals there are because, we argue, the special sciences may disagree about the bounds and status of individuals since they describe the world at different scales. According to the view developed in what follows, real patterns determine what kinds of individuals are legitimate book-keeping devices of the various special sciences.

Once we have considered the complications arising from relations between fundamental physics and special sciences, we return to the claim that the world is made of information and distinguish between two interpretations:

(i) The world is made of a new substance or substantive particular called information ('infostuff').

(ii) The world is not made of anything and information is a fundamental concept for understanding the objective modality of the world, for example, laws, causation, and kinds.

We reject (i) as a further example of domesticating metaphysics, but we have a lot to say about (ii) in the rest of this book. We may entertain the endorsement of information-theoretic fundamentalism in the following deflationary sense: 'it is impossible to distinguish operationally in any way reality and information ... the notion of the two being distinct should be abandoned' (Zeilinger 2004, 219).

---

[102] Landauer's Principle has recently been defended by Ladyman et al. (2007) in a way that depends on explicating what it means for a physical process to implement a logical process in modal terms. For example, to implement the logical operation AND in a physical system, it is not enough that the input states 1 and 0 give the output state 0, it must also be true that had 1 and 1 been the input states, 1 would have been the output state. Information-processing, like information, is modal.

4

Rainforest Realism and the Unity of Science

Don Ross, James Ladyman, and John Collier

4.1 SPECIAL SCIENCES AND DISUNITY HYPOTHESES

We begin this chapter by consolidating the constraints on metaphysics as unification of science that we have articulated up to this point.

First, we require that a metaphysical hypothesis respect the constraint of the PNC. Floridi (2003) criticizes what he calls the 'eliminativist' interpretation of OSR, the view that self-subsistent individuals do not exist, on the grounds that it lets the tail of the quantum-theoretic problems over entanglement wag the dog of our general world-view. But we endorse the Primacy of Physics Constraint (PPC), according to which failure of an interpretation of special-science generalizations to respect negative implications of physical theory is grounds for rejecting such generalizations.[1] Thus Floridi's modus tollens is for us a modus ponens: if the best current interpretation of fundamental physics says there are no self-subsistent individuals, then special sciences had better admit, for the sake of unification, of an ontological interpretation that is compatible with a non-atomistic metaphysics.[2]

The PNC can be invoked again, independently, here. The increasingly widespread use of dynamical systems theory in the behavioural and life sciences has sparked a lively new debate among behavioural and life scientists over

---

[1] We do not mean by this that special scientists are expected to enumerate all such implications and then check to make sure that their conclusions accord with them. Strictly speaking, the negative implications of a theory are infinite. Rather, what we mean is that when a conflict is actually found between what fundamental physics says there isn't and a special science says there is, revisionary work is in order for the special science in question.

[2] Unless the context makes it clear that we mean otherwise, all references to physics in the rest of this chapter are to be understood as referring to fundamental (or putatively fundamental) physics (in the sense explained in 1.4, and further elaborated in 4.5).

individualistic explanation. (See Ross 2005 for a polemical review of these debates; see also Kincaid 1997. Our view of the relationship between dynamical systems and individuals is made clear later in this chapter and in Chapter 5.) We philosophers thus do not need to ask behavioural and life scientists to pause in their work and pay attention to our remarks about their disciplines just because we think they ought to give special attention at the moment to issues of unifiability with physics. Similar remarks can be made about chemistry and chemists (Ponce 2003; Needham 2002).

The following worry now arises. It is easier to give up on self-subsistent individuals in physics than it is in the special sciences because the latter, but not the former, express many (or most) crucial generalizations in terms of transmissions of causal influence from one (relatively) encapsulated system to another (Kincaid 2004). The thrust of methodological individualism in the special sciences can be glossed as follows: isolate the properties of systems that behave *in vivo* just as they behave *in situ*. This is self-subsistence. It is of course controversial to deny that physics identifies and describes causal relations among such systems, but this proposal has at least often been taken seriously; Russell (1913) vigorously defended the claim, and it has recently been revived by Redhead (1990) and Norton (2003), among others. OSR, as defended in the previous chapter, comports naturally with it. In Chapter 5, sections 1–4, we offer a new interpretation of the Russell–Redhead thesis that we reach by applying insights on the metaphysics of the special sciences back to physics. (Although according to the PPC special sciences do not constrain physics, this does not imply that we can't use metaphysical insights derived from special sciences for attempts at unification, so long as all such insights also have some independent PNC-compatible motivations arising from within physics.) Suppose we agree for the sake of argument that physical theory doesn't require or suggest irreducible causal relations amongst encapsulated systems. That still leaves a serious problem for OSR if special sciences do crucially traffic in such relations. Unless the black boxes represented by the encapsulated systems in question all look (from the perspective of current special sciences) as if they're being progressively unpacked in a way that reduces away the causal relations, then these systems are effectively self-subsistent individuals for all practical purposes and we haven't really achieved unification on the terms we've defended.

Another way of approaching the family of issues here is by reference to the literature on 'causal exclusion'. Kim (1998) argues that unless functionally individuated psychological properties are 'functionalized'[3] re-descriptions of physical properties with which they are metaphysically identical, then either local supervenience of the mental on the physical must fail (in Kim's terms, a version of dualism must hold), or mental properties must be epiphenomenal, or one and the same event must have two determining causes, a mental one and a physical one.

---

[3] This is a special term of art defined by Kim. See Kim (1998, 99–101), for the details.

(The latter possibility is referred to as 'causal overdetermination'.) Kim claims that this argument does not generalize to other special sciences that traffic in functionally individuated kinds (ibid. 77–80). However, this claim has persuaded few critical commentators, and we think Marras (2000) decisively refutes it. (See also Bontly 2001.) Now, it follows from the generalization of causal exclusion that when special sciences offer generalizations about functionally individuated kinds, if these kinds are taken to supervene locally on physically heterogeneous tokens (for example, pain-in-a-person and pain-in-a-rabbit, or two monetary transactions where one involves someone handing someone else a dollar and the other involves someone performing an electronic transfer) then the generalizations in question cannot be strictly true (in the sense of supporting the counterfactuals to which they're literally committed). Kim is clear that to the extent that one takes the objects of such generalizations seriously in metaphysics, one is ontologically confused. Ross and Spurrett (2004) then argue as follows. The overwhelming majority of special-science generalizations concern functionally individuated types that do not have physically homogeneous supervenience bases (if, for purposes of argument, one grants that all properties locally supervene on physical properties). It then follows from Kim's argument, if Marras is right, that all special sciences express ontological confusion in their basic assumptions (assuming we take the special sciences ontologically seriously).

As we indicated in Chapter 1, the naturalist must regard this consequence of Kim's argument as a *reductio*; PNC-compatible metaphysics cannot 'discover' that most scientific activity expresses basic ontological confusion. However, this doesn't tell us which of Kim's premises is mistaken; as Ross and Spurrett concede, Kim's argument raises a genuine issue which a sound metaphysical theory must solve, not simply wave away. The premise that Ross and Spurrett argue should be rejected is that which asserts that all functionally individuated kinds, events and properties locally supervene on (respectively) physical kinds, events, and properties. Of course, it is much easier to say that local supervenience doesn't generally hold than to explain how it could fail in a PPC-compatible world. One premise of Kim's we endorsed above is that special sciences generally explain by citing causal relations. If these causal relations do transmit influence from, and only from, one physically constituted object or event to another, then how can the special sciences not be ontologically confused? If, on the other hand, some causal influences propagate among non-physically constituted objects and events, how could this not amount to dualism? If dualism were true then the PPC would be inappropriate and the naturalistic world-view as we described it in Chapter 1 would be in serious trouble.

This way of approaching the problem is directly connected to the issues around ontological reductionism and its denial. The kind of dualism to which Kim thinks one is forced if his version of reductionism is denied is characterized by him as 'emergentism'. In Kim (1999, 19–21) he identifies it with five 'central doctrines' as follows:

(1) *Emergence of complex higher-level entities:* systems with a higher level of complexity emerge from the coming together of lower-level entities in new structural configurations.

(2) *Emergence of higher-level properties:* all properties of higher-level entities arise out of the properties and relations that characterize their constituent parts. Some properties of these higher, complex systems are 'emergent' and the rest merely 'resultant'.

(3) *The unpredictability of emergent properties:* emergent properties are not predictable from exhaustive information concerning their 'basal conditions'. In contrast, resultant properties are predictable from lower-level information.

(4) *The unexplainability/irreducibility of emergent properties:* emergent properties, unlike those that are merely resultant, are neither explainable nor reducible in terms of their basal conditions.

(5) *The causal efficacy of the emergents:* emergent properties have causal powers of their own—novel causal powers irreducible to the causal powers of their basal constituents.

As Batterman (2002) observes, doctrines (1) and (2) express commitment to a world that decomposes uniquely into non-overlapping components, and to a view of the world as dividing naturally into levels related at each step by a composition relation—commitments, we argue, that current physics does not support. Batterman does not quarrel with these commitments, but joins us in doubting the importance of doctrines (1) and (2) for understanding scientific—especially physical—explanation. Batterman also doubts the relevance of doctrine (5), for reasons we discuss and endorse in 5.5. He defends what Kim would regard as a weaker form of emergentism characterized just by doctrines (3) and (4). We substantially agree with Batterman about this, but prefer a different label—'scale relativity of ontology'—for the position, because 'emergent' and all its semantic kin have come to stand for a hopeless jumble of different ideas in different literatures, including popular ones. We discuss this in 4.2.

For now, the key point is that commitment to a world of levels strictly composed out of deep-down little things has played an essential role in leading neo-scholastic metaphysicians to cast doubt on the ontological seriousness of all the special sciences. This is very far-reaching anti-naturalism.

If the basic little elements of reality whose assumption helps to drive these sweeping conclusions actually existed, then surely it would have to be physics that studied them (and their 'causal powers'). We spent the previous chapter arguing that physics rejects them. (In Chapter 5, we argue that physics does not assign 'causal powers' to fundamental aspects of reality either.) But, again, since special sciences *do* seem generally to be about kinds of things and their causal powers, our basis for rejecting neo-scholasticism leaves their metaphysical status mysterious. Kim and those who share his puzzles about 'higher-level' inquiries need to be answered, rather than just brushed off, by an adequate naturalistic metaphysics.

This sets the agenda for the chapter. We need to articulate a metaphysics for the special sciences that (i) respects the PNC by being motivated by problems that matter to special scientists, (ii) keeps the PPC in view by spelling out links to the conclusions about physics arrived at in Chapter 3, and (iii) allows us to explain the ubiquitous appeal to causal relations amongst encapsulated systems in special sciences, in a way that doesn't leave either the causal relations or the apparent encapsulated systems sitting undischarged in the metaphysical principles that do the unifying work.

Before we launch this task, we should re-emphasize that the demand to unify the special sciences with physics is, according to us, the motivation for having any metaphysics at all. In this context, the fact that inquiry into the nature of the world is organized into a multiplicity of distinct disciplines takes on special importance. A naïve naturalism might simply read ontological disunity directly from the institutional disunity, producing a picture like the one promoted by Dupré (1993). This amounts to abandonment of the metaphysical ambition for a coherent general view of the world, and is the obverse of neo-scholastic metaphysics, according to which whatever cannot be forced into the procrustean bed of a neat, compositionally structured hierarchy should be denied 'real' existence. Neo-scholastics accord a priori metaphysical argument too much weight, and advocates of ontological disunity accord a posteriori metaphysical argument too little.

Our respective grounds for resisting these two extreme attitudes towards the role of metaphysics aren't exactly symmetrical. So far as neo-scholasticism is concerned, we can allow that the world might have been compositionally simple. This cuts no ice according to the PNC, which requires positive scientific motivations for taking metaphysical hypotheses seriously; and we have argued that the empirical evidence tells against mereological versions of reductionistic unification. With respect to naïve ontological disunity hypotheses, however, the problem is yet deeper: we are not convinced of their coherence. Is the thesis supposed to be that we inhabit multiple, separate universes at once? We do not understand this hypothesis if, as in Dupré's case, it is intended as a variety of realism. On the other hand, according to the PNC coherence under intuitions doesn't count as metaphysical evidence, so if the PNC motivated both realism and disunity we would just have to adjust our intuitions. Alternatively, perhaps the thesis amounts, Dupré's disavowals notwithstanding, to a version of idealism: the only worlds that exist are anthropocentric ones, and participation by observers constructs multiple, separate ones.[4] Everett's relative state formulation of QM might be used to motivate a PNC-compatible version of Dupré's view.[5]

---

[4] Nelson Goodman (1978) advocates this picture.
[5] We do not think that many physicists take idealist or dualist interpretations of Everettian QM (like that of Albert and Loewer 1988) to be a serious scientific hypothesis. But the PNC just says

However, this would do nothing to explain the disunity that arises *amongst* the special sciences; the idealistic interpretation of Everettian QM is equally compatible with a multiplicity of mereologically unified individual realities[6] (where the mereological relations hold among observer-generated effects rather than traditional little things). Of course, if we're right that the evidence suggests that the world isn't unified by the composition relation, then if idealistic Everettian QM were also well motivated we'd have grounds for believing in two independent senses of disunity.

We discussed the relationship between Everettian QM and our metaphysical thesis in 3.7.3 and 3.7.4. Here, we just note that the disunity of idealistic Everettian QM and disunity of the Dupré type are entirely independent. The burden thus falls on us to provide PNC-compatible motivations for denial of Dupré-type disunity, since by the PNC the institutional multiplicity of science makes our view, rather than the disunity view, the one faced with the burden of argument.

One further point of preliminary clarification is necessary before we proceed. In much of the literature it is customary to use 'special sciences' to refer to all sciences other than physics. This seems to us to be a sloppy habit, since it suggests the obviously false idea that every generalization of physics is intended to govern everything in the universe. The version of unification that we defend in this chapter provides the basis for a non-reductionistic distinction between fundamental and non-fundamental science that does not cut neatly across the divide between physics and other sciences (though no empirical sciences that are not part of physics are fundamental). Thus some branches of physics, for example, acoustics, count as special sciences, according to us. Let us operationalize the concept of a special science as follows: a science is special iff it aims at generalizations such that measurements taken only from restricted areas of the universe, and/or at restricted scales are potential sources of confirmation and/or falsification of those generalizations. For example, only measurements taken in parts of the universe where agents compete for scarce resources are relevant to the generalizations of economics, and no measurements taken in the centre of the Earth are relevant to psychology. According to our view, there are only special sciences and fundamental physics. Thus the overwhelming preponderance of scientific activity is special-scientific activity.

The foregoing discussion has been heavily dialectical, so a recap is in order. Special sciences seem more clearly committed to self-subsistent individuals, and to causal relations holding among them, than do the fundamental theories of

that a metaphysical hypothesis needs to be motivated by a scientific one. So if a philosopher defends a form of idealism by reference to Everettian QM, her procedure is PNC-compatible. We do not mean to suggest that it would necessarily have any other merits.

[6] By this we mean a world in which nothing exists that isn't a strict part of something else, except whole universes. Compositional unification would be a strong version of this in which everything except a whole universe is a component of something else.

physics. If fundamental relationships in physics are described by symmetric mathematical relations (as they may or may not be), the same surely cannot be said of the most important generalizations in biology or economics. Even if quantum entanglement is thought to undermine individualism in fundamental physics, it is generally thought not to 'percolate upwards' into the domains of the special sciences, where this organism is distinct from that one, a selfsame rock can be freely transported from context to context without its integrity being threatened, and two consumers are necessarily separate if either their utility functions or their consumption histories differ to any degree. Denying individualism in fundamental physics while holding on to it as a metaphysics of the special sciences leaves just two options with respect to one's more general metaphysical theory. Either one must be an instrumentalist about the special sciences and deny the literal truth of their generalizations—the reductionist, neo-scholastic option—or one must abandon the project of unification. Both of these options are currently popular among philosophers. But the first asserts philosophical authority to legislate the status of sciences and the second surrenders the metaphysical ambition altogether. Denial of naturalism and denial of the possibility of philosophy have arguably been a recurring Scylla and Charybdis of the twentieth century. We can try to steer between them only by finding a way to deny individualism as an interpretation of the special sciences that doesn't rule out any currently empirically significant special-science hypotheses. Ideally (though, by the PPC, this isn't required) we should try to do this in a way that is second-order PNC-compatible, that is, partly motivated by special-science hypotheses themselves and not just by considerations from physics.

So here goes.

4.2 DENNETT ON REAL PATTERNS

The previous paragraph's Scylla is reductionism (since that is the view that interprets science in terms of an a priori non-naturalistic metaphysics), and Charybdis is instrumentalism about the special sciences (since instrumentalism amounts to giving up on the possibility of metaphysics in our sense).[7] At this point, Scylla is easier to avoid. We will therefore deal with it first before turning to the harder work of steering past Charybdis.

We noted above that reductionism is enjoying a period of popularity among some of the philosophers we have referred to as neo-scholastics. These are *individualistic* reductionists whose programme depends on the idea that at the level of physics everything comes down to microbanging relationships amongst

---

[7] Dupré calls himself a realist, but only because he acknowledges the existence of a mind-independent world. Thus the contrast with his realism is idealism. But in denying that special sciences produce generalizations over classifications independent of specific human purposes, he is an instrumentalist about the special sciences.

little things. As we showed in Chapter 3, this kind of generalized individualistic reductionism is simply refuted by contemporary physics. In addition, as we argued in the previous section, it doesn't even succeed in successfully avoiding instrumentalism with respect to current special sciences, since most of them individuate the kinds over which they generalize functionally, but reductionism tends to 'sunder' functional kinds metaphysically (Kim 1998, Marras 2000, Bontly 2001, Ross and Spurrett 2004). Thus generalized individualistic reductionism is forlorn.

We used the adjective 'generalized' above to mark the distinction between a view according to which *all* kinds reduce to physical kinds from the view that *some* kinds reduce, either to physical kinds, or to other kinds viewed as 'below' them in the traditional model of ontological levels. For example, Bickle (1998) has recently produced a well-received model of mental processes according to which mental kinds reduce to neurophysiological kinds. For reasons that will soon be obvious, we don't agree with Bickle. However, what we want to stress for now is that this is of tangential importance, because as discussed in Chapter 1, nothing in our position entails denial of the possibility of local intertheoretic reductions in the Nagelian sense.

Given the state of play in physics as described in Chapter 3, the only sort of generalized reductionism that would avoid instrumentalism about the generalizations of the special sciences would be a non-individualist variety. Therefore, it would have to be a view according to which all special-science generalizations are special cases of physical generalizations, or combinations of them (where, of course, the physical generalizations do not involve commitment to self-subsistent individuals). One could defend such a view in a PNC-compatible way only by illustrating or producing a sample of derivations of special-science generalizations from fundamental physical ones, and giving reasons for thinking that they were the leading edge of a wave. That is, one would need to provide something like a non-individualist, non-mereological counterpart to the Oppenheim–Putnam argument discussed in Chapter 1. However, far from this looking like the current trend across the sciences, the increasing attention being given to 'complexity' in many special sciences represents precisely the opposite development. Few of the faddish advertisements for 'emergence' (see, for example, Johnson 2001) carry serious burdens of metaphysical argument, but the most interesting and detailed of them are based on the highly plausible idea that derivation of many projectible 'macro-level' generalizations from 'micro-level' ones is computationally intractable. This is a version of Kim's emergentist doctrine (3) from 4.1. We say more about this doctrine below. For now our point is the straightforward one that there is no recent serious defence of generalized non-individualistic reductionism that is motivated by detailed attention to current science.[8] This doesn't entitle us to promote anti-reductionism prior

---

[8] Melnyk (2003) is a possible exception here. See 1.5.

to working out a PNC-compatible version of it. But we needn't also work to overthrow a thesis that has no scientific support.

As we will see, issues related to complexity are among the second-order PNC-compatible[9] motivations for our positive theory of metaphysical unification. But these should be understood in the logical context of the attempt to ward off the instrumentalism about special sciences that seems to threaten when we accept the structuralist interpretation of physics defended in Chapter 3, plus the PPC. Thus we turn to avoiding Charybdis.

Instrumentalism is a slightly elusive target. It arises in two main contexts: radical pragmatism (as defended by, for example, Rorty 1993), which disavows the very idea of objective ontological commitment; and common-sense realism, which allows the objective reality of various kinds of everyday manifest objects, but denies this status to kinds of objects postulated by theories constructed to explain manifest phenomena. (On a broad construal of instrumentalism, van Fraassen's constructive empiricism is perhaps a third variation, which affirms the reality of observable objects according to a quite liberal notion of 'observability', and promotes agnosticism rather than scepticism about the reality of theoretical objects. But the difference between agnosticism and scepticism is plausibly only rhetorical, in which case constructive empiricism is just a particular elaboration of common-sense realism.) We will take instrumentalism in all its possible senses as the contrast class with the realism we mean to affirm. Let us therefore note that we endorse a version of instrumentalism about all propositions referring to self-subsistent individual objects, chairs and electrons alike. But we will be at pains to show that our instrumentalism about propositions referring to self-subsistent individuals in no way implies instrumentalism about theories—which is the sort of instrumentalism that matters, and what we will have in mind whenever we refer to 'instrumentalism' without qualification. Our contrasting realism is realism about the domain of scientific description and generalization.

We may also distinguish global from local instrumentalism. The former has been debated mainly in the philosophy of science literature. A number of thinkers, however, have defended instrumentalism only with respect to particular special sciences or clusters of them—especially the so-called 'intentional sciences' that generalize over kinds of propositional-attitude states. The economist Milton Friedman (1953), for example, argues that firms and consumers and other common-sense economic objects are perfectly real, but not the esoteric types of objects economists use to construct theories about them, such as utility and discount functions (see Mäki 1992). Among philosophers, Alex Rosenberg defends instrumentalism about intentional sciences (1992) and biology (1994), but not physics. Such local instrumentalisms have persuaded far more philosophers over the years than has global instrumentalism. Since our main topic in this chapter is special science, it is thus doubly appropriate that in considering instrumentalism

---

[9] That is, motivated by the contents of special sciences as well as by physics.

we should focus on its local manifestations—especially on its neighbourhood of greatest prevalence and influence, the philosophy of intentionality.

For many years, a lightning-rod figure in debates over instrumentalism about intentional sciences has been Daniel Dennett. He has often been read as an instrumentalist on the following basis. According to him (on this reading), minds are just inferred theoretical fictions, instances of an organizing principle for an inter-defined set of instruments we use in taking 'the intentional stance' to some of the things that the bodies of organisms and some artefacts do. On this reading of Dennett, he shows us that the mind–body puzzle is an artefact of not being thoroughly instrumentalistic, and of taking minds to be common-sense objects, like tables, when they are in fact just elements of predictive models, like quarks. The puzzle results from the assumption of a particular (Cartesian) metaphysical theory, and one dissolves it by dropping the metaphysics altogether rather than by adopting a better metaphysical account.

Dennett (1993, 2000, personal correspondence) says unequivocally that he has never intended to avow general instrumentalism about theoretical kinds (for example, quarks). His early view, in the 1970s and 1980s, seems to have been that intentional attitudes are *sui generis*, and that his 'stance' stance concerning them had no particular implications for the general metaphysics or epistemology of science. However, in a now-classic paper 'Real Patterns' (RP; Dennett 1991a), he emerged from this neutrality to frame his view of mind in the context of what Haugeland (1993) rightly regards as a distinctive metaphysical thesis. According to RP, the utility of the intentional stance is a special case of the utility of scale-relative perspectives in general in science, and expresses a fact about the way in which reality is organized—that is to say, a metaphysical fact. The fact in question is what we (but not Dennett) call the *scale relativity of ontology*.

We take it that scale relativity of *epistemology* isn't controversial. To borrow an example from Wallace (2001), if you want to predict what a hungry tiger will do when confronted with a deer, you should study whole behavioural patterns of whole tigers, not individual tiger cells or molecules. It is clearly motivated by any thesis to the effect that models of complex systems are scientifically useful. It entails Kim's emergentist doctrine (3), and also his doctrine (4), if by a property's being 'explainable' we mean that its specific instantiations, and not just its possibility of being instantiated somewhere, are explained by 'lower-level' theories. This kind of scale relativity is pragmatic, makes no metaphysical hypotheses, and so is consistent with instrumentalism, or with mereological reductionism, or with the denial of reductionism. It is, however, useful for quickly saying what scale relativity in general (that is, applied either to epistemology or ontology) *is*: it is the idea that which terms of description and principles of individuation we use to track the world vary with the scale at which the world is measured. We gave some examples of this in Chapter 3. Here are some more: at the quantum scale there are no cats; at scales appropriate for astrophysics there are no mountains; and there are no cross-elasticities of demand in a two-person economy.

Scale relativity of *ontology* is the more daring hypothesis that claims about what (really, mind-independently) exists should be relativized to (real, mind-independent) scales at which nature is measurable. As we saw in 3.7, Wallace and Saunders endorse something very close to it as their way of handling the measurement problem in QM. Saying that some—typically, macroscopic—objects exist only at certain scales can easily slide into being a tendentious way of saying that the objects in question aren't real at all (as we will see in detail in a moment). However, there is a point to it if one can motivate two principled distinctions: first, between scale-relative objects whose identification contributes to descriptions of real structures and purely fictitious or constructed objects; and, second, between structures *observable* (by any physically possible observer) at (and only at) some restricted-scale parameters and structures that merely are, in point of fact, tracked by particular observers at particular scales. These distinctions are implicitly motivated by Wallace, as we saw, and are explicitly motivated in our account. Wallace draws his version of scale relativity of ontology directly from Dennett's RP.

Though, as noted above, Dennett denies that he is an instrumentalist, in RP he waffles inconsistently between scale relativity of ontology and of epistemology, and at one point goes so far as to explicitly associate his view with Arthur Fine's (1984a, 1986) desperately inconclusive 'Natural Ontological Attitude'.[10] Ross (2000) draws attention to this, and argues that Dennett's (1991a) argument can only be made to work consistently against both of its main targets, classical computational reductionism and eliminative materialism about mind, if it is read as promoting an ontological interpretation of scale relativity.[11] Dennett (2000, 359–62), replying directly to Ross, seems to go along with this, and has not since said anything in press that is inconsistent with it. Therefore, let us here cut through issues about Dennett's self-interpretation, and move straight to an account of his application of scale relativity of ontology to the case of *dynamic patterns*.

Let us begin with an example. Consider a large community of people, most of whom don't interact directly with one another, but who face a collective

---

[10] We find it impossible to determine from Fine's various formulations of the NOA whether it is intended as a version of realism or instrumentalism. It may be that, like Rorty, Fine means to deny the sense of metaphysical questions altogether, in which case his position is instrumentalist by our lights. On the other hand, he may intend only the weaker view—one we share—that there are no specifically philosophical arguments for realism that should be taken seriously on naturalist grounds, but that many scientific realists mistakenly think there are. (We owe this point to Harold Kincaid.) One belief of Fine's we don't accept is that any variety of realism commits its adherent to taking Einstein's side in the debate over hidden variables in QM. This leads Fine to what seems best described as an easy-going realism about both common-sense objects and theoretical kinds, minus belief in hidden variables. This amounts to philosophical quietism about the issues that interest us here.

[11] The 'waffling' just referred to consists in the fact that Dennett responds to computationalist reductionism with arguments appropriate to a scientific realist, but responds to eliminative materialism strictly by reference to what is practical given human epistemic limitations—which is an instrumentalist strategy. See below, but especially Ross (2000), for further details.

decision—say, a national election. Their behaviour might be influenced by a set of national focal points tracked by public information media that try to summarize the dynamics of their opinions, where these focal points themselves organize and stabilize the dynamics in a feedback relationship. Attending to these focal points may be essential for predicting and explaining the election result. However, they are of little predictive efficacy with respect to any particular person's opinions at any particular time, and they may not be perturbable by shifts in view among even relatively large subsets of the population. Such phenomena will often be invisible to an observer who looks for them at a measurement scale that is too fine, because they cannot be inferred by any systematic aggregation procedure over individuals' opinions, and the boundaries of their extensions at micro-scales will tend to be indeterminate. Kelso (1995) provides further and more detailed examples of many such phenomena from a diverse range of scales. Following his (and widely accepted) usage, we will refer to them as 'dynamic patterns'.

There is a large recent philosophical literature on dynamic patterns. A good deal of this concerns itself with puzzles about whether and how such patterns can have causal efficacy over and above the aggregated interactions of the causal capacities of whatever the patterns are 'made of'. Consider Dennett's own leading example, John Conway's 'Game of Life' (Poundstone 1985).[12] This is based on a simple implementation of cellular automata that makes a particular range of stable dynamic attractors highly salient to people. A person using the system naturally book-keeps its state sequences by reference to a typology of 'emergent' objects—'gliders', 'eaters', 'spaceships', etc.—that have only virtual persistence. (That is, two successive instances of 'the same' glider share only structure, and common participation in structures larger than themselves. A glider is clearly mereologically composed of a small number of illuminated cells. However, its successive instances are composed of different cells, and successive instances a few steps apart have no cells in common.) Once this descriptive stance is adopted towards Life, almost all users spontaneously track the dynamics in terms of causal interactions among instances of these types—for example, a glider will be caused to disappear through interacting with an eater. That is, Life users naturally begin logging causal generalizations about the types of virtual objects, and thereby seem to commit themselves to their objective existence.

---

[12] One should be cautious in using the Game of Life as a metaphysical model of the universe. It is useful for the purposes to which Dennett puts it, because it shows how patterns can emerge at grains of analysis coarser than the grain at which what is analogous to the fundamental microphysical level is studied, even when all causal processes governing the latter are non-complex, known, measurable, and deterministic. Life is thus a good antidote to romantic interpretations of emergence. However, because in Life there *is* an unambiguous fundamental level composed of the aggregation of a finite number of 'little things', and because no higher-level object types cross-classify the dimensions of any models of the game relative to classifications in terms of cells, Life differs greatly from the universe with respect to the kinds of reductionism sustainable in it. Life admits of complete decomposition; the universe does not.

Conservative metaphysicians regard this spontaneous response, if it is taken as anything more than a practical heuristic, as a philosophical mistake. In the example of the generalization just given, they would complain, the eater is a redundant causal factor, since the program underlying Life, which in its declarative representation quantifies only over cells, is strictly deterministic. We are reminded that an eater or a glider is, at any given time, 'made of' cells and nothing else. Then we are invited to agree that a thing cannot have causal efficacy over and above the summed causal capacities of the parts with which it is allegedly identical. The result is supposed to be reductionism, and instrumentalism about gliders and eaters.

In our view this whole debate is poorly motivated, because both sides in it presuppose that questions about ontology must be parsed as questions about *which individuals* are the 'real' ones: gliders and eaters (etc.) plus cells, or just cells? We argue below that a main contributor to the confusion is a pre-scientific conceptualization of causation that is undermined by PNC-inspired considerations. We work up to this diagnosis by first discussing Dennett's liberating though incomplete insight concerning the debate, on which our view will then be an elaboration informed by OSR. What we mean by calling Dennett's insight 'incomplete' is that the argument of RP sometimes relies on claims about the independent causal efficacy of dynamic patterns, and thereby mistakenly accepts the false terms of argument established by the conservative critics. However, in other respects the paper breaks through this shopworn dialectic. That is, it succeeds in finding a means of abstracting from rival philosophies of causation that are themselves less secure than the phenomena they are used to analyse.[13]

Dennett recognizes that there is a class of observer-independent facts of the matter about dynamic patterns that is *directly* related to questions about pattern-individuation, rather than just indirectly related by way of issues around causal efficacy. In particular, there are facts derivable from computation theory about which patterns can and cannot be generated by compressible algorithms. These facts permit us to address questions about pattern reality that do *not* reduce to the questions allowed by instrumentalists about which patterns are useful to anthropocentric predictive projects. A 'real' pattern, Dennett argues, must admit of capture using a smaller number of bits of information than the bit-map transcription of the data from which the pattern could be computed. Thus, to refer to our earlier election example, we would not be locating one of the electorate's focal points if we specified a region in 'opinion-space' that could only be identified based on a conjunction of descriptions of each voter. The crucial point about Dennett's version of this requirement is that he does not relativize what 'could' be computed to any special reference class of computers. Instead, he says, 'a pattern exists in some data—is real—if *there is* a description

---

[13] For evidence that Dennett at least fleetingly recognizes this, see n. 22 of RP.

of the data that is more efficient than the bit map, whether or not anyone can concoct it' (1991a, 34). Thus there are (presumably) real patterns in lifeless parts of the universe that no actual observer will ever reach, and further real patterns whose data points are before our eyes right now, but which no computer we can instantiate or design will ever marshal the energy to compact. This is emphatically not instrumentalism.

Nor, obviously, is it reductionism. Dennett is clear in committing himself to the idea that order at a given scale need not depend on any *particular* basis for order at any finer scale from which the cruder-scale order could be inferred.[14] Thus, for example, it is a fact that natural selection sculpts the dynamics of populations even though no individual animal was ever brought into being or killed by natural selection, or by anything that is a vector-component of natural selection conceived as a quasi-mechanical causal force (see Godfrey-Smith 2001; Ross 2005, ch. 8). Natural selection will thus be invisible to an observer who doesn't ascend to a different scale from that on which individual animal biographies are tracked. (Such scale ascendance was Darwin's greatest intellectual achievement, as just before him a similar kind of re-scaling of a special science was Lyell's great contribution.) Invisibility, of course, doesn't imply non-existence, but the former may be diagnostic of the latter. The natural selection example is a case in point: in histories of lineages at small enough temporal scales there is no natural selection, because natural selection requires a substantial minimum number of reproductive events. By similar logic, countries only have GDPs over scales of time substantial enough to produce tradable products (Hoover 2001). Furthermore, Dennett stresses that there may often be multiple patterns in the same data as measured on finer scales, and yet we may not definitively pronounce for one of these patterns as superior to another merely because the former predicts the future value of some micro-scale data-point that the latter gets wrong (47–8). Dennett here anticipates Hüttemann's point, discussed in Chapter 1, about the failure of general microgovernance.

The significance of Dennett's contribution to anti-reductionist metaphysics has been appreciated by a number of philosophers who have recently been working on that project. The reader is referred to Wallace (2001, 2004), Shalizi and Moore (2003), Shalizi (2004) and Floridi (2003). It should be emphasized that none of these authors is motivated by concerns about intentionality or by problems in the philosophy of mind more generally. Floridi is interested in general metaphysical questions, while Wallace, Shalizi, and Moore are explicitly focused on the foundations of physics. All maintain that Dennett's version of scale relativity helps us make sense of the ontology of physical nature as physicists currently model it—that is, as incorporating observer-independent and theory-independent properties tracked by the concepts of superposition

---

[14] Note, with reference to n. 8 above, that this idea does *not* apply to the Game of Life.

and entanglement. (See 3.7.) Nottale (1993), though not citing Dennett, takes elaboration of scale relativity to be the basic preferred strategy for understanding the relationship between GR and microphysics. Since Shalizi, Moore, and Nottale are all physicists who use the scale relativity of ontology in constructing specific physical hypotheses, the proposition that the existence of the objects described by current physics is relative to specific scales is thus well motivated within the constraints of our PNC.

Recent work by Batterman (2002) is also of direct interest here. Though he does not couch his position in terms of either scale relativity or real patterns, he can be seen as providing further PNC-compatible motivation for these ideas. According to Batterman, physics explains why many systems undergo phase transitions at the singular limits of certain system parameters. By means of detailed attention to specific physical theories—not A-level chemistry—he shows that these explanations do not proceed by reference to decompositions of the systems in question into parts that undergo changed relations at critical points; the explanations in question are given in terms of global properties of the whole systems. Specifically, the explanations show why particular *universalities* apply to the systems in question—that is, why they obey phase-transition laws that are independent of their microstructural details.[15] Batterman argues that the properties and kinds picked out by universality classes (for example, renormalization groups in mechanics) are 'emergent', despite the fact that Kim's emergentist doctrines (1), (2), and (5) do not apply to them. The existence of the physical explanations for the universalities must, for a naturalist, block any temptation to try to reduce away the emergent kinds and properties through the introduction of 'metaphysical hidden variables'. Thus, by Batterman's argument universalities are real patterns and his emergentism is ontological rather than merely epistemological. Now, the universalities on which he focuses would be invisible to inquirers who confined their attention to the scale on which microproperties (relative to them) are measured. Thus Batterman's 'emergentism'—which we endorse, while considering its label semantically unwise—is a special case of scale relativity of ontology. We take Batterman's argument as establishing that scale relativity of ontology is PNC-compatible. He also argues that we should expect this to be the basis on which to account for the autonomy of the special sciences in a world where (in our terms) the PPC holds.

Batterman (2002, 134) doubts that we'll often be able to explain rigorously special-science universalities in the way we explain physical ones, because 'the

---

[15] See Belot (2005b), and Redhead (2004) for a critique and Batterman (2005) for a response. The crux of the disagreement between them is that Belot and Redhead think that Batterman underestimates the extent to which the mathematics of the less fundamental theory is recoverable from the more fundamental theory, whereas Batterman thinks that they neglect the fact that the boundary conditions that must be used to do this are only motivated by considerations from the less fundamental theory.

upper level theories of the special sciences are, in general, not yet sufficiently formalized (or formalizable) in an appropriate mathematical language.' We are more optimistic. There are ongoing efforts both to discover and to explain universalities in the most formalized of the social sciences, economics. Ball (2004) provides a popular survey of the considerable scientific literature that has recently sprung up here.

Having defended its PNC-compatibility, let us now return to Dennett's account of real patterns. There are, as noted above, problems with it. Though it offers the major achievement of the compressibility requirement—that a real pattern cannot be a bit-map of its elements—as a *necessary* condition on real-pattern-hood, Dennett never provides an explicit analysis of *sufficient* conditions. (Obviously compressibility isn't sufficient, or every redescription of anything would conjure a new real pattern into being.) Thus he only gestures at an implicit background metaphysic without actually providing one. This leads to some backsliding in the direction of instrumentalism towards the end of his paper, when he imagines arguments among some actual (as opposed to possible) observers over how many patterns there are in a given series of data. Each observer, in Dennett's discussion, is allowed to determine her own idiosyncratic answer based on her preferred level of tolerance for noise in producing predictions. If this were a basis for adjudicating the reality of patterns, rather than just indicating the grounds on which someone might defend a pattern's practical significance, then the principle would self-evidently generate an infinity of real patterns, since there are as many possible preference profiles over tolerable levels of noise as there are ratios of noise to signals, and only pure randomness is always predictively useless. 'Everything that is not randomly organized exists' is not a potentially helpful basis for unifying science; treated as an ontological principle, it is an invitation to embrace disunity to a degree that would make Dupré look conservative.

Dennett identifies 'common-sense' patterns as being those that are products of pragmatic trade-offs in favour of high compression with high error levels, whereas 'scientific' patterns tolerate less noise. This is unobjectionable in itself—indeed, it is surely right. But then, in the case where the scientist's pattern predicts everything that the common-sense pattern does and more besides—though at the cost of more computing—a realist should regard the common-sense pattern as ontologically redundant. Dennett wanders away from this clear consequence of the scientific realism manifest through most of his paper because, toward the close of it, he is anxious to see off eliminative materialism about the mind, but recognizes that whether intentional psychology is a redundant pattern in light of other ones that science may hit upon is a strictly empirical matter. He should therefore be content to simply say that intentional psychology is certainly practically useful, and that we don't presently have good reasons for thinking it is in fact redundant. He does say exactly this at the very end; muddles in a few penultimate paragraphs arise from his first trying to get eliminativism into a deeper level of purely philosophical trouble than his own naturalism really allows.[16]

The origins of these problems lie in RP's roots in Dennett's earlier defences of his theory of the 'intentional stance' in the philosophy of mind. In one of his founding presentations of that idea (Dennett 1971), he argued that one *can* usefully assume the intentional stance towards a house thermostat (for example, 'It prefers the room to be 68 degrees and believes it's now 64 degrees, so it decides to turn on the furnace'), but that one *must* take the intentional stance towards a fancy chess-playing computer if one wants to maximize predictive leverage with respect to it. The reading of Dennett as an instrumentalist that has prevailed over the years under-emphasizes the distinction he has always drawn between these two cases. In the case of the thermostat, use of the intentional stance is purely pragmatic and optional, because the physical stance is potentially available given more effort, and throws away no information the intentional stance tracks. By contrast, where the artificial chess-player is concerned, failure to reference its structural dispositions to the reasons furnished by its problem space—chess—will cause one to miss some true counterfactual generalizations about its behavioural tendencies. This is due to facts about the world: the chess-player can adapt its behaviour to environmental contingencies that *no physically possible* computer could exhaustively specify (because writing out the game tree for chess would require more energy than exists in the universe), whereas the thermostat cannot.

We said that Dennett's crucial insight in RP consists in his identification of a necessary condition on the reality of a pattern. The condition in question is just that which makes the intentional stance non-optional without sacrifice of information: that the pattern cannot be compressed by any physically possible computer. However, we also noted that Dennett fails to specify sufficient conditions on a pattern's reality consistently, suggesting at one point that any non-compressible pattern that any observer of some data finds useful relative to a given level of epistemic laziness—that is, tolerance for error—should be regarded as real. This also replicates slack left in Dennett's earlier discussions of the intentional stance. In these papers, there is as yet no sign that he is thinking of 'information' in a rigorous or generalizable way. Thus his 'could not' in the formulation of non-optional intentional stances was originally relativized to a *sui generis* enterprise of predicting rational behaviour. The insight of

[16] We can express the distinction in the terms we have set up in this book as follows. A metaphysical thesis like eliminative materialism is not ready for scientific acceptance if it *is* motivated within the constraints of the PNC but the balance of present empirical considerations don't favour it. This is how we think the situation stands for eliminativism, and that is also what Dennett says (in other terms) at the very end of 'Real Patterns'. The hypothesis would be in one of two kinds of *philosophical* trouble if it were (i) motivated according to the PNC but not unifiable with hypotheses from other domains, or (ii) not motivated according to the PNC. The parts of Dennett's discussion we are calling 'muddled' arise from his flirting with the suggestion that eliminativism is in one of these kinds of philosophical trouble. His discussion doesn't allow us to tell which kind he's imagining; we suspect he doesn't anticipate the distinction.

RP, by contrast, precisely consists in the recognition that which patterns are compressible by physically possible processes is something about which there are non-instrumental facts of the matter; but then Dennett confuses himself by hoping (at least for a few paragraphs, before coming back to his consistently naturalistic attitude at the end) that the existence of patterns of rational behaviour is something in which we can have a priori confidence, so that we can reject eliminativism about mental and rational patterns from a purely philosophical vantage point. However, this is inconsistent with naturalism; and if Dennett has any flag of first allegiance it is, as with us, naturalism. We indeed believe that there are patterns of rational behaviour, but not because we found this out while sitting in our armchairs. It is a potentially revisable belief for which we think there is good empirical evidence.

Ross (2000) argues that Dennett's equivocation as just discussed expresses itself in a second way in RP in the form of a poorly motivated distinction between two different *kinds* of patterns. Dennett refers to these kinds, following Reichenbach, as 'illata' and 'abstracta'. This distinction is based on one way of trying to preserve the unity of physics with the special sciences: one maintains that physics describes some structures that are independent of interpretation, the illata, and then interpreters perform re-classifications of the elements of these patterns for special purposes, giving rise to abstracta. (In RP patterns in rational behaviour are the main examples of abstracta, while physical patterns are held to be illata.) Reichenbach had not assumed that the distinction speaks to any first-order ontological facts, since for conventionalists like Reichenbach there are no such facts we have any serious business trying to talk about. However, Ross argues that Dennett derives his interpretation of the illata/abstracta distinction from Quine (1953), for whom it is based on our adopting, for pragmatic reasons, 'Democritean faith' that we will someday be able to decompose everything over which we want to quantify into elementary particles and relations between them. Like Oppenheim and Putnam at around the same time Quine believed, for broadly PNC-compatible reasons, that the science of his day motivated this 'faith'—which in this case is just an extremely unfortunate word for what we have been meaning by a unifying metaphysic.

We have of course been arguing for two chapters now that science no longer motivates this metaphysic; and it is nearly the opposite of the non-reductionist metaphysic that is otherwise promoted in RP. The source of Dennett's incompatible mixture of assumptions here is fairly obvious, we submit. On the one hand, he is trying to defend the real, non-reducible existence of intentional patterns. On the other hand, he doesn't want to risk dualism by letting the intentional patterns float altogether free of the physical. The problem is that he doesn't have a clear or consistent eye on what naturalistic ontological unification requires and doesn't require. The idea that intentional patterns are re-classifications of the 'basic' patterns—that is, abstracta on physical illata—won't work, because this idea only has coherent content if we suppose the reverse of what Dennett mainly argues, viz, that what is necessary for making some abstractum *real* is that it can't be computed from all the facts about the illata. Dennett is a verificationist in the sense we defended in Chapter 1, so he can't suppose that it makes sense to talk about re-classifications that no possibly observable system could compute.

These reflections help to show us what is needed for a consistent account. First, we need some empirically motivated story about how the physical patterns constrain the intentional patterns without, even just in principle, reducing them. Furthermore, if this is not to come out as instrumentalism about the intentional patterns after all, the failure of implication can't be relativized to the inferential capacities of some arbitrarily distinguished computers in arbitrarily restricted observational circumstances (like groups of humans doing philosophy of psychology in 1991).

A number of special sciences traffic in intentional concepts, but let us for the moment concentrate on intentionality's 'home' domain, psychology. We will stipulate that 'psychology' here refers to a province of the sciences that study evolved living systems. Its generalizations thus need not answer to counterfactuals about hypothetical computers that share no common ancestors with organisms. One does not discover facts about native human inferential capacities by attending to the capacities of, for example, quantum computers.[17] Such limits on the scope of psychological generalizations must be identified from outside psychology. But this does not entail that we have to identify them directly from the domain of metaphysics. Psychologists do not believe that humans can bend distant spoons just by concentrating. They believe this, in the first place at least, not on the basis of attending to what brains (*qua* brains) can and cannot do, but on the basis of attending to physics, which tells us, in general, what sorts of channels can carry what sorts of information and what sorts cannot. Thus physics constrains psychology.

Now, physics also constrains the domain of the computable. Computation is a kind of flow of information, in at least the most general sense of the latter idea: any computation takes as input some distribution of possibilities in a state space and determines, by physical processes, some other distribution consistent with that input. Physics can tell us about lower bounds on the energy required to effect a given computation, and since it also sets limits on where, when, and how physical work can be done, it can show us that various hypothetical computations are (physically) impossible.[18]

We will have more to say about the details of this constraining relation between physics and computation later in this chapter, but for the moment let us keep our eye fixed on the broader metaphysical picture. What makes a given special

---

[17] Litt et al. (forthcoming) argue that there are no grounds for supposing that the brain is a quantum computer.
[18] As mentioned in 3.7, Landauer's Principle is a constraint of the kind to which this paragraph refers (see Ladyman et al. 2007).

science, like psychology, the particular *special* science that it is, is its scope. By 'scope' we refer to limits around the classes of counterfactual generalizations it is responsible for supporting. Thus if some psychologists say that unaided animal brains cannot perform a certain class of computations, one does not furnish an objection to their generalization by showing that a quantum computer, or a supercomputer, or a language-equipped person embedded in a culture, could perform some member of that class. This relationship between psychology and computer science is not symmetrical. If one shows that no physically possible machine could perform a given inference, then one thereby shows that brains cannot perform it (nor quantum computers nor supercomputers nor enculturated people either). We do not know about this asymmetric constraint relationship on the basis of a priori reflection. If Cartesian dualism had turned out to be a fruitful research hypothesis then this might (or might not) have amounted to the empirical discovery that, *inter alia*, the relevant constraint in one direction didn't hold. But dualism has been an utter failure as a research programme, so psychologists are professionally obliged to believe that brains process the information they do by means of physical state transformations.

Thus we have learned empirically that computer science has wider scope than psychology. (This does not imply that thinking *reduces* to computation, but only the weaker claim that what cannot be computed cannot be thought.[19]) Again, this means that computer science generalizations are responsible for supporting counterfactuals that apply across a wider set of conditions. The limits on the scope of computer science must come from somewhere other than computer science itself, and as it happens those that don't come from mathematics come directly from physics (that is, not by way of a further mediating special science). Only a mathematical or a physical generalization can tell you what can't be computed anywhere or anytime in the universe by any process.

This fully explains why we should believe, just as Dennett says, that there are real patterns no people have discovered, and still other patterns that no people will ever discover. Our evidence to date tells us that there is computable information out there that we can't compute (and that none of our artefacts will ever compute). Much of this information is just spatio-temporally too distant from us and/or so dispersed as to be practically inaccessible. Other physically possible computations require more input of energy than we'll ever be able to muster.[20] Our evidence doesn't, of course, tell us what these inaccessible patterns

---

[19] Some readers might interpret this as endorsing old-fashioned computational internalism about thought. We intend no such endorsement. A person whose thought extends through processes occurring outside her body, as in the kinds of cases made salient by Clark (1997) and Rowlands (1999), participates in a computational process performed by a coupled organism-environment system.

[20] It is in this light that our extension of our information-processing capacities through recent and continuing massive increases in machine number-crunching power is every bit as portentous for the future of science (and for the philosophy of science) as Humphreys (2004) argues in a recent book.

are. But we violate no verificationist principle in supposing that they're out there; this just follows from what we've empirically learned about the relationship between computers in general and specific classes of computing machines.

This all says something—far from the whole story, of course—about how we can say that physics constrains psychology in a way that is not instrumentalist or reductionist but that does not require us to invoke a metaphysical principle incompatible with the PNC. Some might think that the third part of this achievement is so far trivial, on the grounds that we haven't yet made any metaphysical moves at all: we've merely *described* some working assumptions of current science. But this isn't in fact true, according to the understanding of metaphysics we've promoted. One goes a bit beyond psychology and computer science, respectively, in pronouncing on their scopes. To comment on the scope of a special science from the domain of another special science with which the scope of the first is asymmetrically related is to engage in unification, and that is what metaphysics is about, according to us.

4.3 CONCEPTS OF INFORMATION IN PHYSICS AND METAPHYSICS

Our only example so far of a practical belief disallowed to psychologists by the constraining hand of physics—brains can't bend spoons at distances—isn't interesting by the lights of the PNC. If the kind of unification at which we aim isn't to be trivial, the constraining relation had better be given enough additional content to do more substantial work. We will develop this further content by continuing to exploit Dennett's insight about the importance of an objective concept of information. This will lead us to something nascent but not realized in RP: an explicit theory of ontology, which is about as strong a device for metaphysical unification as one can request. A full section of digression is necessary before we proceed directly to this, however. The reason is that the central concept in our theory of ontology, 'information', has multiple scientific interpretations (and goodness knows how many philosophical ones). We cannot do any clear work with this concept until we have reviewed these uses with a view to regimentation for the purposes ahead of us.

Fortunately, in the context of the main topic of the present chapter, the discussion to come is not entirely a digression. As we noted at the top of the chapter, special sciences are incorrigibly committed to dynamic propagation of temporally asymmetric influences—or, a stronger version of this idea endorsed by many philosophers, to real causal processes. Reference to transfer of some (in principle) quantitatively measurable information is a highly general way of describing any process. More specifically, it is more general than describing something as a *causal* process or as an instantiation of a *lawlike* one: if there are causal processes, then each such process must involve the transfer of information between cause and effect (Reichenbach 1956, Salmon 1984, Collier 1999); and if there are lawlike processes, then each such process must involve the transfer of information between instantiations of the types of processes governed by the law. Much of the next chapter will be devoted to fleshing out each of these ideas. However, we must avoid commitment to the idea that causal processes are fundamental, since we must leave room for our promised variation of the Russell–Redhead thesis about the role of causation in physics. (As noted earlier, that thesis is motivated in a PNC-compatible way, which doesn't show it to be true but disallows us from rejecting it on the basis of its implications for the metaphysics of special sciences.) And if causal processes are taken to require actualization of effects after actualization of their causes, then there may be an important sense in which nothing is a causal process at the level of abstraction suitable for metaphysical unification, since physics motivates the hypothesis that we may live in a block universe (as we argued in Chapter 3), and then an acceptable metaphysics will have to be set within such a universe. For now, in conditions where some serious physical models seem to imply a block universe and others seem to imply irreducibly tensed processes, we will need to find a way of talking about information transfer—if we think that is metaphysically useful—in a way that makes sense in either sort of universe.[21] Evidently, this objective is more challenging in the context of the block, since talk of information *transfer* seems to imply process on its face. Accommodating special sciences to a block universe in a way that doesn't imply instrumentalism about them thus requires some work.

Some accounts of physics embrace exactly the sort of radical dynamicism that has become fashionable in the philosophy of the life sciences. There is a substantial literature in physics and around its fringes (for example, in complex systems theory of the sort associated with the Santa Fe Institute) that in different ways substantivalizes information 'flow'. (See, for a highly representative sample, the papers in Zurek 1990a.) Workers in this tradition typically conceptualize information as 'order', and thus as the complement of entropy (or, in terminology due to Schrödinger 1944, as 'negentropy'). So long as the whole universe isn't disordered, so that boundaries between 'systems' can be distinguished, energy input to a system can then be viewed as ordering it, and thereby carrying information. The Second Law of Thermodynamics is often interpreted as implying that systems cannot endogenously self-organize except by exchanging information from outside for entropy. In that case, it is sometimes said, all information found anywhere must have 'flowed' to where it is from somewhere else. It is natural to turn at this point to cosmology. If all information everywhere has flowed there from elsewhere, and if information can only flow from states that are out of thermodynamic equilibrium (another common interpretation of the

---

[21] Henceforth, talk of processes in the context of fundamental physics is to be understood as referring to irreducibly tensed features of reality.

implications of the Second Law), then it might be supposed that all information in the universe must ultimately flow from the Big Bang. A host of very strong philosophical interpretations of physics thus lurks around this literature: we seem to have a generalization of traditional causation, with an emphasis on temporal asymmetries as fundamental. Everything is process according to this account, and the block universe appears to be excluded.

By the lights of the PNC, we as philosophers must take these ideas just as seriously as some physicists do, but no more than that. That is to say, we mustn't rule out Santa Fe-inspired metaphysics, but we also must not exceed our station and take a process perspective for granted while rival conceptions—including conceptions suggesting the block universe—remain on physicists' tables. In Chapter 3, we discussed some grounds for concern about reconstructions of physics in terms of information theory: sometimes it incorporates philosophical idealism and sometimes it doesn't, but in both cases on the basis of little that a philosopher would regard as a serious argument. To convey some idea of how far this sort of thing can go, let us provide a quotation from no less a figure than J. A. Wheeler: 'Every item of the physical world has at bottom—at a very deep bottom, in most instances—an immaterial source and explanation; that which we call reality arises in the last analysis from the posing of yes–no questions and the registering of equipment-invoked responses; in short, ... all things physical are information-theoretic in origin and this is a *participatory universe*' (1990, 5). A philosopher might be inclined to tremble at the boldness of this, but for the fact that it can be interpreted to suit any taste along the spectrum of metaphysical attitudes from conservative to radical. As we said above, the root of the problem is that the concepts of 'information' and 'entropy' as we find them used across physics are highly flexible. Let us briefly canvass some of the disagreements.

On the one hand, one finds in the literature some forthright statements to the effect that there is ultimately *one* concept of information to which all others reduce *because* (i) information is negentropy and (ii) all concepts of entropy likewise univocally collapse to one—so information, on any reading, is just the complement of *that*. For example, the physicist Asher Peres (1990) defends this view. What seems to be necessary to the position is the idea that thermodynamics is as fundamental a part of physics as GR and QM.[22] Peres also asserts that all three members of this triad rely on a 'primitive' idea of time-ordering:

In thermodynamics, high-grade ordered energy can spontaneously degrade into a disordered form of energy called *heat*. The time-reversed process never occurs. More

---

[22] Some philosophers will boggle immediately at this because stock discussions in the philosophical literature on intertheoretic reduction have told them since primeval times that thermodynamics reduces to statistical mechanics. However, this is an instance of 'philosophy of A-level chemistry'. Classical statistical mechanics could only conceivably be a part of fundamental *classical* physics, and in any case the reduction of thermodynamics to classical statistical mechanics has not been achieved (see Sklar 1993). The contemporary version of the reducibility question concerns the relationship between thermodynamics and QM. That issue features prominently in the literature we discuss below.

technically, the total entropy of a closed physical system cannot decrease. In relativity, information is collected from the *past* light cone and propagates into the *future* light cone. And in quantum theory, probabilities can be computed for the outcomes of tests which *follow* specified preparations, not those which *precede* them. (1990, 346)

This strong metaphysical hypothesis allows Peres to construct a three-part argument. First, he provides a formalization of physical entropy, as defined for application to quantum theory by von Neumann, on which it is syntactically identical to Shannon and Weaver's (1949) entropy of communication channels. He then argues that the concept expressed by this formalization reduces the traditional thermodynamic notion of entropy. On the basis of this conclusion he finally argues that if Copenhagen-interpreted QM is tampered with (if non-orthogonal quantum states are distinguished, or if non-linear 'corrections' are allowed in Schrödinger's equation) then the Second Law is violated. This is intended to defend the Copenhagen interpretation by *reductio*.[23]

We are not very interested in this as an argument for avoiding hidden-variables models in QM, since the overwhelming majority of physicists is persuaded of this conclusion anyway on other grounds. Rather, we have described Peres's argument in order to offer it as an exhibit of physical reasoning within a framework that relies on strong *metaphysical* dynamicism. We say 'metaphysical' here to indicate that the view that thermodynamics is as fundamental as quantum theory and GR is not an empirical result. However, in the context of Peres's argument it is also more than an intuition or prejudice. It is pragmatically motivated: if it is supposed, then some formal similarities between thermodynamics and information theory can be regarded as non-coincidental, and strong constraints on physical hypotheses can be derived. Thus Peres's assumption is metaphysical but its justification is not appeasement of metaphysical curiosity.

The eminent physicist Wojciech Zurek (1990b) agrees with Peres in identifying so-called 'physical entropy', the capacity to extract useful work from a system, with a more abstract entropy concept—in this case, the idea that entropy is the complement of algorithmic compressibility, as formalized by Chaitin (1966).[24] Zurek also regards the formal similarity of von Neumann entropy and Shannon–Weaver entropy as reflecting physical reality, so in his treatment we get formal reification of *three* entropy/order contrasts. Let us refer to this as 'the grand reification'. Zurek expresses the physical entropy of a system as the sum of two parameters that sets a limit on the compressibility of a set of measurement outcomes on the system. These are the average size of the measurement record and the decreased statistical entropy of the measured system. He then argues

---

[23] We noted in 3.7.1 that Peres's claim about the temporal irreversibility of QM is highly contentious. Note also that the assimilation of the von Neumann entropy in QM to the thermodynamic entropy has recently been questioned by Shenker (1999), and defended by Henderson (2003).

[24] This is the same idea of compressibility as Dennett appeals to. Many writers (for example, Seife 2006) defend the idea that patternhood is the complement of entropy.

that if the latter is allowed to be greater than the former, we get violation of the Second Law. In regarding this as a constraint on fundamental physical hypotheses, Zurek thus implicitly follows Peres in treating thermodynamics as logically fundamental relative to other parts of physical theory.

We have referred to concepts of information and entropy from communications and computational theory as 'abstract' because they measure uncertainty about the contents of a signal given a code. Zurek substantivalizes this idea precisely by applying it to the uncertainties yielded in physical measurements and then following standard QM in interpreting these uncertainties as objective physical facts. However, biologist Jeffrey Wicken (1988) contests the appropriateness of this kind of interpretation.[25] Following Brillouin (1956, 161) and Collier (1986), Wicken (1988, 143) points out that Shannon entropy does *not* generalize thermodynamic entropy because the former does not rest on objective microstate/macrostate relations, whereas the latter does. Entropy, Wicken argues, doesn't measure 'disorder' in any general sense, but only a system's thermodynamic microstate indeterminateness. He claims that it increases in the forward direction in time in irreversible processes, but that this cannot be presumed for 'disorder' in general. 'Everything that bears the stamp

$$H = -k\sigma P_i \log P_i'$$

says Wicken, 'does not have the property of increasing in time. Irreversibility must be independently demonstrated' (144). He argues that algorithmic complexity as formalized by Chaitin is the proper target concept at which Shannon and Weaver aimed when they introduced their version of 'entropy'.[26] Wicken's general view is that both thermodynamic information and what he calls 'structural' information (as in Chaitin) are relevant to the dynamics of biological systems (as leading instances of 'self-organizing' systems), but that they are essentially different concepts that play sharply distinct roles in the generalization of the facts.[27]

Notice that Wicken's argument depends for its force on not taking for granted that physics incorporates irreversible processes, since if it did then one would not need to 'independently demonstrate irreversibility' in particular cases.

---

[25] Note that the argument isn't directly over the possibility of substantivalizing Shannon–Weaver information in *any* context. Rather, it concerns the appropriateness of trying to do so *at the level of macrosystem description*. Thus Wicken's objections to be canvassed raise no issues for quantum information theory.

[26] This point is entirely independent of the linkage between thermodynamic and computational reversibility embodied in Landauer's Principle.

[27] The curious reader will want to know what he thinks these roles are. He says: 'As remote-from-equilibrium systems, organisms have thermodynamic information content; maintaining their non-equilibrium states requires structurally informed autocatalytic processes. In evolution, the dissipation of thermodynamic information drives the generation of molecular complexity from which the structural information can be selectively honed' (ibid. 148). Collier (1986, 18) makes the same point. See also (Layzer 1990).

Irreversibility and governance by the Second Law are strictly independent of each other. But this ceases to be true if the Second Law is interpreted by way of the grand reification. Wicken thus implicitly questions the framework in which Zurek treats the Second Law as fundamental. Since Wicken is a biologist, it is perhaps unsurprising that the ubiquitous grip of the Second Law doesn't strike him as the first premise to pull down from the shelf in trying to understand complexity. But the physicists work with different background principles for individuating systems than do the special scientists (a point to which we soon return at length). In any case, as metaphysicians interested in unification of science, and upholders of the PPC, we cannot just help ourselves to background assumptions about 'emergence' and thereby wave aside the possible generality of the Second Law. If Peres and Zurek are mistaken, this must be shown within the context of physics. At the same time, Wicken's view concerning appropriate assumptions in biology needs to be explained by any unifying metaphysics, and must not depend on demonstrating that Zurek and Peres are wrong (lest metaphysics devolve into speculative physics). Again, given the current state of play in science, the PNC obliges us to account for incorrigibly dynamic special sciences given either a fundamentally tensed universe or a block one.

Note that it is possible to support the grand reification without taking thermodynamics to be fundamental physics. Physical astronomer David Layzer (1982, 1988) furnishes an example. He argues that the physical information is the difference between the system's potential entropy and its actual entropy, where 'potential entropy' refers to the largest possible value the entropy can assume without the system's boundary conditions becoming ill-defined.[28] Thus Layzer produces one of the plethora of functional equivalence conditions on information and entropy that we find in the literature: $H + I = H_{max} = I_{max} \equiv J$, where H denotes actual entropy, I denotes information, and J denotes potential entropy (Layzer 1988, 26). On this basis, Layzer rejects the common claim that the growth of thermodynamic entropy in the universe drives the growth of order in self-organizing systems, on the grounds that nothing rules out J increasing faster than H as the universe develops, in which case, by his definitions, information is generated. (See also Collier 1986 for a more general account of the same point.) The relevance of this argument depends in turn on Layzer's general view that the Second Law is not fundamental because it holds only given specific initial or boundary conditions. He defends a very strong symmetry principle as essential to a law's being fundamental. This is the 'Strong Cosmological Principle', according to which 'in a complete description of the physical universe all points in space, and all spatial directions at a given point, are indistinguishable' (Layzer 1982, 243). From this he derives the conclusion that a complete quantum-mechanical

---

[28] This value is guaranteed to be uniquely defined for any system, since it is a basic theorem of thermodynamics that the order of removal of constraints does not affect the value of state variables at equilibrium.

description of the universe cannot give a complete description of individual physical systems (since this would imply preferred positions in space). While this is again an argument based in speculative physics, it offers a more general form of the view discussed in Chapter 3 of the place of theories of dynamic systems in a block universe.

The foregoing tour of a few perspectives is intended to show how apparent conceptual anarchy around ideas of information and entropy mainly arises from divergence of topic emphases among different theorists. Some are mainly concerned with applications to physics, others with problems in communication and computation theory, and still others with explaining growth in biological and ecological order. It is neither Zurek's nor Wicken's 'job' to unify the sciences, and one will misread both of them—and therefore misdiagnose their disagreement—if one reads them as trying to do this job. This draws our attention to the potential scientific value of metaphysics as we have defended it. Working in their separate problem areas, different scientists have developed a range of interpretations of concepts that might be identical but might alternatively just be analogous in some respects. When, as in applications of the Second Law and its implications to biology, the PPC plays a direct role in evaluation of the plausibility of specific research hypotheses, the result is considerable conceptual uncertainty. Here is where metaphysics can help—but only stringently naturalistic metaphysics that is done for the sake of its contribution to the larger scientific project.[29]

Unification, the goal of the metaphysician, does not in this context require defending the grand reification. Though it is not our proper station as philosophers to advance our hunches as evidence one way or another for such empirical questions, we are sceptical about the basis of the reification. Is the syntactic identity of von Neumann and Shannon–Weaver entropy really evidence of anything physical? In both cases the topic is reduction of uncertainty. How surprising is it that this would be addressed in both cases by a logarithmic function? We think people should find it worrying that from one Santa Fe application of information theory to another, one finds little commonality of state variables. It is thus difficult to investigate research programmes in complex systems theory by attention to the kinds of structures that, according to OSR, furnish the basis and evidence for scientific progress. Rather, unification depends on showing how special scientists can pursue their explanatory aims without tripping over the PPC, as Wicken implicitly does by, in effect, presupposing that Zurek and Peres are wrong for reasons not motivated from within physics.

One of us, Collier, has over the years had illustrative adventures as a philosopher in these information-theoretic conceptual pathways that run between physics, computation theory, and biology. Collier has on several occasions (for example, Collier 1990, 1996) defended views that casual readers might find

[29] That is, we do not think that analysis of the *concept* of information just for its own sake is a worthwhile kind of inquiry, because the enterprise lacks stable success criteria.

hard to distinguish from Zurek's. However, Collier's aim in this work is not to follow Zurek in doing speculative physics, but to construct a conceptual fusion of material information ('bound information', in Collier's terminology) and abstract information that he can use to make sense of the contested issues in theoretical biology that concern Wicken and Layzer. On other occasions (for example Collier 1990, 2000; Collier and Burch 1998), where he discusses implications of information theory for the Maxwell's Demon thought experiment, his topic is computation theory rather than fundamental physics. We will digress a bit more on the topic of Collier's motivations, not for the sake of autobiography but to draw some lessons for the business of philosophical unification.

As described earlier, a basic idea in 'biological thermodynamics' is that self-organizing systems exchange entropy for information (conceived as negentropy). In direct contrast to Wicken's stance, this takes governance of the Second Law at all scales for granted. Collier's work on biology is likewise representative in this way. However, he has not had to take sides on the question of whether fundamental physics includes thermodynamics. His position is open to easy misunderstanding in this respect, because his leading interest has been in philosophical unification of Brooks and Wiley's (1986) project in theoretical biology[30] with wider domains of functional explanation, and Brooks and Wiley *do* assert that the physical world is fundamentally time-asymmetric. Thus a casual reader might assume that Collier is stuck with the same assumption—which violates the PPC. However, careful reading of Collier's (1986, 2003) defence of Brooks and Wiley shows that his way of reifying the information and entropy concepts is specifically restricted to contexts in which both reproductive dynamics and hierarchical structures—exactly as govern the biological domain, but *not* that of physics—yield physically instantiated, objective codes, and thus justify particular interpretations of abstract Shannon–Weaver information *in that context* as bound. This has no direct implications at all for the parallel reification in physics (for example, Zurek's).

Hopf (1988) in effect scolds all parties to this tangle *except* Collier when he reminds them that 'Thermodynamic entropy does not compel irreversibility. Irreversibility arises from kinetics. Kinetic processes must be reversible on microscopic scales. Otherwise the kinetic description would erroneously predict a final state in thermodynamic equilibrium that differs from the state of maximum

---

[30] We apologize for here having to mention a theory the reader may not have heard of, but which we will not pause to explain. The Brooks–Wiley theory concerns issues too deep inside the philosophy of biology for treatment in a book about metaphysics. All that is relevant here is that Brooks and Wiley were intensely criticized (for example, by Bookstein 1983, Lovtrup 1983, and Morowitz 1986) for, in effect, being confused about physics. Thus battle was joined along the seam of interdisciplinary unification, the territory for metaphysics as we understand it. Then, as we will see, a very high-level gloss of Collier's defence of Brooks and Wiley consists in saying that he shows how to read them as, despite themselves, saying nothing with implications for physics after all.

entropy' (265). Hopf goes on to diagnose the argument between Layzer and others over whether applications of dynamic systems theory require distinctions between 'initial' and 'boundary' conditions as 'pointless' (266). 'Kinetic systems have constraints', Hopf says; 'one wants to know what the consequences of those constraints are—end of story.' We gratefully concur with this PNC-friendly puncturing of a metaphysical bubble. Collier's defence (with corrections) of Brooks and Wiley in the philosophy of biology is independent of basic questions in the foundations of physics.

Collier has also applied his ideas on information and entropy to attempt resolutions of some problems in the foundations of computability theory (Collier and Hooker 1999, Collier 2001). This constitutes speculative physics only if one insists on viewing the universe as a kind of computer, which seems to imply denial of block universe hypotheses.[31] It is possible to discuss formal relations between thermodynamic and structural information as topics in computation theory in ways that involve no presumptions about fundamental physics (including the presumption that thermodynamics is part of fundamental physics). Caves (1990) provides a clear example. Landauer's (1961) seminal work furnishes another as we have mentioned before. That Collier does not confuse applications of information theory to (respectively) physics, computation theory, and biology with one another is indicated by the following observation. In his work on computation, he emphasizes *logical depth* (a measure of computational complexity) as the criterion for object individuation, rather than *thermodynamic depth* (the minimum amount of entropy that must be produced during a state's evolution). He appeals to the latter only as an individuation criterion for biological objects (and then, by extension, other 'special' objects), not (what we will call) fundamental physical ones. We of course doubt that there are fundamental individual physical objects at all; but we will need to show how to recover, in our metaphysical unification, the apparent commitment to self-subsistent individuals that is ubiquitous in biology and other special sciences (including non-fundamental physics). Thus we can (and will) maintain Collier's practice of adverting to thermodynamic depth in discussions of special systems, while reconstructing Dennett's idea of Real Patterns in terms of logical depth. Sketching possible empirical relationships between the domains of these two ideas, without presupposing their reification, is how we respect the PPC in this instance.

We can simultaneously show that we respect the PNC here, since we can cite instances of leading physicists who motivate our metaphysical approach from strictly physical considerations. First, Zeh (1990) shows that *if* macroscopic descriptions of the physical world describe real processes in time—including, when macroscopic instruments take quantum measurements, real collapses of

---

[31] That pragmatically isolated *aspects* of the universe can instantiate computers is obvious. Quantum information theory and quantum computing depend only on this non-controversial truth—not on any metaphysical identification of physical structure with computational structure.

real wave packets—then a consistent quantum re-description of these processes must yield irreversible processes, and, therefore a 'quantum arrow of time' and a failure of determinism in the forward direction. Alternatively, if we opt for the Everett interpretation (which Zeh interprets as being based on assuming that all time evolution is in accordance with the Schrödinger equation with a time-independent Hamiltonian), the physical facts all become fixed and indeterminism transfers to identities of observers and relative states of universal wave functions correlated with them. Then all the usual notions of entropy (information-theoretic and thermodynamic) become observer-relative. Finally, Zeh considers how the account he gives for the Everettian universe can be extended to the universe as described by a version of canonical quantum gravity where the Wheeler–DeWitt equation seems to eliminate time. Thus Zeh shows us how to understand the relationship between irreversibility and entropy-increasing measurement in both a universe with collapse as a physical (tensed) process, and a block universe; so we can theorize about variations in both logical and thermodynamic depth without presupposing either the tensed or the block universe.[32]

As a second case, consider Gell-Mann and Hartle (1990). They are interested in explaining the existence of measurement domains in which observers can gather information by making crude approximate measurements of a density matrix and a Hamiltonian, and then applying QM to (just) the decohering variables that appear in the resulting models. Their account is quantum-cosmological, and appeals to the time-asymmetric adaptiveness of such observation. For such observers, the grand reification of information concepts is natural (and useful) at their scale of observation. They will thus model the universe, at this scale, as time-asymmetric. However, Gell-Mann and Hartle are careful to note that this does not require them to model it as fundamentally time-asymmetric (though it might be, and the authors perhaps guess that it is). All that is required in the quantum-mechanical formalism is a knowable Heisenberg density matrix, concerning which 'it is by convention that we think of it as an "initial condition", with the predictions in increasing time order [readable from the syntax of the equations describing it]' (439). They go on to note that when we physically interpret the formalism we render the physical universe 'time'-asymmetric, but they are careful to formulate this in conventionalist tones (440) of a kind that would have pleased Reichenbach. Once again, then, we are shown a way—in the context of physics, not philosophy—of using

---

[32] Leggett (1995) and Stamp (1995) consider 'reversible measurements' by which they mean the coupling of macroscopic and microscopic systems without decoherence thereby producing macroscopic quantum behaviour. Leggett claims that time asymmetry in QM 'is in some sense only a special case of the more general problem of the so-called thermodynamic arrow of time' (99). Stamp agrees and thinks that, in the light of contemporary thinking about decoherence, 'the irreversibility in "quantum measurements" becomes no more fundamental, and no less subjective, than in any other thermodynamic process' (108).

concepts of information and entropy without presupposing a fundamentally tensed universe.[33]

None of this yet shows that when we set out to unify the special sciences with a physics interpreted following OSR (and thus devoid of little things and microbangings), we can achieve a non-instrumentalistic interpretation of the former in the context of a block universe. The discussion in this section may be summarized as motivating two more limited points. First, even accounts of special sciences that are most strongly dynamicist on their face need not presume tensed physics. Second, in our positive account to come our caution about speculating on physics will be reflected in our observation of the distinction between logical and thermodynamic depth. (Both concepts will be explicitly defined before we put them to work.) If future physics ratifies the grand reification in information theory (and if our scepticism about the prospects of this turns out to have been misplaced), then perhaps there really is just one notion of 'informational depth' to which the distinct concepts reduce. In the meantime, the required institutional modesty of the philosopher entails keeping them apart.

4.4 RAINFOREST REALISM

We now offer our theory of ontology—that is, a theory which makes explicit the metaphysical claim made by saying of some x, in a context where metaphysics governs the intended contrast class, 'x exists'—which makes explicit and consistent the scientific realist aspect of Dennett's reasoning in RP. The theory is stated in terms of four concepts that need special interpretation: information, compressibility, projectibility, and perspective. We provide our interpretations of these concepts before stating the theory.

On the basis of the discussion in the previous section we can now stipulate that the relevant sense of information as it appears in our theory of ontology is the Shannon–Weaver notion from communication theory. However, Shannon and Weaver provide only a theory of the capacities of channels for transmitting information in which the quantitative measure is relativized to initial uncertainty in the receiver about the source. To obtain an objective measure of informational content in the abstract (that is, non-thermodynamic) sense, one must appeal to facts about algorithmic compressibility as studied by computer science. The important measure for our purposes will be *logical depth*. This is a property of structural models of real patterns. It is a normalized quantitative index of the execution time required to generate the model of the real pattern in question 'by a near-incompressible universal computer program, that is, one not itself computable as the output of a significantly more concise program' (Bennett 1990, 142).

---

[33] Recall that we mentioned in 3.7.2 how Saunders and Hartle attempt to reconstruct the arrow of time for observers by considering them as information-processing systems.

In Chapter 3 we understood the modal structure of the world as manifest in regularities. We say more later about why regularities are the basis for something fit to be called modal structure. For now, however, let us connect our talk of regularities with Dennett's account of patterns by endorsing the spirit—but not quite the letter—of Paul Davies's (1990, 63) remark that 'The existence of regularities may be expressed by saying that the world is algorithmically compressible.' Davies glosses this by explaining that 'Given some data set, the job of the scientist is to find a suitable compression, which expresses the causal linkages involved' (ibid.). As we will discuss extensively in Chapter 5, reference to 'causal links' at the level of fundamental metaphysics presupposes a time-asymmetric universe instead of just leaving that possibility open. We will therefore substitute 'information-carrying' for 'causal'. Note that Davies's own immediately following example in fact makes no appeal, not even implicitly, to causal structure: 'The positions of the planets in the solar system over some interval constitute a compressible data set, because Newton's laws may be used to link these positions at all times to the positions (and velocities) at some initial time. In this case Newton's laws supply the necessary algorithm to achieve the compression' (ibid.). Those regularities that Newtonian theory correctly identified, stated as they are relative to just one allowable foliation of spacetime, are themselves further compressible by SR, which is similarly still further compressible by GR.

As scientific realists we understand the foregoing as referring to compressibility by any physically possible observer/computer—that is to say, given our verificationism, compressible *period*, rather than compressibility *by people*. We explain our restriction to 'physical', as opposed to logical or mathematical, possibility later in this chapter.

Now we introduce *projection* and *projectibility*. We will make use here of our previously introduced concept of *locating* (see 2.3.4). Consider a mathematical structure S that is used to model some aspect of the world. As soon as S is minimally physically interpreted by a function that maps some of its elements onto real patterns then those elements of S are locators. Recall that we claimed that such a locating function will have a dimensionality.

The notion of dimensionality here is not in general that of the dimensionality of a vector space, nor of a differential manifold, although these are paradigm examples of structures that are used to locate real patterns. What both these kinds of dimensionality have in common is that they pertain to the number of parameters that must be specified to pick out an element of the structure in question, by specifying the weights attached to a set of basis vectors in a linear combination of them, or by specifying the values of coordinates respectively. When a real pattern is located relative to some structure that models patterns of the relevant kinds, enough parameters must be specified to distinguish it from other real patterns in the same domain. As we mentioned in 2.3.4, in practice it is a pragmatic matter how precisely such distinctions must be made. Precision may merely be a matter of resolution, or it may be a matter of how fine-grained the distinctions between patterns are required to be. The former case is exemplified by grid references on maps that only resolve locations to (say) the nearest hectare. The latter case is exemplified by the case of the locater 'eagle' in contexts where the ambiguity of the locator with respect to eagle species is not significant, even though the more precise distinctions among such patterns could be made *in the same structure*.[34] In general, the dimensionality of a locating function for some structure is the maximum number of parameters needed to make the distinctions necessary in the most precise contexts in which real patterns are studied using the structure.

We will provide two examples, one involving locators that are precisely specified in a formal theory, the other regimented only by practical experience. Consider, first, the 'normal form' (matrix) representation of a so-called 'inspection game' as defined by classical game theory and used to model a variety of social situations (monitoring of workers, discouraging athletes from using steroids, reducing drunk driving, etc.[35]). Picking this game out from all the other games in 'normal-form-game space' requires specifying parameters along three dimensions: a definite number of players, a definite set of available strategies for each player, and a set of payoffs (one per player) at each possible combination of strategies. Now suppose I want to consider the extensive (tree) form of the inspection game. This object lives in a space of higher dimensionality since now two new ranges of parameters must be specified: the order of play and the distribution of information among the players about the game's structure. Thus we say that the locating function for extensive form games has higher dimensionality than the locating function for normal form games. In addition, since a normal form game will generally map onto an equivalence class of different extensive form games, modelling the world in terms of the latter involves examining it on a finer grain of resolution. (In general, we can only talk meaningfully about comparative grains of resolution in cases where there are mapping relations of this kind.)

Our second example will illustrate other aspects of the locating function and its dimensionality. Both biologists (Wheeler and Meier 2000) and philosophers

---

[34] It is not pragmatic what patterns are real. 'Eagle' is a real pattern and so is 'golden eagle'. The former is not redundant because there are projectible generalizations about eagles that are not captured by the conjunction of all the generalizations about particular species of eagles. For example, most eagles have feathers on their feet, hence we can predict that were some new species of eagle discovered its members would probably have feathers on their feet; but this is not predicted by the conjunction of the generalizations about each of the known eagle species because the crucial information on which the generalization is based is that both they, and the new species, are all *eagles*—that is, a class over which there's independent biological reason to project inferences to new members.

[35] A social situation instantiates an inspection game wherever the behaviour of an agent X in a set of n agents, $n \geq 1$, will conform to the preferences of another agent Y whenever X is sure that Y will monitor X's behaviour, but will conform to Y's preferences with a probability $p < 1$ whenever X's estimate of the probability of being monitored by Y is also <1.

(Wilson 1999) have repeatedly drawn attention to the plethora of species concepts (phylogenetic, ecological, morphological, with alternative partitioning schemes available within each of these principles). Suppose we call a species concept 'well defined' if there is some pre-specified range of information I associated with the concept such that, whenever scientists have I with respect to any set of organisms, the concept will partition the set into species. Then I will in each case be a function that maps sets of organisms onto an n-dimensional 'species space'. This is a locating function. n will have different values for different species concepts. The higher the value of n the larger the number of species there will be into which sets of organisms may be partitioned. Suppose that two biologists differ over whether two similar songbirds are members of the same or of different but closely related species. If they have the same genetic and morphological data, and are using the same species concept and agree on its dimensionality, then this implies that they disagree concerning some putative fact in the history of evolution. However, they might instead be applying locators of different dimensionality.[36] As we develop our analysis of real patterns below, we will suggest a basis for resolving such disputes, where they are amenable to rational resolution. In general, we may distinguish two kinds of cases of comparative relationships among species concepts. There are cases in which one scheme of larger n is a resolution of another scheme of smaller n, but the schemes do not cross-classify any sets of organisms relative to one another. There are other cases in which two species concepts (whether of same or different n) cross-classify some sets of organisms relative to one another. Then *pluralism* about species (as defended by Dupré 1999 and many others) is true just in case at least two cross-classifying schemes locate real patterns. (Masses of empirical evidence suggest that pluralism *is* true.) The general points illustrated here are that the dimensionality of locators is motivated by the empirically determined complexity of the spaces in which systematic discriminations serve scientific purposes (about which we will be more specific in a moment); and that attending to the dimensions of locators allows us to compare the ontological implications of their successful use without commitment to general reductionism or even supervenience.

Let us understand the idea of projection on S as requiring computation of the value of one or more measurable properties of an element y_L located at a point on S specified in terms of dimensionality D, given input of a measurement taken by a physical computer M of another element x_L located on S under D. Computation is necessary but not sufficient for projection. Projection requires not only that

---

[36] In cladistics, disagreements are usually about relative parsimonies of alternative phylogenetic trees. Sometimes disputants assign different conditional probabilities to relationships between priors and predictions based on rival conceptions of how evolution generally works. But other arguments that look very similar or identical on the surface are really about the appropriate dimensions of analysis for discrimination, and these different judgements are encapsulated in competing software packages for generating trees from data.

the value of a parameter of y_L be output given measurement of x_L, but also that the probability of that value being correct within a scale of resolution R (on D) given the measurement of x_L be greater than the probability that the value M would give as output in the absence of the measurement would be correct within R. Let $x_L \to y_L$ denote a *projection* from x_L to y_L, where projection refers to any effected computation that yields y_L as output given x_L as input. Then we will say that $x_L \to y_L$ is *projectible* if (i) there is a physically possible M that could perform the projection $x_L \to y_L$ given some R and (ii) there is at least one other projection $x_L \to z_L$ that M can perform without changing its program. (Condition (ii) is necessary to avoid trivialization of projectibility by reference to an M that simply implements the one-step inference rule 'Given input x_L, output y_L'. Projectibility, unlike projection, is a modal notion and so has stronger conditions for applicability.) Intuitively, then, projectibility is just better-than-chance estimatability by a physically possible computer running a non-trivial program. Estimatability should be understood as estimatability in the actual world that science aims to describe, not as estimatability in some class of merely possible worlds.

Our notion of projection is compatible with that discussed by Barwise and Seligman (1997, 47).[37] Projection is a particular kind of 'information flow', in their terms. However, their primary concern is with properties of computers that make them possible mediators of (more or less efficient) information flow. Here, we are interested instead in properties of information structures that warrant inference to real patterns. 'Projectibility' is the concept of information-carrying possibility—applied now not to channels but to models of real patterns and ultimately to real patterns themselves—that we will use.

Projection is related to counterfactual-supporting generalization by means of a special concept of *perspective*. Consider an observation point X from which x_L is located by M. The model of X may be either coincident with a region on S, or may be external to S and so referenced from somewhere else in S_p, the superset of all structures endorsed by current physics.[38] (The reader is advised not to take the idea of locations in spacetime as her exemplar of what we mean by 'a region on S' despite the fact that most regions on S that will be referred to in theoretical generalizations in science are spacetime locations. An example of a region on S that is not a region of spacetime is the vantage point from which a historian considers Napoleon. This will include the historian's spatio-temporal relation to Napoleon since such specification will be relevant to which information about Napoleon the historian cannot access. However, it will also include facts about Napoleon that are included simply because this historian

---

[37] Specifically: 'projection' in our terms denotes the extensional equivalent of the class of transformations for which Barwise and Seligman give intensional conditions.

[38] S_p is the maximally consistent conjunction of all theories currently endorsed by physics. A way of expressing the PPC in the context of the concepts now being developed is to say that no science, except physics, concerns itself with perspectives that lie outside of S_p.

happens to know them, and they therefore enter into inferences about Napoleon that the historian may make given new information she receives.) We will refer to such points of observation as 'perspectives'. By 'what is located from X by M' understand now 'whatever is measured at X by M as input to some computation going on or initiated at X by M'. A 'computation' need not involve a mind, where by mind we mean a non-optional intentional-stance characterization of patterns of behaviour. Any physical system that a mind could use for processing information, if there were any minds around, is to count as a computer. X is distinguished, in the first place, by reference to the projections that can be performed from it by M when M is a universal computer running a program that asymptotically approaches incompressibility. Thus X given a coarse-grained R on S_p will be an equivalence class of perspectives $\{X'_1, \ldots X'_n\}$ indexed relative to a finer-grained R on S. Necessarily, the upper bound on the logical depth of patterns projectible from X will be lower than the upper bound of patterns projectible from each of the $X'_1, \ldots X'_n$.[39] ('Coarse graining' and 'fine graining' will always be relative to physical interpretations that it is the job of physics, not metaphysics, to justify. For the relevant physical sense of differential 'graining' see Gell-Mann and Hartle 1990, 432.) This captures Dennett's version of scale relativity of ontology. Then projectibility implies counterfactual generalization as follows. Suppose there is an x_L such that M can *simulate* x_L by modelling S under R from the perspective of X. (That is, M need not 'go to' X physically. It need merely enter as input some parameters that it would transduce if it were at X.[40]) Then if $x_L \rightarrow y_L$ is projectible, any simulated input of an estimation of x_L to M implies simulated output of an estimation of y_L with higher probability than obtains when x_L is not simulated. If there are exceptionless laws of nature, these will be the special cases of counterfactual-supporting generalizations in which the probability of the output estimation given the input estimation is 1; this then also holds in simulation.[41] (We will consider more general issues about laws in 5.5.)

Note carefully that this last conditional about exceptionless laws isn't a theorem we imagine deriving from some model of M. It is an analytic stipulation that incorporates an aspect of our version of verificationism. We are asserting that what it is to say that something is a counterfactual-supporting generalization is that it is projectible by the maximally efficient computer that is physically capable

---

[39] If this doesn't hold in some particular case, then the $X'_1, \ldots X'_n$ are redundant and should be reduced to the model of X. We will justify this claim later in the chapter.

[40] Given what we have just said X is, the reader may well wonder how to distinguish physical from virtual occupancy of X by M. The answer is that virtual X is a model of physical X, suppressing some of the information available at physical X. We intend the initially counterintuitive consequence of this: a perfect simulation of a perspective (namely one that has all the same information content as the target) thereby just *is* that perspective and *physics* should not be expected to distinguish them.

[41] M might or might not 'know' that this is so. In general, M might not 'know' that it 'knows' anything. But, more specifically, if it does 'know' that it 'knows' some things, it might 'know' a law without 'knowing' that what it 'knows' is a law.

of performing the projection in question (given the actual physical limitations on computers). The motivation for this stipulation will be obvious in a moment, when we put it to work.

Now we have introduced the special concepts we will need for our theory of pattern-reality. Its first, crude, version was given by Ross (2000). Here is the original statement:

To be is to be a real pattern; and a pattern is real iff[42]

(i) it is projectible under at least one physically possible perspective; and

(ii) it encodes information about at least one structure of events or entities S where that encoding is more efficient, in information-theoretic terms, than the bit-map encoding of S, and where for at least one of the physically possible perspectives under which the pattern is projectible, there exists an aspect of S that cannot be tracked unless the encoding is recovered from the perspective in question.

We will now formulate this theory more precisely in light of the technical notions we introduced in the present section.

First, with respect to what precedes the biconditional. Recall from Chapter 2 that x_L is a mapping from the formal to the material. x denotes the relevant material-mode entity here; and now our theory says that x must be a real pattern. If the whole universe were just (*per impossible* given the capacity of anyone in such a universe to theorize about this) one black box, we would be in the Kantian situation with respect to the one real pattern: all we'd be able to do is locate it.[43] In the actual case, we understand the 'x in itself' (if you like—but see n. 43) recursively, by reference to $x_L \rightarrow y_L$. So $x \rightarrow y$ is the schema for an 'epistemically basic' real pattern.[44] Now: let the models of all x and y in the schema $x \rightarrow y$ be instantable by any sentences of first-order logic. Suppose that reality is closed under transitivity of the projectibility relation among models of it,[45] so that if $x \rightarrow y$ and $y \rightarrow z$ satisfy clause (i) above for pattern-reality, then so does $x \rightarrow z$. Then without loss of generality we may modify what precedes

---

[42] In the original formulation in Ross (2000), the conditions stated only necessary, rather than necessary and sufficient, conditions on pattern-reality—just as we complained is the case in Dennett's account. We will not go into details here about why we think that Ross's earlier caution was unwarranted, except to say that Ross then had non-verificationist scruples about in-principle undiscoverable mathematical structures that he has since got over.

[43] Then, perhaps, Kantian idealism would be in order and we should drop talk of real patterns altogether. Well, perhaps. But this enters the territory of precisely the sort of philosophy against which we inveighed in Chapter 1. We introduced the idea of the universe as one black box to motivate the syntax of our model pedagogically, *and that's all*. We cannot usefully think, in a PNC-compatible way (hence, cannot usefully think *period*), of a world in which all that exists is one black box and yet somebody has a philosophical theory about it.

[44] We use 'basic' in this book *only* to designate epistemic status; so the reader should not conflate this with our use of 'fundamental' as in 'fundamental physics', which reflects ontology.

[45] If it isn't, we have disunity *à la* Cartwright (1999) and/or Dupré (1993).

the biconditional to read 'To be is to be a real pattern; and a pattern $x \to y$ is real iff ...'

Next, the definition of projectibility allows clause (i) to be simplified. Both projectibility and perspectives were defined above in terms of physically possible projectors. We have yet to address the philosophical warrant or implications of this. However, modulo this coming explanation, clause (i) is evidently now phrased redundantly, and should be modified to read simply 'it is projectible'.

The original formulation of clause (ii) is clearly insufficiently explicit. It remains useful, we believe, as an intuitive gloss of what the theory of ontology aims to do. However, we must now try to state it more transparently.

First, what can it mean to say that a real pattern 'encodes information' about a structure? In *one* sense this formulation is indeed backwards given the new battery of concepts we've introduced here. From the point of view of someone working with a scientific theory, it's the structure (formal mode) that 'encodes' information about a real pattern (material mode).[46] However, we don't want therefore to surrender the idea that real patterns bear Collier's 'bound' information. For the reasons canvassed in 4.3, they *might* do so in the sense supposed by Peres and Zurek (which might in turn be the sense intended by Wheeler, for all we can tell), but we don't want to presuppose this. Nevertheless, if someone claims to have no warrant for regarding real patterns as bearing bound information in *any* sense, then they deny *minimal* realism. That is, they deny that the projectibility of $x_L \to y_L$ is based on any sort of transduction of properties of x; for if transduction isn't itself a form of projection then no one who isn't a follower of Berkeley has the slightest idea what it might be.[47] (And Berkeleians need an alternative idea, viz. the mind of God, that the PNC excludes straightaway.) Fortunately, we have PNC-compatible reasons for thinking that some real patterns might carry bound information directly: quantum information theory is often so interpreted, with some justification as we noted in 3.7.5. Thus we have PNC-compatible motivations for this *very* minimal level of realism. (Notice that it's even minimal enough for van Fraassen to allow, since without it he'd have no reason to be epistemically impressed by observation as he surely is even if he resists the idea that empiricism be defined as the claim that 'experience is our sole source of information about the world' (2002).)

Thus we want, in the new formulation of clause (ii), to go on saying *something like* what that clause now says about the information 'in' real patterns. But 'encode' is surely misleading; and since we have now adopted 'structure' in its technical guise from OSR, we must no longer say that real patterns carry information *about* structures. So what does a real pattern carry information about? The answer can only be: about other real patterns. As discussed in 4.2, the

---

[46] Unless S models the whole fundamental universe, in which case we should talk of real pattern*s*.

[47] Note, for completeness, that an occasionalist like Malebranche has the same idea of transduction as everyone else, but denies that it actually goes on.

core intuition we have extracted from Dennett is that, his own wayward remarks notwithstanding, the distinction between illata and abstracta has no scientific basis. This just means, to put matters as simply and crudely as possible: it's real patterns all the way down.[48]

This slogan could summarize the whole conclusion of this book, so we can't ask the reader even to understand it immediately, let alone believe it. For the moment we are mainly concerned with *formulating* part of our view, but this effort can't be separated from explicating its point. Let us thus pause in our technical reconstruction for some motivating meditations.

Self-evidently, a statement of the necessary and sufficient conditions on the reality of patternhood cannot make first-order reference to real patterns. It can, however, make reference to 'patterns' simpliciter, because it is as easy as can be for us to say what a pattern is: it is just any relations among data.[49] That there is such a distinction between real patterns and mere patterns (OSR), and how it is to be drawn (the subject of this section, 'Rainforest Realism'), and that nothing else about existence in general should be said (PNC), is the content of the metaphysical theory defended in this book.

Consider, as an example, the many curves that can be drawn through the past price data on a stock market. All of these curves, assuming they link accurate data points, are patterns—that is, records of past values and whatever mathematical relations hold among them. What the financial economist or stock analyst wants to know, however, is which of these curves can be reliably projected forward—which ones generalize to the unobserved cases. This is of course a bland and familiar point. (A standard example of a non-projectible functional relation is Elliot Sober's (2001) hypothetical case of reliable covariation between Venetian sea levels and the price of bread.) But why do we want to associate the distinction between projectible and non-projectible patterns with the distinction between what's *real* and what isn't? Is this not just an eccentric semantic preference on our part?

The semantic preference may indeed be unusual, but we think it well motivated. The main ontological implication of OSR is that reality is not a sum of concrete particulars. Rather, individual objects, events, and properties are devices used by observers (when these observers aren't making mistakes)

---

[48] Ross suggested this slogan to Dennett in correspondence in 1999, and Dennett assented enthusiastically. Since this exchange wasn't public, we're sure that Wallace's (2003a) promotion of the slogan is independent. (Although in 3.7.3 we argued that his own view may not be consistent with it.) In fact, we suggest that someone would have to be poorly educated in twentieth-century philosophical rhetoric if, after a careful study of RP and the critical discussion of it in Dahlbom (1993), this slogan did *not* occur to them. (Cf. similar slogans we quoted from Saunders (3.7.3) and Stachel (3.4), the idea of 'dynamics all the way down' introduced in 3.7.2, and the idea of 'correlations all the way down' in 3.7.5.)

[49] Social scientists often refer to 'mere correlations'. However, this usage is not suitable for us since we use the notion of 'correlation' in Chapter 3 to refer to any circumstance in which $P(A/B) > P(A)$, where $P(.)$ is a probability and not merely a finite relative frequency.

to keep cognitive books on what science finds to be sufficiently stable to be worth measuring over time, viz. some but not other patterns.[50] Some of these patterns are indeed conceptualized as individuals in some special sciences, while simultaneously *not* being so conceptualized by other special sciences making projections at other scales of resolution. For example, from the point of view of a behavioural economist measuring choices, a person is an extended pattern over a run of data, whereas at the scale where 'common sense' is most comfortable but also at which social psychology is done, a person is the paradigm case of an individual. When we deny that individuals are 'real' we do not mean that, for example, Napoleon is *fictitious*. Observers tracking him in 1801 could get lots of highly useful leverage projecting the pattern forward to 1805; so (sure enough) Napoleon is a real pattern. What OSR denies is that real patterns resolve 'at bottom' into self-subsistent individuals. Furthermore, we must put 'at bottom' inside scare-quotes, because we find the levels metaphor misleading. *The single most important idea we are promoting in this book is that to take the conventional philosophical model of an individual as being equivalent to the model of an existent mistakes practical convenience for metaphysical generalization.* We can understand what individuals are by reference to the properties of real patterns. Attempting to do the opposite—as in most historical (Western) metaphysical projects—produces profound confusion. (Witness the debates about identity over time, identity over change in parts, and vagueness, none of which are PNC-compatibly motivated problems.) Or, at least, it produces such confusion in the context of scientific theorizing—which is to say, in precisely the institutional context in which we achieve objectivity.

Thus, saying that some patterns are real in a way that individuals are not is intended to express the idea that individuals are resolved out of patterns rather than vice versa. Now let us return to considering 'mere' (non-projectible) patterns in light of this point. From the ontological point of view, a non-projectible pattern exactly resembles the traditional philosophical individual. There it stands, in splendid isolation. One can point at it: lo, there is a curve drawn through this finite set of observations. Abolish the rest of the universe, and it remains as ostended, all its properties intact.

To live as a cognizer is to de-contextualize, to foreground selected elements from the network of reality and build referential structures around them. Peter Strawson (1959) brilliantly illuminated the impossibility of practically effective thought that doesn't organize experience into individuals. But the moment we forget that we do this for practical reasons grounded in our teleological circumstances, we encourage ourselves to forget that objectivity consists in finding reliable *generalizations*. 'Reliable' here means: prediction-licensing and counterfactual-supporting. Tracking an individual such as Napoleon is

---

[50] We recognize that the significance of stability requires both further motivation and further explication. This will come in 4.5.

epistemically helpful precisely in so far as doing so supports some predictions and counterfactual claims about him. The ontological truthmaker for the epistemic fact is thus that he is a real pattern. But then a non-projectible pattern through a finite data set is exactly the paradigm case of a chimera. By exact semantic symmetry, we find it appropriate to say that non-projectible patterns are epistemically useless because they are not real. Common-sense usage by no means always disagrees with ours here. 'Don't bet on that trend in the market data,' your broker might tell you—'It's not real.' The broker of course doesn't mean that what the data record didn't actually happen; she means that the pattern in them isn't projectible.

Hoping that the reader now has a decently clear idea of what's at stake, let us return to a consideration of clause (ii). The first part of its formulation is 'it [the would-be real pattern] encodes information about at least one structure of events or entities S'. We have now explained the basis for replacing this by the following: 'It carries information about at least one pattern P'. Notice that this allows that the pattern P about which the real pattern carries information might be itself. Our leaving this loophole open anticipates some issues we will later confront in connection with fundamental patterns describing the whole universe. It is *not* motivated by concessions to philosophers who believe in 'ineffable' patterns tracked by special sciences (that is, 'qualia').

Since many or even most philosophers might have jumped to the latter interpretation of our new formulation had we not explicitly blocked it, we can illuminate another aspect of our view by explaining the point. Philosophers' favourite examples of putatively incompressible special-science patterns are so-called 'sensory qualia'. Suppose, to give these philosophers as plausible an example as we can allow without ignoring the PNC, that there is an aroma someone can reliably re-identify, but which can't be compressed in a linguistic description or tagged by correlation with a neural signature because its verification (as *that* odour) requires multi-modal integration of environmentally distributed, multiply realized information. Then we must ask: does re-identification of the odour in question facilitate inference about *anything else* or not? If it does ('Mom baked here recently!' or even just 'I smelled this in the old country'), then the (real) odour carries (and compresses) information about at least one *other* pattern P. If it doesn't facilitate any such inference, then our verificationist instincts are aroused. How could one justify, just to oneself, one's confidence that it re-identifies any more general pattern at all? 'I could swear I've smelled this somewhere before' does not constitute *evidence* of anything, since neither the subject nor the recipient of the report could tell whether it indicates re-identification or confabulation. This is so *in principle* by virtue of our stipulation that the aroma can't be described or be independently detected by a neuroscientist. (And if we don't stipulate that then the example ceases to be interesting for the imagined purpose.) Distinctions that can't be marked in principle are of no interest to science even in the limit, and what is of no interest to science in the limit doesn't exist. (As Dennett has insisted over and over again—but now putting his point in our terms—pure metaphysical epiphenomena are not only not real patterns, they are not even *mere* patterns.)[51]

It isn't necessary to block epiphenomena in clause (ii) because they're already excluded by clause (i): epiphenomena aren't projectible, even if people mistakenly try to project them. Again, we allow some real patterns to carry information only about themselves because we would otherwise disallow physically fundamental patterns. In particular, the whole universe would be declared not to be real.

The main point of the rest of clause (ii) is to capture Dennett's insight from his work on the intentional stance (the insight on which he equivocates in responding to the eliminative materialist in RP), that to be real a pattern must be informationally non-redundant, that is, must be *required* if some counterfactual-generalization-supporting information is not to be lost.[52] Clause (ii) is thus restrictive, aimed at blocking the implications of Dennett's flirtation with instrumentalism that encourages an ontology made anthropocentric by reference to possible degrees of epistemic laziness, and made infinite by the infinite number of possible degrees of such laziness. Furthermore, there are two kinds of redundant pattern we want clause (ii) to exclude: those that are mere artefacts of *crude* measurement that bring no informational gain—the cases Dennett wrongly lets in—but also putative patterns that are just arbitrary concatenations of other patterns. To borrow an example of the latter from Ross (2000), the object named by 'my left nostril and the capital of Namibia and Miles Davis's last trumpet solo' is not a real pattern, because identification of it supports no generalizations not supported by identification of the three conjuncts considered separately. All these considerations conspire to show that it is clause (ii) that is supposed to make our account a variety of realism.

The concept that captures informational non-redundancy was introduced earlier: logical depth. We respect Occam's razor with respect to patterns by regarding as real only those that minimize the overall logical depth we attribute to the world while nevertheless acknowledging all projectible correlations. We achieve this by replacing the original wording 'where that encoding is more efficient, in information-theoretic terms, than the bit-map encoding of S, and where for at least one of the physically possible perspectives under which the pattern is projectible, there exists an aspect of S that cannot be tracked unless the

---

[51] A mere pattern is a locatable address associated with no projectible or non-redundant object.

[52] Notice that we do *not* refer here to counterfactual-supporting *or* prediction-supporting information. The only way in which a pattern could inform non-redundant predictions from a perspective without supporting non-redundant counterfactuals from that perspective would be by reference to epistemic limitations of some particular group of predictors. This kind of case is relevant to philosophy of science, but not to metaphysics. Consider the case of the non-projectible pattern in past market data. Someone ignorant of the fact that the pattern isn't real might use it to make predictions, and might luckily be right. However, the non-reality of the pattern implies that it doesn't support counterfactual claims about what would have obtained had some data in the pattern been different.

encoding is recovered from the perspective in question' by the following: 'in an encoding that has logical depth less than the bit-map encoding of P, and where P is not projectible by a physically possible device computing information about another real pattern of lower logical depth than [the one at hand, viz.,] $x \to y$'. Note that the reference to real patterns in this formulation does not refer to the instance being analysed; therefore this is a case of recursion rather than vicious circularity as we pointed out above.

The above formulation, like its less precise ancestor, blocks both kinds of redundant patterns from being taken as real. If a pattern is crudely demarcated and permits of refinement, then the requirement of minimal logical depth excludes it. Thus, for example if eliminative materialists about intentional patterns turn out to be right, and all projectible patterns tracked by intentional psychology along with others it doesn't track are projected by a mature cognitive-behavioural neuroscience, then by clause (ii) intentional patterns aren't real. The instance of the pattern cobbled together out of three real patterns is not itself a real pattern because it increases logical depth in exchange for no gain in projectibility.

Note also that clause (ii) does *not* exclude traditional so-called 'high level' patterns (for example, traffic jams, bird migrations, episodes of price inflation, someone's belief that Tuesday precedes Wednesday) that arise only on broad grains of resolution *if* these pay their way in terms of projectible information. Only those patterns whose individuation is motivated by nothing other than epistemic limitations or pragmatically driven epistemic laziness are refined out of reality as *crude* approximations.

Finally, note that the clause makes reference (here following Dennett) to 'the' bit-map encoding of P. A bit-map encoding is a zero-compression encoding, a one-to-one mapping from each bit of information in the pre-encoded representation of P to a distinct bit in the encoding. Thus one can meaningfully talk of 'the' bit-map encoding of P only relative to some background structuring of P. Such background structures are always presupposed in scientific descriptions; they are the locating systems introduced in Chapter 2. For an advocate of disunity clause (ii) will have no ontological bite, since she supposes that we are always free to choose new locating systems in describing reality, thus leading to alternative encodings of P that are all bit-map encodings relative to the different locating systems. Hence the importance of our reference to 'regarding as real only those patterns that minimize the overall logical depth we attribute to the world while nevertheless acknowledging all projectible correlations'. In the context of commitment to unifiability of science as a regulative norm, the modal force of 'projectible by a physically possible device' has teeth. For any given real pattern $x \to y$, there will be a locating system that attributes just enough logical depth to the world to allow projection of all data projected by $x \to y$. The claim that the world is unified is equivalent, under our verificationism, to the claim that there is a unique such efficiency-promoting locating system. Of course, we do not know what this is. If we did, and Melnyk's core-sense reductionism (CR) holds (see 1.5),[53] then this would imply that we had a complete and final physics; if we knew it and CR does *not* hold, then this would imply a complete and final unified science going beyond a merely complete and final physics. However, our inability to identify *de re* the target of the regulative ideal against which we acknowledge existence is not embarrassing to realists.

Just one further tweak is necessary to complete the new statement of the theory. We explained above why it is appropriate to attribute structural information directly to real patterns, and not just to representations of them. However, in cases where we are wondering about the reality of a pattern, actual observers may have access only to a model of the pattern in question. (Consider, for example, our access to the real pattern that is the interior of the Sun.) We don't want to legislate applicability of the analysis as out of reach in such cases. On the other hand, we also want the analysis to apply potentially to patterns over which no model has been constructed (though in such cases no one will typically know that the analysis applies to the case). If we stipulate as sufficient for real patternhood that the candidate pattern has a model that carries information about another pattern, and meets the rest of the sufficiency condition as stated immediately above, then this satisfies both desiderata, since every pattern is a model of itself. However, the second invocation of logical depth should apply only to real patterns directly, since here we want to restrict existence to patterns that are not redundant given the actual universe—we mustn't rescue some patterns from exclusion just because their logically shallower surrogates haven't yet been modelled, on pain of repeating Dennett's instrumentalist equivocation.

Thus the more precise statement of our theory of ontology is as follows:

To be is to be a real pattern; and a pattern $x \to y$ is real iff

(i) it is projectible; and

(ii) it has a model that carries information about at least one pattern P in an encoding that has logical depth less than the bit-map encoding of P, and where P is not projectible by a physically possible device computing information about another real pattern of lower logical depth than $x \to y$.

Ross (2000) called the original analysis of pattern-reality 'Rainforest Realism' (RR). We will continue here to refer to it by that name; and explaining the motivation for the label will help the reader to place its motivations and implications in a wider philosophical context. As noted, clause (ii) of the analysis expresses the principle of Occam's razor, in restricting ontological commitment

---

[53] 'All nomic special- and honorary-scientific facts, and all positive non-nomic special- and honorary-scientific facts, have an explanation that appeals only to (i) physical facts and (ii) necessary (i.e. non-contingent) truths' (Melnyk 2003, 83).

to what is *required* for a maximally empirically adequate science. The razor has sometimes been interpreted in a stronger way, as suggesting that ontologies should be not just restricted but *small*. Quine frequently expressed this idea by appeal to metaphors involving a preference for deserts over jungles. Thus at the end of *Word and Object* (1960), after the scientist has been assigned sole responsibility for populating our collective ontology (just as our PNC requires), the philosopher is depicted as the janitor of the enterprise, coming into the workspace left by the scientist and 'smoothing kinks, lopping off vestigial growths, clearing ontological slums' (275). Elsewhere, Quine says that 'cognitive discourse at its most drily literal'—that is, in his terms, the voice in which quantification is intended with metaphysical force—'is largely a refinement ... characteristic of the neatly worked inner stretches of science. It is an open space in the tropical jungle, created by clearing tropes away' (Quine 1978, 162). As mentioned earlier, Quine thought that the state of science in the 1950s and 1960s motivated speculative ontological reductionism and atomism, according to which all that ultimately exists are physical particles. However, our analysis of projectibility, incorporated into the theory of ontology, is (following Dennett's lead) deliberately made compatible with the scale relativity of ontology. If this speculation is better motivated by contemporary science than Quine's now is, as we have argued, then ontology should exclude many of the things people (including scientists) talk about for practical purposes, but is substantially larger than what people can ever be expected to be in a position to specify. Ours is thus a realism of lush and leafy spaces rather than deserts, with science regularly revealing new thickets of canopy. Anyone is welcome to go on sharing Quine's aesthetic appreciation of deserts, but we think the facts now suggest that we must reconcile ourselves to life in the rainforest.

Some metaphysicians will complain that our forest isn't verdant enough. RR restricts existence to patterns that support informational projection under some *physically*—as opposed to logically or mathematically—possible perspective. This expresses our Peircean verificationism. No empirical science is responsible for counterexamples drawn from just anywhere in the possibility space allowed by currently accepted logic and mathematics. Empirical sciences thereby institutionally incorporate a verificationist rather than a Platonist conception of reality. For a Platonist, the boundaries of the real are the limits of what is mathematically coherent. The verificationist objects that there is no perspective from which the limits on possible mathematical patterns can be identified. Metaphysical speculations on such limits (to be distinguished from formal demonstrations of the limits on what is implied by particular ways of writing out axiomatizations) are not empirically verifiable. This, we suggest, is the most plausible version of the conviction that some philosophers—for example, positivists—expressed with the idea that logic and mathematics are not *empirical* sciences.

Following our usual approach, we define 'empirical' recursively and by reference to institutional practices of scientists. Empirical sciences are sciences that motivate hypotheses by measurements of contingent magnitudes. (By this we mean contingent as contrasted with logically necessary; many are not contingent relative to physics, and perhaps, for all we know now, none are.) No empirical science explains patterns that are not empirically motivated. The only empirical science that no other empirical science may contradict, as a matter of institutional fact, is physics (and the scope of fundamental physics is also defined recursively, as whatever is measured by those parts of physics that institutionally constrain every empirical generalization). This of course just re-states the PPC. A pattern that could not be tracked from any perspective that current physics tells us is possible is a pattern whose existence cannot be empirically verified as far as we now believe. Such patterns are therefore not possibilities that empirical scientists should take seriously. By the PNC, these are therefore also patterns that metaphysicians should not take seriously. A non-idle example here would be the pattern some people try to refer to when they talk about God: this is a physically uninteresting idea, so it is also metaphysically uninteresting. (It is of course anthropologically interesting.)

The above reasoning rested on the supposition that limits on mathematical patterns are not empirically verifiable. What is supposed to justify this idea? Did Gödel, among others, not identify such limits by an entirely non-speculative procedure? Gödel, like many mathematicians much of the time, did two things at once. He indeed showed limits on what patterns can be tracked from physical perspectives that can be (and are) occupied by certain machines. These limits are contingent, as Hogarth (1992, 1994, 2001, 2004) showed: some functions that are non-computable by machines that satisfy Gödel's explicit model in what appears to be *actual* spacetime would be computable in alternative hypothetical spacetimes. Patterns in the input–output behaviour of certain sorts of physical machines in certain sorts of physical circumstances are indeed real patterns in our sense. But Gödel's theorem was not motivated by the failures of particular computing machines to halt. Being derived from, and thus being about, a system of formal axioms, it applies in its intended formal domain universally and independently of any measurements of restricted parts of the world. But the conviction that a given range of phenomena should be modelled in a given formalism must always itself be empirically, rather than logically, motivated. Scientists sometimes—indeed, often—bind their generalizations within certain formalisms, in the sense of being suspicious of data that can't be modelled in those formalisms, because alternative formalisms are not yet available and it is typically more productive to be precisely wrong than to be imprecisely right.[54] But recalcitrant measurements hold trumps in these cases, and motivate efforts aimed at developing more flexible formal systems.

---

[54] This of course reverses a populist slogan often aimed at people who value quantitative over purely qualitative knowledge. Like populism generally, it is a bad meme so far as human welfare is concerned.

These are all familiar sorts of things for philosophers to say. Our only innovation here, against the recent climate in metaphysics, consists in expressing them in the voice of our version of verificationism. The sole source of *verifiable* limits on the scope of the formal is our collective mathematical ingenuity to date, whereas the limits on empirical theories are set by the physical possibilities for measurement. Non-verificationists about *metaphysics* think it is appropriate for metaphysics to be constrained only in the way that mathematics is constrained, that is, by nothing more than our collective ingenuity. We have been arguing that such constraints are too loose; there is nothing in metaphysics that corresponds to the alternative constraints provided in mathematics by the requirements of formalization. We think there is much attractiveness in the idea that good metaphysics can best ensure its liberation from anthropocentric myths by accepting a requirement for formal articulation, leaving natural language behind. But if this ambition is in principle achievable, we won't be trying to rise to it here.[55] If we thus grant that metaphysics can throw off the shackles of formalization, at least provisionally, then it is all the more essential that it restrain itself by respecting the PNC, which, given the actual recent history of science, entails the PPC. This is why we say that a pattern, to be real, must be projectible from a perspective that physics tells us could be physically occupied.

Let us emphasize that our motivating principle here is not dogmatic empiricism but a concern for discipline in objective inquiry. That is why we don't think that we should worry about the possibility of irresponsible formal logic and mathematics in the way we worry about actually irresponsible metaphysics. We are open to alternative interpretations of the ultimate force of RR that emerge from rival traditions in mathematics. A constructivist about mathematics might want to urge on us the view that mathematical generalizations are empirically verifiable after all (in so far as whether or not anyone has actually produced a constructive proof is an empirical fact), so RR is a *complete* theory of ontology. In that case, since physicists certainly live within modelling constraints set by whatever mathematicians are in consensus about, the PPC should perhaps be dominated by an analogous PMC. On the other hand, a Platonist about mathematics might ask us to recognize that RR is a theory of only *informal* existence. If all the Platonist thereby wants to make ontological room for are numbers, and perhaps sets and other mathematical structures, we are happy to be mere spectators to this enterprise. We will be interested only in PNC-compatible

[55] It is possible that quantum logicians are right to think that physics demands a more flexible formal apparatus than the existing fully developed ones. If this is true, then a fully formal metaphysics is something that is not yet achievable given the state of formal inquiry. We should not give the impression that we find such arguments terribly compelling; we just don't have a firm view about them. In any case, since our metaphysical theses in this book mainly aim to address the unification of the special sciences, and since many special-science generalizations are based firmly in empirical measurement but aren't presently modelled in any accepted formalism, the ambition for a purely formal metaphysics could at best be a limiting ideal for now. Perhaps it will always be; but such speculation is not itself motivated within the constraints of the PNC.

motivations for Platonism (as for any of its rivals), but those might be defensible (J. Brown 1999). One distinct, and very interesting, possibility is that as we become truly used to thinking of the stuff of the physical universe as being patterns rather than little things, the traditional gulf between Platonistic realism about mathematics and naturalistic realism about physics will shrink or even vanish. The new wave of structuralism in the philosophy of mathematics, which has a number of supporting arguments in common with OSR discussed in 3.6, adds substance to this speculation.

Meanwhile, however, it *is* among RR's distinctive objectives to deny reality to putative aspects of the informally modelled universe about which physics tells us no information can be obtained. Suppose, for example, that cosmological models according to which the Big Bang is a singularity are correct. This is of course not a settled question, but the speculation that the Big Bang is a singularity is obviously motivated compatibly with the PNC. In this case it would follow that there is no physically possible perspective in the universe from which information about what lies on the other side of the Big Bang could be extracted. RR would then say that, if this is how physics plays out, no serious metaphysical theory should try to describe reality in a way that relaxes the boundary conditions imposed by the Big Bang. Furthermore, although some current physical theories provide grounds for not *asserting* that there is a Big Bang singularity, none, so far as we know, motivates any particular description of any aspects of the universe that involve positive breaching of singular boundary conditions. By RR, it would thus not constitute a significant objection to a metaphysical theory if a philosopher dreamed up a counterexample that depended on the suspension of such boundary conditions. States of the physical universe that are merely logically possible are not physically interesting. Therefore, they are not metaphysically interesting either (modulo our caution about mathematics above).

RR, as we explained, allows room for the scale relativity of ontology. It thus allows for unity of science without reduction. There are various possible ways in which real patterns at different scales can carry information about one another without inter-reducing. However, RR does not legislate scale relativity; according to it all issues about reduction (and ontological elimination) are empirical. As we have discussed, Oppenheim, Putnam, and Quine in the 1950s thought that the course of science up to that point suggested a progressive reduction/elimination of the entities and properties identified by special sciences. They thought there would go on being special sciences, but only for reasons of practical anthropocentric convenience. (Thus, in terms of the discussion in 4.2, they steered towards Scylla.) By contrast, the widespread turn to dynamic-systems theories in rigorous special sciences over the past few decades suggests that information will be lost if we refuse to project patterns that cross-classify data relative to the classifications of physics. (Thus, again with respect to 4.2, over the past few decades ever more boats, especially those with pilots from the Santa Fe Institute, have headed for Charybdis.) Batterman (2002) provides examples of physicists explaining phenomena by appeal to singular asymptotic limits, and on the basis of this argues that reduction (both intertheoretic reduction and micro-reduction) is blocked *within* physics. (What Batterman regards as the non-reducing parts of physics will all be special sciences, in our conceptual terminology.) These facts about the recent history of science then motivate, compatibly with the PNC, speculations about what kind of universe we might live in such that this fact about the need for cross-classification—that is, for the scale relativity of ontology—is true. We will thus offer such speculations, but insist that each particular one must be motivated by science.

We thus face a double task in the remaining sections of this chapter. In showing how it is possible that special sciences track real patterns, we have to say more about the kind of scale relativity we think is currently motivated within the constraints of the PNC. Another way of expressing this is: we must explain how to draw the distinction between physics and special science given the conjunction of OSR and RR. This sub-task must in turn serve the main goal of the chapter, explaining how some of the patterns identified by special sciences can be real even if their dynamical character turns out to have to be reconciled with a physics that posits a tenseless and even time-symmetric universe.

4.5 FUNDAMENTAL PHYSICS AND SPECIAL SCIENCE

From early in the book (1.4), we have distinguished between fundamental physics and parts of physics that are special sciences. This is likely to be thought odd for avowed anti-reductionists, particularly as disunity advocates such as Cartwright have lately stigmatized 'fundamentalism' as the most tenacious residue of ontological reductionism. Disunity theorists can regard *every* science as 'special' without embarrassment. But this option would undermine the whole point of our project. In 1.4 we operationalized what we mean by 'fundamental physics' by fiat, and have used it in this operationalized sense ever since. But we have not yet said why, in a world where ontological reductionism appears to fail, we find both fundamental and special sciences.

In fact, we will now argue, a natural interpretation of the distinction between fundamental and special science falls directly out of the conjunction of OSR and RR. We will henceforth refer to the latter as 'Information-Theoretic Structural Realism' (ITSR). A crucial basis for thinking that ours is the right interpretation of the distinction is that it explains the PPC as scientific practice actually observes it. Thus our account of the distinction provides additional PNC-compatible evidence for ITSR itself.

Our distinction must obviously avoid identifying the 'fundamental' with the study of 'ultimate building blocks' out of which everything else is composed. Special sciences seem somehow to focus usefully on distinctive interrelated subsets of the real patterns that there are. Might there then be some other parts of science—the fundamental parts—that study *all* the real patterns that there are, in some sense? Micro-reductionism provides one possible such sense: if all the real patterns were made out of arrangements of a finite number of kinds of atoms, then the science that studied the properties of these atoms would study everything, even if all the facts about everything couldn't be discerned at that fine a grain of resolution. We must, obviously, seek an alternative sense. But this pursuit, like the micro-reductionist's one, will have to begin from the most basic fact about special sciences: that they observe boundaries on their objects of study. Boundaries seem to imply individuals, or at least groups of objects that can be objectively arranged under sortals. So far in this book, we have introduced two concepts to play, respectively, the parts of the traditional metaphysical role of individuals and the parts of the logical role of sortals. *Locators* are representational devices used by observers to fix attention on addresses in some kind of structured background. Some locators successfully pick out *real patterns*, and these are the objects of genuine existential quantification. This all leaves two crucial requirements of the account dark, however. First, just what is this background? Second, what is the metaphysical basis for the boundaries between real patterns such that we can regard these as objective rather than nominal without reintroducing traditional individuals? Notice that these two questions together just amount, in the context of naturalism, to the components of the main question of this section: the first question concerns the basis for fundamental science and the second one concerns the basis for special sciences.

We will begin with the second question. In keeping with the PNC, let us start with an institutional fact about science. All scientific disciplines except mathematics and, arguably, some parts of physics, study temporally and/or spatially bounded regions of spacetime. By this we mean that the data relevant to identifying real patterns are, for most disciplines, found only in some parts of the universe. Biology draws data only from regions in which natural selection has operated. Economists study the same region as biologists (Vorbeij 2005, Ross 2005)—since natural selection depends on competition created by scarcity—but at a different scale of resolution, since only significant aggregates of many biological events are relevant to economic generalization. All psychological measurement is confined to subregions of those examined by biologists and economists, namely, locations in the neighbourhoods of central nervous systems. Anthropologists and other ethologists study a tiny sliver of the universe: specific populations of organisms during the very recent history of one planet.

Let us reflect on the epistemological and metaphysical implications of this institutional fact. What accounts for the specific selection of spacetime regions, at specific scales, to which disciplines are dedicated is obviously, to a very great extent, a function of practical human concerns. Scientific institutions organized by dolphins wouldn't devote more than 40 per cent of their total resources to studying human-specific diseases. There is no puzzle as to why far more scientific attention is lavished on the dry parts of the Earth's surface than on the waterlogged parts. To some extent this same consideration explains why more information is gathered about the Earth than about the Moon.

Self-absorption and convenience, however, account only partly for the distribution of scientific activity. It is also important, in the third instance above, that the Earth is a great deal more complex than the Moon. This statement has a number of interpretations, not all of which matter here. The one that interests us in the context of RR is the following: there are far more real patterns to be discovered by isolating specific spacetime regions, and scales of resolution on those regions, and treating them as relatively encapsulated from other regions and scales, on Earth than on the Moon. This was true before life on Earth began, and is the main part of the explanation for why life did begin on the Earth but not on the Moon; but the extent to which the ratio of real patterns to possible physical measurements on Earth has increased relative to that on the Moon has been made staggeringly great by the progress of first biological, and then social and cultural, evolution. If the Moon were as complex as the Earth then scientists driven by non-self-interested curiosity would indeed have reason to devote as much attention to the former as to the latter. But this is a counterfactual.

These claims about complexity are matters of objective fact, not perspectival foregrounding. Though real patterns are defined in terms of logical, not thermodynamic, depth, a highly general fact manifest across many sciences is that the kinds of processes fecund with real patterns are those that exchange thermodynamic entropy for information at high rates. Indeed, the 'real pattern production rate' in a region O of spacetime is a non-linear function of the thermodynamic depth measurable at whatever scale of O has greatest depth. (The higher the minimum entropy produced by a state's evolution, the greater is the thermodynamic depth of that state. On Earth now, perhaps this trench is found within the highly encapsulated innards of our largest and fastest virtual-reality producing computers. Or perhaps it's in human brains.)

Over a series of papers Collier (1988, 2003, 2004, Collier and Muller 1998) has developed principles for understanding individuation in terms of the production of thermodynamic depth. The central concept in this work is that of *cohesion*, a notion introduced by Wiley (1981) to explain the coherence of species in phylogeny, but subsequently generalized by Collier. The idea has been developed in a context that takes the existence of individuals for granted and aims partially to analyse individuality. Here, we precisely do *not* want to take the existence of individuals for granted, since we hold them to be only epistemological book-keeping devices. We will thus put the cohesion concept to a different sort of work than it does in Collier's earlier papers. Rather than thinking of cohesion as a property that helps to explain what individuals *are*, we will introduce it with the aim of saying which kinds of properties of nature make epistemological book-keeping of real patterns by reference to hypothetical individuals useful for observers like us.

We understand 'cohesion' as a *notional-world* but not *folk* concept that turns out to be useful in metaphysics despite not referring to any real pattern identified by physics. Before we can say more about cohesion and use it in our account of the relationship between physics and special sciences, we will need to develop the idea of a notional-world object and indicate what ontological status these have in the context of RR.

No observer ever has access to the complete extent of a real pattern. This obviously applies to real patterns that are represented in models by epistemically indeterminate sets (from our perspective), like 'oxygen' or 'monetary inflation' or 'aardvarks', but the general point becomes obvious only when we recognize that it applies even to real patterns of the kind that are taken to be paradigmatic cases of individuals. There is a great deal of information about Napoleon that none of us considering this passage will ever be able to recover. He had a definite number of hairs on his head at the moment when the order for the Old Guard to advance at Waterloo was given. There are (or were) a number of perspectives in the universe from which information identifying this number is (or was) available. But we cannot occupy these perspectives, whether we live in a block or in a tensed universe. All sorts of inferences about the state of Napoleon's hair at other times during his life could be made from the inaccessible information if we had it, so there are aspects of the real pattern that is Napoleon—projectible, non-compressible regularities—we are missing and can't get. Such is the fate of observers.

This compels us to be pragmatic. We track all sorts of real patterns that we can approximately locate, but we can't draw their complete or precise objective boundaries in so far as they have them. In this circumstance, we fall back on a second-best epistemological device by focusing on, or aiming to focus on, diagnostic features of real patterns that we can treat as 'core'—that is, as very reliably predicting that our attention is still tracking the same real pattern through any given operation of observation (and reasoning). These are Locke's 'essences of particulars'. Or, once one recognizes that 'natural kinds' are like individuals, they are the 'hidden structures' that Hilary Putnam (1975) imagines we know we mean to pick out—even when we might not know how to—when we refer to, for example, 'tigers'.

It is of course one of the main theses of this book that one makes a metaphysical mistake if one reifies these essences and imagines that they are the real constituents from which the world is fashioned. We can illustrate the difference between real essences and essences as book-keeping devices by way of the following homely analogy. Imagine you are driving in the south-western USA and you want to know, at any given time, which state you're in—this matters with respect to the speed you drive etc.—but, let us further suppose, you can't get direct information about the location of the borders. You might discover that the following second-best epistemic policy works well: assume that whichever state's licence plates appear on the greatest numbers of cars you've passed over the past

5 kilometres is the state you're in now. Of course, it would be a philosophical mistake to think then that you'd stumbled on a fact about the universe to the effect that dominating 5-kilometre stretches with its licence plates is what it is for some part of the world to *be* (for example) Texas. Whether this also represented a scientific mistake (that is, a mistake that matters) would depend on whether the philosophical mistake led you to waste resources on silly experiments, or caused you to fail to notice some good experiments you could try, where 'silly' and 'good' are spelled out in terms of which procedures help to find and measure real patterns.[56]

Individual things, then, are constructs built for second-best tracking of real patterns. They are not necessarily *linguistic* constructions, since some non-human animals (especially intelligent social ones) almost certainly cognitively construct them. However, all questions about the relationship between real patterns and the individuals that feature in special sciences concern individuals constructed by people. These constructions are performed communally, and represent negotiated agreements on applications of a communally controlled resource for book-keeping, namely, a public language.[57] Thus individuals that feature in generalizations of special sciences are community representations.

Many special-science generalizations concern these very representations. Thus, if special sciences in general track real patterns, and sciences that discover generalizations about communal representations are not specially deviant and defective as sciences, then the communal representations that have projectible properties (which might be all representations, communal and otherwise[58]) are real patterns. As naturalistic philosophical under-labourers to science we must accept both antecedents in this conditional. So communal representations are real patterns, which are studied, qua real patterns, by behavioural and cognitive sciences.

It would be consistent with widespread philosophical usage to refer to such real patterns as 'second-order', with the real patterns that are not, in the first place, representational structures being called 'first-order'. We will use this terminology, but only after first being careful to delimit its metaphysical significance. We do not suggest as a metaphysical claim that there are two *kinds* of real patterns, first-order ones and second-order ones. To take this view would be to repeat Dennett's mistake, which we criticized, of dividing reality into illata and abstracta. However, one can harmlessly use the first-order/second-order distinction to say

---

[56] In Chapter 1 we argued that neo-scholastic metaphysics is silly in just this sense.

[57] Since we think that some non-language-using animals (and pre-linguistic human infants) construct individuals, we should not be read here as suggesting that individuals can only be constructed communally. Rather, the point is that we are interested in individuals as these appear in generalizations of special sciences, and special sciences *are* necessarily communal constructions (with essential linguistic aspects).

[58] This alludes to semantic issues about what to count as representation that we doubt are of scientific interest. Thus we will pass over them with the weasel formulation 'might be'.

something about contingent relationships among real patterns: some of them, those that are representations, depend for their genesis and maintenance on their utility to observers as devices for tracking other real patterns. Then a real pattern R2 that exists because it is used to track another real pattern R1 may be said to be 'second-order' just with respect to this special type of genetic dependence on R1. 'Being second-order' is not a property of a real pattern that makes it 'less real'; calling a real pattern 'second-order' merely says something about its historical relationship to some other designated real pattern, and so 'is second-order' should always be understood as elliptical for 'is second-order with respect to pattern Rx'. Then let us say that a real pattern is 'extra-representational' if it is not second-order with respect to any other real pattern. Real patterns that are not extra-representational will be called 'representational'. The overwhelming majority of real patterns that people talk directly about are (we will argue) representational. This is the not-very-exciting but true point at the heart of the exciting but false idea that people think only about 'phenomena' while what really exist are 'noumena'. According to us, people *can* think and communicate about extra-representational real patterns but usually don't try to; scientists often try and *succeed* in so thinking and communicating (which doesn't mean they're making philosophical mistakes when they're being practical communicators and aren't trying); and metaphysicians should be trying harder than they usually have.

All individuals, we will argue, are second-order real patterns with respect to some other real patterns. Some individuals (for example, Tarzan) are second-order only with respect to other representational real patterns. These are the 'fictitious' individuals that form the usual contrast class with those regarded as 'real' in atomistic metaphysics. Such metaphysics treat 'real' individuals as (in our terms) first-order real patterns with respect to their representations. By contrast, according to the view we will articulate, no individuals are extra-representational real patterns. (This is a less startling version of our slogan that 'there are no things'.) However, some real patterns that are targets of first-order property assignments by special sciences are second-order real patterns with respect to some extra-representational real patterns. Things that are treated as real individuals by traditional metaphysics are, according to us, so treated to the extent that they *approximate* extra-representational real patterns (and approximate the 'right' ones[59]). This is why we can understand what a standard individualistic metaphysics is getting at, even if we don't endorse such positions.

'Things' are locators for correlations holding across less than the whole universe that manage not to fall apart instantly. The last part of this statement should strike the reader as unusual, and so we pause to emphasize its import. Correlations too fleeting to be registered from *any* physically possible perspective are not real patterns—this is an expression of our basic verificationism. However,

---

[59] Clarifying the meaning of this restriction is the task of a theory of reference, a topic we will bypass.

sound scientific inference assures us that there are real patterns too fleeting to be registered by *people*. People will not construct individuals to try to track such real patterns. Traditional metaphysics might infer from this that there are individuals out there whose existence we miss; our view denies this. The limits of existence are not determined anthropocentrically but the limits of thinghood *are*, because 'thinghood' denotes no scientifically motivated extra-representational real pattern. It is in this sense—and this sense only—that as realists we deny the existence of individual things.

Each locator constructed around the idea of a thing gives users of the locator a network of reference points which they can compare from limited perspectives. Everyday things are at best merely *correlated* with the extra-representational real patterns to which they stand in a second-order relationship, rather than being identical to them, because the everyday ontology is built for many purposes in addition to, and that partly divert it from the job of, giving true descriptions of the world.

Things as individuated by reference to the practical purposes for which a community has constructed them are elements of these communities' 'notional worlds' (Dennett 1991b). A notional world is a network of background assumptions that permits, but also restricts, projections of objects by reference to their utility to some set of practical purposes. Following Dennett's own leading example, consider the assumptions about Sherlock Holmes's environment that a reader of Conan Doyle must make in order to judge that the question 'Is Holmes's small left toe some definite size?' has the answer 'yes' even though there is no definite size the toe is (because Conan Doyle never said, nor did he say anything from which a definite size can be inferred). Since no extra-representational pattern that has this sort of logical profile is physically possible, no one who is not deluded or severely ignorant imagines that Holmes is identical to or even an approximate description of any extra-representational real pattern. Nevertheless, construction of the roughly correct notional world is essential if a reader of the Holmes stories is to make the projections necessary for appreciating their plots.

The actual world (including its representational real patterns) constrains human goals and purposes, and these in turn constrain the notional worlds that people use. This is obviously compatible with there being generalizations about the actual world that don't apply to notional worlds, and vice versa. In addition, there are constraint relations between notional worlds and representational real patterns that go in both directions: notional-world construction always includes representational real patterns among its basic elements, and many representational real patterns arise just in and for the construction of notional worlds.

These two points give sense to the following questions. Are there facts about the actual world as described by physics that, together with specifically *epistemic* purposes (that is, purposes relevant to science), constrain the notional world that must be constructed for building special sciences? And then might we be able to understand the basis for the representational real patterns—specifically including the individual objects—that feature in these sciences by reference to these constraints? If we can justify positive answers to these questions then we will understand why special-science objects don't reduce to extra-representational real patterns (for example, as described by models in physics); but, so long as the purposes served by special sciences are genuinely epistemic, and cannot be performed by physics, we can conclude that the special sciences track real patterns, necessarily using individuals as constructed epistemic book-keeping devices, but without identifying the real patterns in question with the individuals.

How might our local actual (physical) circumstances be such that constructing individuals is necessary for tracking some extra-representational real patterns? Let us approach this question by trying to imagine counterfactual ways our local circumstances might be in which these constructs would *not* be scientifically helpful. Suppose that every measurement we took of everything around us, sampling randomly (as far as we could, from our perspective, tell) in the measurement space made available by our range of physical probes, was consistent at all scales of resolution with a field model of our data that gave us the same gradient on every potential. In this circumstance, there would be no benefit to be had in trying to go beyond mere location and to sort groups of locators into individuals. The environment would be uniform and homogeneous. Now let us make the surroundings slightly less boring. Imagine that each measurement is still consistent with a single field representation, but gradients on potentials varied from location to location. We would now have evidence about some distinguishable real patterns—'variable forces', we might call them—but there would be no obvious temptation to model these 'forces' as individuals with 'essences' we might use to track them. We could flesh out the structure of the field as we took more measurements, and that would be that.

Next let us add the following additional structure. Suppose that (i) we have some independent asymmetric relations with which we can correlate sets of measurements (we anticipate our hair falling out but never remember it growing back), and (ii) we have reason to believe that the Second Law of Thermodynamics governs our region. That is, we believe that, *ceteris paribus*, the gradients *should* all tend towards an equilibrium set of values when we compare suites of measurements that observer X at rest at spatial location L took when X was hirsute with those X at rest at L took when X was bald—and yet, for *some* special parameters clustered in *some* locations they didn't. We will want to explain what is special about these clusters. Note that nothing introduced to this point requires us to seek our explanation by looking for special properties *of the clusters themselves*; we might instead look for global properties of the whole local field that predicted a statistical distribution of clusters consistent with our measurements. But we might find, if we have limited access to the whole field or to measurements of its global properties, that we could get decent predictive and explanatory leverage by hypothesizing that some clusters of measurements are individuals, which work against entropy to maintain their cohesion for lifespans.

After enough habituation in thinking of this sort, some might even come to believe that all of reality is the mereological sum of such individuals.

The situation we have imagined is underspecified, in all sorts of ways, as a thought experiment in the philosophy of science, let alone physics, so who knows what we'd do? But this question isn't the point of the exercise. The point is instead just to imagine inverting the metaphysical relation between background and foreground against which modern science arose. We of course didn't start with models of fields and forces and then build individuals into them; the conceptual order went the other way around. Once we have constructed things we can construct other things as parts of them. Then 'forces'—what binds or pushes some things together and pulls other things apart—can be correlated with locators in mathematical field representations, and therefore (sometimes) with first-order real patterns (in this case, with gradient potentials in, historically, first Laplaceans and later Hamiltonians). A naturalist should infer no metaphysical conclusion from this one way or another (even if natural selection couldn't, as may be, design beings that conceptually synthesized fields and forces before they conceptually synthesized individuals). We no more think that individual 'forces' are PNC-compatible metaphysical entities than are individual little things. But there is one (scientifically demonstrable, not conceptually asserted) relevant fact lurking in this neighbourhood: we can't get the information we'd need to predict everything we can measure about the distribution of entropy-resistant clusters just from global properties of the universe. (Again, this reports a fact about our perspective of observation, not a first-order claim about the real patterns that physics will find.) As we will see shortly, various ideas in fundamental physics are all consistent (as far as we know now) with this fact. But a fact it is: our measurements to date tell us that the Second Law governs at least our local surroundings, and yet different inputs of work systematically delay entropy increases at different rates in different kinds of applications.

Thus the notional worlds we must construct to track most local real patterns are irreducibly dynamic and time-asymmetric. By contrast, the actual world occupied by all observers we know about is *either* truly tensed, or else merely seems to us to be so given the measurements we can obtain, but is in fact a block universe. A PNC-compatible metaphysics should avoid presupposing either side of this disjunction.

If things and forces are epistemically essential representational real patterns for some observers—for example, us—then we can understand why cohesion is a useful idea in the construction of the notional world we use for projecting them. Collier's first formulations introduced cohesion as the closure of the causal relations among the dynamical parts of a dynamical particular that determine its resistance to external and internal fluctuations that might disrupt its integrity. But 'cause' is another notional-world (and folk) idea that has caused no end of confusion in metaphysics (as we will discuss in the next chapter), so let us avoid using it for the moment. Collier (2003) offers an alternative metaphorical framework for the relevant notional space in saying that an object coheres when the balance of intensities of 'centrifugal' and 'centripetal' forces and flows acting on it favours the 'inward' ('centripetal'). In fact, we can allow for a bit more sophistication in the notional world than this, and say that, owing to fluctuations, it may be enough in a given case if there are propensities of forces and flows that show statistical distribution in space and time (or along other dynamical dimensions thought to be relevant to an application) such that propensities of centripetal forces and flows are favoured over propensities of centrifugal ones.

Talk about 'centripetal' and 'centrifugal' forces is obviously and irreducibly metaphorical if interpreted as proto-physical theory. In this respect, it is exactly on all fours, according to us, with talk of little things and microbangings by philosophers when *that* is intended as proto-physical theory—as indeed the philosophers we criticized in Chapter 1 intend it.[60] That individual forces acting in definite directions, like individual things with particular essences, find no representations in mathematical physical theory is just what we mean when we say that they don't exist. (Similarly, Sherlock Holmes has no representation in the measured biography of the human genome; he is correlated with no real pattern that was a human being.) However, as with the example of Napoleon earlier, it is misleading to think in terms of a simple dichotomy between real elements of physical reality and purely empty fictions. People can do useful work with their notional concepts of things because the stability of their kind of observation platform—relative to the independent stability of the objective world they're trying to talk about—provides them with stable (enough) exemplars (like Napoleon, conceived as an individual). Similar exemplars of centripetal and centrifugal forces can be offered, for the same general reason. Applied physicists often talk of 'energy wells'. Evolutionary theorists say that natural selection plus the fitness-damaging character of most mutations 'holds species together' over time. Biochemists model 'proton pumps' (devices that move hydrogen ions across membranes) as 'doing work against potential gradients' or as altering potentials so that protons 'flow differently'. They will then say, without thinking that they are being poetic, that proton pumps 'hold organisms together' by producing stored energy potentials for use in biochemical processes (such as conversion of ATP to ADP).

In fact, we *do* think that description of reality using notional-world background assumptions is more like poetry than it is like objective modelling, in a quite precise sense. Exemplars of things and forces, qua things and forces,

---

[60] Thus these philosophers should be as willing to invoke centripetal and centrifugal forces in their metaphysics as they are to talk about ultimate particles and classical hydrogen atoms. They aren't; and we can think of no explanation for this asymmetry but the fact that 'centripetal' and 'centrifugal' smell like scientific mildew in a way that classical concepts still useful in mechanical engineering idealizations don't. This suggests that our opponents implicitly accept our major critical premises.

can't be identified with extra-representational real patterns because their status as exemplars is perspective-dependent. (This is at least part of the reason why poetry doesn't go over without loss into prose.) This doesn't mean that the exemplars themselves aren't real patterns—they indeed always are when scientists take them seriously. Napoleon is a real pattern and so are the processes modelled by proton pumps. But when we treat Napoleon as an individual, or say that proton pumps create centripetal forces, we add nothing to the correct objective description of reality. Instead, we help to keep our audience's (and our own) attention focused on a consistent approximate location. That is what notional-world objects are for.[61]

In this pragmatically governed context, the key conceptual job performed by cohesion is that it is an equivalence relation that partitions sets of dynamical particulars into unified and distinct entities. As Perry (1970) argues, construction of such a unity relation is a necessary basis for being able to say how some Ga, Hb, a, and b might all be the same F, while a and b are not the same G or H where F, G, and H are sortals.[62] Cohesion is this unity relation for dynamical objects and their properties. It does not apply to things that are not taken to be unified by internal dynamical relations, for example, rainbows or uncontained gases. It is invariant under conditions that do not destroy it, so that breaking the cohesion of a thing is equivalent to destroying its status as a thing. When a star explodes it ceases to be that star. When a species splits into distinct descendent species there are no longer any members of the original species. These examples risk causing readers to think of cohesion as only synchronic when it is also diachronic. Obviously, it is hard to give examples of *direct* loss of diachronic cohesion. But diachronic cohesion *can* be lost as a result of synchronic changes elsewhere. The set of male dominance displays would fall apart if, as a result of ecological and morphological changes, identification of a bit of behaviour as a dominance display ceased to predict or explain anything.

Attention to concepts like cohesion shows the need for a more sophisticated category space for concepts than the simple binary opposition of 'scientific' and 'folk'. It would be misleading to call cohesion a folk concept, since it is not clear that any such concept is needed, or often used, outside the special sciences. In the binary picture, the temptation to think of cohesion as a folk concept arises from the fact that it regiments folk concepts of things and forces when these are extended to apply in the context of dynamical systems. However, the job the concept performs is a scientific one: it gives individuation conditions for dynamical entities, constructed out of measurement data by observers for second-best keeping of epistemic books. This activity is (at least often) genuine science

---

[61] So, 'Sherlock Holmes' keeps a reader's attention focused on a region of spacetime where English-ness and detective-ness and thin-ness and friend-of-Victorian-doctor-ness (etc.) all make sense together in one model.
[62] For example, a cub and an alpha male may be the same lion, but they are neither the same cub nor the same alpha male.

(where 'genuine', as always here, is benchmarked by reference to institutional practices and controls). The cohesion relations that specify the closure conditions for individuation of a given object might be energetic, kinetic, or organizational (in cases of so-called 'self-organizing' systems). They may be taken to be internal, or routed through the external environment. Cohesion may occur as a sum of local effects (as in the case of molecular bonds in a crystal) or else as a non-localizable property (as in the case of the organizational closure of an organism or species). Because cohesion determines the appropriate levels for dynamical analyses, cohesive entities of the kind the folk tend not to recognize as individuals because they are distributed in space and time (for example, again, species) are sometimes confused with classifications. Thus biology texts often remind students that phylogenetic trees are historical processes, not systematic classifications; the former are abstractions and need not be cohesive, while the latter model particulars and depend on cohesion if the models are to have biological applications. We emphasize this example because it is an instance where scientists must *correct* the (student) folk in the use of a concept for organizing notional-world objects[63]—thus illustrating the inadequacy of the simple folk/scientific opposition over concepts.

Where cohesion, like little things and microbangings, will *not* figure is in *fundamental* science—or, therefore, in first-order metaphysics. We are busy explaining it now because it turns out to be important for the second-order metaphysical job of accounting for the difference between special and fundamental science and, thereby, in turn, for explaining the nature of scientifically driven metaphysics.

We now offer a generalization about institutional practice in special science that we claim is empirically motivated by a range of specific scientific hypotheses and contradicted by none. All special sciences stabilize location of the real patterns they aim to track by constructing particular role-fillers to exemplify the cohesion relation. This construction activity is often contested. Indeed, it is the standard happy hunting-ground of methodologists and philosophers of individual special sciences. For example, philosophers of biology have been preoccupied for years now with the question of whether natural selection reduces all other cohesive forces needed in applications of evolutionary theory; this is the so-called 'adaptationism debate' (see Orzack and Sober 2001 for a representative sample). Economic methodologists obsess over how much mileage economists can squeeze out of individual consistency in choice as the source of both synchronic and asynchronic cohesion in markets (see Ross 2005). The reason why these debates are typically resistant to clear resolution is precisely that 'cohesion', in general, refers to no definite real pattern against which specific proposals can be rigorously tested. Cohesion is a locating tool and location is irreducibly pragmatic and perspective-dependent.

[63] This describes most of what goes on in a typical philosophy classroom.

Saying that cohesion *in general* picks out no real pattern is consistent with invoking objective features of the world to explain why special scientists rely so heavily on *this* locating tool.[64] Special scientists assume, at least implicitly, that all real patterns they aim to track must be stabilized by *something* against entropic dissolution. (So, exactly as per Collier's use, the phrase 'cohesive forces' just gestures at the indefinite class of conceivable 'somethings'.) By a pattern's 'stability' they understand its probabilistic disposition to resist background perturbations normally present in its normal kind of environment. In our terms, 'resistance' can be interpreted here as referring directly to the extent to which a pattern supports projectibility by physically possible observers. Continuing thus to interpret institutional practice in special science by the terms of RR, we can identify the following general background hunch of the special sciences: as a pattern's stability goes asymptotically to zero, it ceases to be real. To the extent that the world is governed by the Second Law of Thermodynamics, to be (in this context, to persist) is to resist entropy.[65]

The qualification at the beginning of the previous sentence is crucial. Physics does *not* presently tell us that the Second Law is part of fundamental physics, although some current approaches to grand unification, as we saw in Chapter 3 and in 4.3, make the assumption that it is. Others, such as superstring/M theory, make it derivative, and allow that whether it holds may depend on boundary conditions. Furthermore, it is especially clear that thermodynamics cannot be basic in any fundamental physical theory that is time-symmetric (and as we discussed in Chapter 3, this may be all of them).

It is on this basis that we propose to identify the promised distinction between fundamental and special science. The explanation for the fact that theories and research programmes in special sciences assume background cohesion relations is that they confine their attention to what seem to us observers, unable to obtain information from outside our backward-projecting collective light cone, to be dynamic patterns. We do not know whether the universe is asymmetrical in a way that mirrors the epistemic asymmetry we face with respect to (speaking loosely, just for brevity) 'the past and the future'. Physics is trying to discover this. It cannot do so by gathering information from outside our collective light cone, Instead, it aims to do so by (i) investigating the accessible parts of the universe at ever larger scales of logical resolution (which often requires measurements at ever *smaller* scales of physical resolution) and (ii) betting on extrapolation of a trend from the history of science to the effect that as we broaden the grain of

---

[64] Note carefully that *particular* cohesion relations discovered by special sciences are themselves real patterns, since there are physically possible (actual) perspectives from which they non-redundantly compress information.

[65] As is usual with 'implicit' and 'background' assumptions in science, this hunch will be found at varying distances from surface presentations in different theoretical exercises. Two cases in which it is unusually close to the surface are Brooks and Wiley (1986) in biology and Durlauf (1997) in economics.

resolution we converge on 'universal real patterns'—the modal structure of the universe.

We have put 'universal real patterns' in quotes because this phrase is only intended to connect our proposal with more traditional philosophical usage. We will not use it unquoted, because we don't endorse most of its traditional associations. Let us thus be immediately explicit about our surrogate idea. The hypothesis that the universe is unified rather than dappled amounts, in the terms presented here, to the suggestion that there are some real patterns for which measurements are maximally redundant. That is, there are some real patterns about which measurements taken *anywhere in spacetime at any scale of measurement* carry information (in the logical, not thermodynamic, sense). Fundamental physics is that part of institutional science responsible for trying to discover maximally redundant real patterns.

We can identify a number of theories that have been intended during the history of science as fundamental ones. Aristotle's magnificent generalization of four-element theory by the hypothetical dynamic mechanism of turning spheres was perhaps the most fully psychologically satisfying, until extensions of measurement technology tore it apart. Cartesian mechanism was the next serious candidate, though much less well worked out in detail. Newtonian mechanics was of course treated as fundamental physics by many. Here at the beginning of the twenty-first century we find ourselves with several proposed fundamental theories among which evidence cannot yet decide. We reviewed the leading contenders in 3.7. Any of these theories, to be regarded as adequate, faces the task of explaining why the Second Law of Thermodynamics governs the regions of the universe in which we live, at least at all scales of resolution above a certain threshold, and faces this task regardless of whether they are based on block or fundamentally tensed foundations. The theories just mentioned are all plausible candidates now because we can see at least in general how they might provide such an explanation, along with the explanations of other measured data they supply.

Special sciences are, by pragmatic fiat, not responsible for finding generalizations that support counterfactuals arising outside of their institutionally evolved and negotiated scope boundaries. By contrast, fundamental physical theory is (analytically, *given* the historically supported and contingent PPC) responsible for supporting counterfactuals across the entire actual universe, whether this is taken to be infinite (as by some fundamental theories), or finite and limned by spacetime singularities (as in other theories). Saying that a measurement from *anywhere in spacetime* is a potentially relevant counterexample to fundamental physics is then the analytic extension of—the philosopher's logical generalization of—the institutional PPC. This is in turn just equivalent to the unity hypothesis in our framework.

OSR is the hypothesis that science provides a unified account of the world by modelling structures that modally constrain inferences from measurements. RR is the metaphysical theory of the relationship between these structures and extra-representational real patterns. ITSR is their conjunction. ITSR then accounts for the partial appearance—and the appearance is only partial, since the balance of evidence from the history of science suggests otherwise—of disunity as this expresses itself in the plurality of special sciences. There is room for special sciences because ITSR is compatible with the scale relativity of ontology. Indeed, RR just *is* a regimentation of the idea of the scale relativity of ontology, and that is the primary metaphysical explanandum for which recent science furnishes direct motivation in the spirit of the PNC. Special sciences are free to hypothesize any real patterns consistent with the measurements they accumulate so long as these do not contradict what physics agrees on. (Thus they cannot hypothesize forced motions or entelechies or instantaneous perceptions across space or acts of divine intervention.) Were the world fundamentally disunified, this institutional norm—the PPC—would be mysterious. The hypothesis that there is a true fundamental physics explains our observation of the PPC: every measurement of some real pattern on a scale of resolution appropriate to a special science that studies real patterns of that type must be consistent with fundamental physics.

Finally, because all the special sciences take measurements at scales where real patterns conform to the Second Law of Thermodynamics, all special sciences traffic in locally dynamic real patterns. Thus it is useful for them to keep epistemic books by constructing their data into things, local forces, and cohesion relations. As we will discuss in the next chapter, causation is yet another notional-world conceptual instrument ubiquitous in special science. The great mistake of much traditional philosophy has been to try to read the metaphysical structure of the world directly from the categories of this notional-world book-keeping tool. The effects of this mistake are often subtle, which no doubt explains why we have had so much trouble recognizing it as a mistake. The *exemplars* of thinghood are precisely those entities that are most stable under tracking in the terms of the notional world. These are 'particulars' such as the table beside the wall and Napoleon. What explains the stability of the exemplars is that they *are* real patterns—in general, just the kinds of real patterns that native human cognition *is* well adapted to track. In this context, reductionistic scientific realism has the odd consequence of denying their existence—thus we get Eddington's dissolution of the table, Kim's sundering of 'overdetermining' higher-level entities, and Merricks's sweeping elimination of everything except atoms (and, in his special strange twist, people). According to ITSR, tables and chairs *are* real patterns—they just do not have fundamental counterparts. At the fundamental level, where all proper talk about the entities whose status fuelled the twentieth-century scientific realism debate goes on, reliance on the notional-world idea of cohesive things is just completely misleading.

This last remark is apt to remind readers of Eddington's famous ruminations on his two tables:

One of them has been familiar to me from earliest years. It is a commonplace object of that environment which I call the world. How shall I describe it? It has extension; it is comparatively permanent; it is coloured; above all it is substantial ... Table No. 2 is my scientific table. It is a more recent acquaintance and I do not feel so familiar with it. It does not belong to the world previously mentioned—that world which spontaneously appears around me when I open my eyes, though how much of it is objective and how much subjective I do not here consider. It is part of a world which in more devious ways has forced itself on my attention. My scientific table is mostly emptiness. Sparsely scattered in that emptiness are numerous electric charges rushing about with great speed; but their combined bulk amounts to less than a billionth of the bulk of the table itself. (Eddington 1928, ix–x)

The question Eddington addresses in these passages is how to reconcile the scientific and manifest images of the world. Ontological fundamentalism maintains that the fundamental part of the scientific image is an accurate portrait of reality and that common-sense image is an illusion.[66] Hence, we find philosophers such as Merricks and van Inwagen denying the existence of the everyday table. On the other hand, for philosophers such as van Fraassen, the manifest image is epistemologically primary. In this book we have argued that to be is to be a real pattern, and that science is engaged in describing the modal structure of the world. It is an advantage of our view that it makes it possible to understand how both the scientific image and the common-sense image can capture real patterns.[67] The everyday table is probably a real pattern. Strictly speaking there is no scientific table at all because there is no single candidate aggregate of real microscopic patterns that is best suited to be the reductive base of the everyday table. We deny that everyday or special science real patterns must be mereological compositions of physical real patterns, and we deny the local supervenience of the table on a real pattern described by physics. Hence, we reject the dichotomy between reductionism and eliminativism about everyday objects.

Horgan and Potrc (2000, 2002) are eliminativists about everyday objects because they regard ontological vagueness as impossible. Since they think that if there were tables their boundaries would be vague, and since there are too many candidate sets of particles for their composition, they conclude that there are no tables. Another problem for reductionism is that composition is not identity. We are happy to agree that reductionism about everyday objects is hopeless. However, we are sanguine about ontological vagueness in the sense that (like Wallace 2003a, 95) we think it makes no sense to ask questions about cats over very short timescales or about mountains over very short-length scales. There is no fact of the matter about the exact boundary of Mount Everest because it

---

[66] Note that the scientific and manifest images are not totally disjoint. Entities like stars, or a macroscopic sample of a gas, are both observable and theoretical.
[67] In advocating his NOA, Fine (1984a, 1986) recommends that we accept scientific and homely truths as on a par, yet he does not explain how they can be compatible despite appearances.

is not a real pattern at all length scales. We regard puzzles about where exactly mountains stop and start, and whether or not the table is the same if we remove a few particles from it, as pseudo-problems *par excellence*, of no scientific or factual relevance.

We share Horgan's and Potrc's suspicion of the idea that there is a direct correspondence between sentences such as 'the book is on the table' and the world, if what is meant by that is the claim that there are individuals that act as the referents of the singular terms 'book' and 'table'. So with them we conclude that we need a different semantics from that of direct reference and correspondence to explain how everyday utterances can be true despite there being no self-subsistent individuals. They defend the claim that there is only one concrete particular, namely the whole universe. We saw in Chapter 3 that there are research programmes in physics that reject the idea that whole universe can be considered as a single individual. This leads us to a position even more austere (in a sense) than that of Horgan and Potrc, since we claim that there is *at most* one contingent individual. However, the theory of ontology we have defended in this chapter means that, unlike Horgan and Potrc, we do not have to deny the existence of, for example, tables, corporations, or symphonies. On the contrary, according to RR, there are plenty of such complex real patterns.

Our relationship with common-sense realism is not straightforward. On the one hand, our ontology makes room for everyday objects and treats them on a par with the objects of the special sciences. On the other hand, we attribute no epistemic status to intuitions about ontology derived from common sense, and in particular we deny that scientific ontology is answerable to common sense, while insisting that common-sense ontology is answerable to science. We take it to be an empirical question for any particular common-sense object whether it is a genuine real pattern, and so eliminativism about, for example, tables or mental states, cannot be ruled out a priori. If cognitive science concludes that mental concepts do not track any real patterns then the theory of mind will have to go. We do not at present think that behavioural and cognitive sciences are tending in this direction (though we think that the *folk* theory of mind is clearly false in all sorts of important ways), but the issue is not up to philosophers. In any given instance, it may be that our concepts and intuitions about the real patterns we are tracking may be widely mistaken, as with intuitions about qualia and phenomenal conscious states (Dennett 1991b). So, for example, everyday talk of time and space tracks real features of the universe, but our intuition that there is absolute simultaneity may be wrong, our everyday talk of tables tracks real patterns but our concept of tables as individual substances is in error, and finally our concept of the self may turn out to track a real pattern, but it may also attribute properties to selves that the real patterns lack. To build on what was said in Chapter 1 now that we have more elements of our ontological framework in place, it is important to distinguish between the question of whether our concepts and intuitions track anything real, and the question of whether the real patterns they track have all the properties that their intuitively familiar surrogates appear to have.

According to Rainforest Realism, what grounds the claim that some real pattern exists is its projectibility. Induction is therefore part of the manifest and the scientific images in the sense that we individuate objects in the light of inductive knowledge of what is relatively invariant. The objects that we identify are ontologically secondary to the modal structure of the world. In his classic paper 'Identity, Ostension and Hypostasis', Quine (1953, ch. IV) offers a similar account of how we arrive at our everyday ontology. The idea is that we posit identity between ostensions as a method of hypostasis. Our hypostases have survival value because engaging in the pretence that there are enduring objects makes things much simpler in 'manifest image' contexts even though they may be superfluous in 'scientific image' contexts. On his view, there is no fact of the matter about what things there are: 'objects' shrink or expand to their most convenient size.

Induction is any form of reasoning that proceeds from claims about observed phenomena to claims about unobserved phenomena. If we think of induction as simply the generalization from observed cases to a universal claim about observed and unobserved cases, and then the derivation from that of claims about unobserved cases, then it may seem entirely without justification because it appears to be merely question-begging. What we are really interested in are the unobserved cases and we arrive at a conclusion about them by simply assuming it. Consider the argument

1. All of a large number, n, of observed cases of Fs have been Gs.
2. All Fs are Gs.
3. All as yet unobserved Fs are Gs.
4. The next observed F will be a G.

Even though we may only be interested in the next cases we will observe, we need to know 3 because we don't know in advance which of the unobserved cases we will observe and which we won't. So we need 2, which is what we are supposed to infer by induction from 1. But 2 is just the conjunction of 1 and 3. (This is a massive idealization of course, but the essential point carries over to more complex renderings of inductive inference.)

David Armstrong (1983, ch. 4, sect. 5) argues that if that were all there was to it induction would indeed be irrational. However, like Peirce he thinks that induction is really grounded in inference to the best explanation. As such it is to be regarded as genuinely inflationary in that it requires the existence of genuine lawlike or necessary connections among events. Mill thought that the principle of the uniformity of nature amounted to the same thing as determinism. We certainly don't need as much as that; objective potentialities and probabilities will do.

It is supposed to be the existence of things, whether everyday or scientific, that explains the phenomena, but what really does the explanatory work is not things but their stability as part of the world's modal structure. Tables are particularly robust real patterns because they allow us to make predictions about many other real patterns, for example, about light distributions and about mass distributions. We only know if at all about unobservable objects in virtue of our theories about them. Since we only understand the entities as we do because of the way they are supposed to relate to our observations, belief in the unobservables entailed by our best scientific theories entails belief that those theories are at least approximately empirically adequate, in other words that there really are the relations among the phenomena they attribute to the world. If those relations were merely randomly correlated to the objects then the claim that the objects exist could not count as an explanation of what is observed. It is noteworthy that scientific realists often invoke thick anti-Humean notions of causation in their attempts to convince us that unobservables exist. We need not go so far as belief in objects, observable or not. The history of science undermines not only materialism and classical views of space and time, but also the claim that science describes the individuals that lie beyond the phenomena.

So, to reiterate a point made a few pages above, at the fundamental level, where all proper talk about the entities whose status fuelled the twentieth-century scientific realism debate goes on, reliance on the notional-world idea of cohesive things is just completely misleading.

Such, at any rate, is the most general metaphysical upshot of ITSR. In stating it now, we have run somewhat beyond what we have established a basis for saying. That special science objects, along with tables and chairs and Napoleon, exist as at least representational real patterns should now be clear enough. By RR, representational real patterns exist. But how does this amount to more than just saying we metaphysically affirm them? Isn't second-order real patternhood still, in all *substantive* senses, second-order existence—just what we criticized Dennett for allowing? Are we not, in fact, effectively Kantians where they are concerned, letting them exist only as constructions set against hypothesized noumenal real patterns modelled by fundamental structures? We will confront this question directly at the end of the next chapter.

First, we must elaborate further aspects of our position by considering, at length, an issue from recent debates in the philosophy of science that is closely connected with the question just posed. Harold Kincaid (1990, 1996, 2004) has sought to vindicate special sciences as being metaphysically on all fours with physics by defending the claim that special sciences regularly discover true causal laws. John Earman and John Roberts (1999) have directly contested this, arguing that the need to include *ceteris paribus* provisos in most special-science generalizations that otherwise have the logical form of law statements suggests that most special sciences are at best on the road to finding laws, but might in general not be in the business of doing so. Earman and Roberts write that:

If laws are needed then only laws will do, and ceteris paribus laws will not. The point of this slogan is not that special sciences cannot be scientifically legitimate. Rather, the point is that when only 'ceteris paribus laws' are on offer, then whatever scientific purposes are being fulfilled ... do not require laws. Hence there is no need to try to rescue the special sciences by finding a way to minimize the differences between 'ceteris paribus laws' and laws. To do so is to try to stuff all good science into the pigeon hole modelled on fundamental physics, which ... does articulate strict laws of nature. This can only obscure what is important about the special sciences, as well as what is important about fundamental physics. (1999, 470)

In the next chapter we show how to arrive at a conclusion complementary to this in the framework of ITSR. This will in turn lead us to the basis for our explicit view on the metaphysical status of special-science generalizations (and thus of the real patterns that special sciences discover). This will come with a twist from Earman and Roberts's point of view, however. We will argue that Kincaid is also right about something crucial in this debate: the idea of causation, as it is used in science, indeed finds its exemplars in the special sciences, and it is a presently open empirical question whether that notion will have any ultimate role to play in fundamental physics. This argumentative tour through differences in the kinds of generalizations furnished by, respectively, special sciences and fundamental physics will then enable us to complete our account of the metaphysics of the former.

5
Causation in a Structural World

Don Ross, James Ladyman, and David Spurrett

5.1 RUSSELL'S NATURALISTIC REJECTION OF CAUSATION

Microbangings, a vaguely characterized class of events that, according to us, should not be taken seriously by a naturalistic metaphysics, are typically the paradigm instances of causal connection in modern Western philosophy and, as we argued in Chapter 4, continue to have this status in contemporary neo-scholastic metaphysical pictures such as Kim's. In denying that there are microbangings, then, are we denying that there are such things as causes?

We will argue that the idea of causation has similar status to those of cohesion, forces, and things. It is a concept that structures the notional worlds of observers who must book-keep real patterns from perspectives that involve temporal and other asymmetries they cannot ignore on pain of discarding information. Appreciating the role of causation in this notional world is crucial to understanding the nature of the special sciences, and the general ways in which they differ from fundamental physics. However, we will devote deeper and longer attention to causation than we did to cohesion. This is for two reasons. First, causation, unlike cohesion, is both a notional-world concept *and* a folk concept.[1] It has, as we will explain below, at least two 'centres of semantic gravity'—one in science and one in everyday discourse—that have usually been in tension with one another in the history of metaphysics. Indeed, in attempting to resolve this tension, philosophers managed to construct a third centre of semantic gravity of their own, a bit of mischief which has greatly compounded the difficulties in

---

[1] We explained the idea of a notional world concept in 4.4. By 'folk concept' we mean what is standardly meant by this term in the philosophical literature.

achieving clarity on the subject. Second, causation, unlike cohesion, is a basic category of traditional metaphysics, including metaphysics that purports to be naturalistic but, according to us, falls short of this ambition. We cannot claim to have shown how to supplant the tradition with a rigorously naturalistic but unified world-view unless we articulate a clear and careful story about how the various tasks performed by causal concepts in pseudo-naturalistic metaphysics get displaced and redistributed in the real thing.

We will argue that causation, just like cohesion, is a representational real pattern that is necessary for an adequately comprehensive science. Thus, we will not urge its elimination. However, although we will argue that—again, as with cohesion—fundamental physics and metaphysics can together explain why causation plays an essential role in science, the correct account of fundamental physics might not itself incorporate any causal structures, mechanisms, or relations. Therefore, a PNC-compatible metaphysic cannot go on making causation fundamental.

Instead of beginning our account with metaphysical pictures in which causation is fundamental and then showing how to delimit its role, we will follow the opposite procedure. That is, we will begin with an account that eliminates causation altogether on naturalistic grounds, and then show, using PNC-mandated motivations, why this outright eliminativism is too strong. The eliminativist argument we will discuss is due to Bertrand Russell. As we will see, Russell's view has some important contemporary adherents among philosophers of physics. We are basically sympathetic with him and them, because, as we will argue, their appreciation of the problem with causal ideas in physics is sound, and the motivations driving Russell's argument are gratifyingly PNC-friendly. (Indeed, Russell's paper on causation is among the clearest anticipations of PNC-based metaphysics to be found in the philosophical canon; this alone would justify our lavishing attention on it in this book, since an enhanced understanding of Russell's position entails an enhanced understanding of the commitments of genuine naturalism.) Russell goes too far in his across-the-board eliminativism. However, we will argue, he does so only because of his assumption—made for PNC-compatible reasons—of a version of reductionism that we have argued should be decoupled from naturalism. Close consideration of Russell's paper thus keeps all of our main themes at the focus of attention.

Russell (1913 [1917])[2] characterized what he called 'the law of causality' as a harmful 'relic of a bygone age', and urged the 'complete extrusion' of the word 'cause' from the philosophical vocabulary. His main premises are the descriptive claim that practitioners of 'advanced' sciences, particularly physicists, don't seek causes, and the normative claim that it is improper for philosophers to legislate on whether they should. This is why we said above that Russell's argument is based

---

[2] All references to 'Russell' in what follows are to 'the Russell of 1913' unless otherwise stated. Citations are to 'On the Notion of Cause' as reprinted in *Mysticism and Logic* (Russell 1917).

on exactly the sort of naturalism we are promoting in this book. His contention that causation is not a significant concept in metaphysics is based on the idea that it is an artefact of an anthropocentric perspective that science supersedes.

In pointing out that Russell's argumentation here is naturalistic, we do not mean to imply that Russell consistently observed naturalism as his first allegiance. He was first and foremost a Platonist. But as we pointed out in 3.5, 3.6, and 4.4, there are versions of Platonism that are compatible with naturalism; and Russell's Platonism was motivated by facts about mathematics and its relationship to science, so was PNC-compatible. In any case, Russell's main basis for argument in the essay on causes is description of the general content of science. Such appeals have no force without an accompanying normative thesis to the effect that science should have authority over philosophy. Thus Russell cites James Ward as complaining about physics on the grounds that 'the business of those who wish to ascertain the ultimate truth about the world...should be the discovery of causes, yet physics never even seeks them', to which Russell objects by saying that 'philosophy ought not to assume such legislative functions' (1913, 171). Later, having found Bergson attributing an event-regularity principle of causation to scientists, Russell scolds that 'philosophers...are too apt to take their views on science from each other, not from science' (1913, 176). Towards the end of his essay, Russell diagnoses commitment to a law of causation as stemming from projections into the metaphysical interpretation of science of 'anthropomorphic superstitions' based on the practical human predicament with respect to the asymmetry of past and future in our memories and capacities for control. Finally, at one point (1913, 184) he suggests the very radical naturalist response to the conflict between everyday and scientific conceptions for which we have also expressed attraction: the idea that sound metaphysical insights cannot be stated in natural language, but require expression in a formalism from which anthropocentric distortions have been purged.

The Russell of the 1913 essay also closely anticipates our stance in supposing that the only legitimating point of metaphysics is unification of our world-view—which, given the accompanying naturalistic theses, means unification of otherwise separated branches of science. Russell denies (1913, 185 f.) that scientists make inferences from the hypothesis of the 'uniformity of nature', because such a procedure would treat this hypothesis as at least implicitly a priori. However, he then goes on directly to argue (1913, 186 f.) that science advances generalizations about 'relatively isolated systems' only for practical purposes: its guiding aim is to achieve generalizations of maximum scope. Thus 'it should be observed that isolated systems are only important as providing a possibility of discovering scientific laws; they have no theoretical importance in the finished structure of a science' (1913, 187). Despite anticipating Cartwright's criticism of the event regularity interpretation of laws, then, Russell is not led from this towards the view that the world is dappled. As sciences widen the scope of the generalizations they produce, he says, they abandon simple regularities among 'causes' and 'effects' which they used as ladders to discovery in their immature phases, and identify progressively 'greater differentiation of antecedent and consequent' and a 'continually wider circle of antecedents recognised as relevant' (1913, 178).

Where more recent work in naturalistic metaphysics has been preoccupied with questions about unification, this has usually been in the specific context of trying to understand the relationship between causation and explanation (Friedman 1974, Kitcher 1981, Kitcher 1989). These issues are in turn closely related to debates around reductionism. For example, Kincaid (1997) takes as his central problem the challenge to unification of the sciences raised by the failure of reductionism, and by the fact that (he argues) different sciences display a proliferating range of explanatory strategies distinguished by reference to special, parochial causal patterns—patterns which indeed provide much of the basis for the locations of the boundaries between the scopes of the sciences in the first place. Because of the priority Kincaid attaches to the empirical evidence against reductionism, he regards the reductionist conclusion of neo-scholastics like Kim (1998) as a *reductio* on their premise that where there are multiple (for example, both mental and neural) causal claims about what happens in some part of the world, all but one of these must be false unless we accept reduction or overdetermination. Of course we agree with Kincaid about this. In later work, Kincaid (2004) makes realism about causes the basis for an argument that special sciences (including social sciences) discover laws of nature, despite the fact that the laws in question must usually be stated with *ceteris paribus* clauses. Thus we find a reductionist champion of the manifest image (Kim) agreeing with an anti-reductionist naturalist (Kincaid) that our world-view obtains such unity as can be had by reference to causation. In the context of this current dialectic, it is interesting that Russell combines the commitments to naturalism and unification shared by us and Kincaid with the latter's reductionistic physicalism, and to the mereological atomism embraced by the neo-scholastic metaphysicians and rejected by us. In Russell's case, unlike that of the neo-scholastics, this attitude was PNC-consistent. The new physics coalescing into being in 1913 made the late nineteenth-century impasses that had inspired Mill's emergentism about chemistry (and, by implication, the life and behavioural sciences) look much less likely to remain pressing problems. Specifically, Einstein's introduction of photons made it seem likely that wave–particle duality with respect to light was about to be dissolved in a way favourable to atomism. That the duality would instead be generalized to all matter by de Broglie—let alone that quantum entanglement would be discovered—was something Russell could hardly anticipate.[3]

---

[3] Russell kept a close eye on the frontier of science, however. By the time of his *ABC of Relativity* a few years later (Russell 1925), he knew that de Broglie's work spelled trouble for metaphysical atomism. We remind the reader that many philosophers writing now still don't seem to know this.

Russell construes cause-seeking as the quest for, or assumption of the existence of, 'invariable uniformities of sequence' (1913, 178) or laws of constant succession. According to Russell, belief in the 'law of causality' leads us erroneously to expect and to prize these uniformities. He blames philosophers for supposing, against the evidence in science, that there exists such a law identifying causal relations with uniformities. Here he anticipates Cartwright's (for example, 1980) contention that laws understood as stating facts about event regularities have numerous counterexamples. Russell agrees with Cartwright that laws stating event regularities don't describe how bodies actually behave. Instead of concluding from this that the laws of nature are false, he maintains that they don't state event regularities, defending this through an account of the features of scientific practice that produce the counterexamples *and* show the inappropriateness of an event-regularity view of laws. First, he argues, the Humean interpretation of laws as regularities depends on a qualitative basis for individuating events that cannot be rendered in objectively quantitative terms; yet a mature science is one that aims at strictly quantitative description in the limit (1913, 183 f.). Second, the idea that uniformities of *sequence* might be 'laws' makes asymmetry between the past and the future fundamental, whereas maturing science aims always at laws that make no reference to time in this sense (though they often refer to intervals, some of which may be rightly called 'temporal') (1913, 185 f.). Russell concedes that in the 'infancy' of a science (1913, 178) the principle of 'same cause, same effect' may prove useful for discovery, but maintains that advance in science consists in moving away from such simple formulae, and towards understanding phenomena in terms of mathematically specified relations of functional interdependence in which past and future are not systematically different,[4] which the 'law of causality' is too crude to embrace, and which Russell therefore thinks should not be called 'causal'.

Those who now look to causation to provide the unifying structure of the world have in mind a notion of cause as in some sense the 'cement of the universe' (see Hume 1978 [1740], 662 for the leading source of this metaphor) or what we elsewhere called the 'glue' holding all objective relations in place (Ross and Spurrett 2004). The causation as glue tradition contains a substantial body of work, developing what are collectively called causal process theories. Such glue would be or would provide the necessity that Hume couldn't find in the impressions and their regular relations. Causal process theories, that is, can be understood as attempts to answer Hume's epistemological challenge to say how anyone could know, by any amount of observation, which links between processes are causal and which are not. They purport to show that we can, at

---

[4] That is, as Field (forthcoming) notes, being '*deterministic in both directions*'. We agree with Field that the (alleged) lack of past–future asymmetrical dependence in fundamental physics is not Russell's most important argument against the relevance of causes to science, and that the failure of the facticity view of laws is.

least, observe something that is precisely *diagnostic* of what we have traditionally meant by 'causation'. Contra Russell's complaints about the law of causality, causality, or the glue, isn't itself an event regularity; it is supposed to be that in virtue of which there are some projectible regularities in the first place.[5]

Salmon's (1984) causal process theory followed Reichenbach (1957) in describing real processes in terms of the transmission of *marks*.[6] On this view a genuine process is one that can be modified (or 'marked') at some stage, and observed to carry the same modification at subsequent stages. Some processes, that is, transmit information about their antecedent stages, while others do not; those which do are genuine processes, the others pseudo-processes. A crucial scientific motivation for Salmon's project of distinguishing causal from pseudo-processes, as with Reichenbach, is the alleged relativistic requirement not to count a faster-than-light process as causal (Salmon 1984, 141; Reichenbach 1957, 147–9). A commonly used illustration of the problem concerns a spot of light moving on some surface because the light source is rotating. For large enough distances between spotlight and surface, the spot cast by the rotating light will move across the surface at a superluminal velocity (Salmon 1984, 143). Even so, it would not be a causal process for Salmon, because it cannot be marked. A filter that turned the beam red, placed just before one part of the surface, would make the light that struck that bit of surface red (diagnosing the process between the filter and surface as causal) but not change the spot elsewhere, showing that the successive stages of the spot weren't causally connected.

The metaphysic suggested by process views is effectively one in which the entire universe is a graph of real processes, where the edges are uninterrupted processes, and the vertices the interactions between them. Thus process views, if correct, would make putatively causal claims by scientists subject to a critical test. Those that pick out real processes could be causal, those that don't, can't. Then, when scientists find causes in their domains of inquiry, whether they put the matter this way or not, they find ways of picking out real information-carrying relations. That is to say, if macroeconomics, for example, has its own notion of 'macroeconomic causation', this notion will turn out to correspond to some information-carrying relations really instantiated on the general graph of real processes.

Russell's 1913 claims about the irrelevance of causation to science and the tradition of causal process theories are, then, in conflict. It is bound to seem at first glance that unless Russell is completely wrong, the process theory tradition must be a mistake. There are at least two central reasons why most contemporary

---

[5] To suppose that unification might consist in the identification of some privileged 'glue' is, then, to suppose that there is a master relation (in this case causation) that is implicated in all projectible generalizations.

[6] Process theories differ on the question of what real processes actually are. For Dowe (for example, 1992) they are possessors of conserved quantities, and for Collier (1999) they are transfers of information.

naturalistic philosophers are apt to regard Russell as wrong. First, because of his reductionism and atomism Russell takes for granted a now quaint-seeming view (among naturalists) of the capacity and right of physics—rather than science more generally—to populate our ontology. As we've explained, attention to the proliferation of non-reducing patterns of causal explanation in special sciences largely defines the contemporary naturalistic (unificationist) metaphysician's mission. For this, process theories are one possible piece of useful ordinance. Second, as also discussed above, Russell relies on the Humean conception of causation in pronouncing for eliminativism. Many, perhaps most, contemporary philosophers of science would agree with him that scientists don't seek causes in that sense, but would then go on to say that that is not the sense in which the idea of causation is either scientifically or metaphysically interesting. For these reasons taken together, Russell's thesis might be written off as of historical interest only.

Yet Russell's thesis is not dead. Some contemporary philosophers of physics continue to endorse and argue for it. Redhead (1990) explicitly acknowledges it when developing his own argument that talk of causes has no place in physics. (We qualify this by saying *fundamental* physics.) He considers an explanation of the fall of a body using Galileo's law, and asks what in the explanation might be regarded as the cause. As he notes, closely following Russell's logic (Russell 1913, 183 f.) the position at an earlier time 'can hardly be cited as the cause', that the acceleration is merely 'defined by the kinematic relationship' expressed in the law, that the law itself cannot be regarded as a cause because it isn't an event, and finally that citing the force as a cause is 'a very anthropocentric notion' (Redhead 1990, 146).[7] Developing his charge of anthropocentrism, Redhead maintains that '[t]o most physicists the old-fashioned idea of cause arises from the idea of our interfering in the natural course of events, pushing and pulling objects to make them move and so on. In modern physics there are just regularities of one sort or another' (147). Batterman (2002) adds further endorsement motivated by taking physics seriously. As noted in 4.1, the way of interpreting the concept of 'emergence' defended by Batterman in our terms is to say that if there is emergence, this means that there are real patterns that do not reduce to physical patterns taken, for explanatory purposes, as networks of causal relations. (The explanations in question will be Salmon-style, rather than Kitcher-style, explanations.) Then, Batterman notes, many philosophers reject emergence (in his sense) on the basis of their claim that attributing causal powers to emergents that are not shared by their 'basal constituents' violates physicalism (on any

---

[7] Compare this charge with Gunther Stent (1971), defending what amounts to a version of idealism as follows: 'Certainly the most basic law projected by man into nature is causality, or the belief that the events he observes in the outer world resemble his own conscious acts in their being connected as cause and effect, rather than occurring haphazardly... Indeed, even the most elementary dimensions in terms of which scientists attempt to describe the very events that causality is supposed to connect, such as time, space, mass, and temperature, are nothing more than projections into nature of man's own physiology and anatomy' (128).

reasonable construal of 'physicalism') (2002, 126). To this Batterman responds that 'I just do not understand the talk of causal powers. I realize that largely this is just an appeal to our intuitions about causes, but they are, as far as I can tell, intuitions that resist codification in a coherent theory or framework' (ibid. 127). We agree with Batterman that the explanatory force of the kinds of physical theories that he takes as exemplary does not consist in appeal to networks of causal relations. Finally, Norton (2003, 4) endorses Russell's thesis for Russell's reasons, and says 'mature sciences ... are adequate to account for their realms without need of supplement by causal notions and principles. The latter belong to earlier efforts to understand our natural world or to simplified reformulations of our mature theories, intended to trade precision for intelligibility' (ibid. 3). In our terms, Norton here claims that reading commitment to causation into 'mature' science—by which, to judge from his examples, he means physics—is a way of trying to domesticate it.[8]

The thought that, somehow, both the process tradition and Russell of 1913 could be correct raises perplexing possibilities. Could process theories unify the special sciences but not unify them with physics? That would be surprising and confusing, not least because one of the best known and most carefully articulated criticisms of Salmon's causal process theory is, roughly, that the opposite is the case, that is, that it works better for physics than for the special sciences (Kitcher 1989; see also Ross and Spurrett 2004). Were the thought that process theories worked for everything except physics instead taken as grounds for doubt about process theories, then perhaps the rebound view might be that Kim's causal exclusion problem and the attractions of Russell's thesis in physics mutually support one another: causal overdetermination is evidence for reductionism—in particular, for doubt that parochial causal patterns in special sciences are metaphysically significant—and then, once we go reductive, causation, absent in the reduction base, drops out altogether. Two overdetermining causes aren't one cause too many but two.

We agree with Russell—and with most current philosophy of science—that scientists do not generally seek 'invariable uniformities of sequence'. We also agree, for PNC-based reasons to be given below, that it doesn't follow from this that they seek nothing worth calling causes. We argue that scientists do seek causes, in an everyday sense we describe, and that they find them. This might look like a straightforward confutation of Russell's thesis. We go on to argue that it is not, since Russell's argument is most interestingly read as attacking the idea that there is a 'master' concept of causation, independent of anthropocentric bias, that should be expected to feature in the metaphysical unification of the scientific world-view. This might be a basis for dissolution of Kim's problem, but also seems to imply rejection of the point of process theories. However, we will skirt that conclusion too. Instead, we will defend the following

---

[8] Glymour (1999, 463) also denies that there are causal laws in physics.

consilience between Russell (and philosophers of physics who agree with him) and Salmon: the naturalist metaphysical project of seeking universal glue is well motivated; process theories are on the right track in this search; but 'causation' is a semantically unfortunate name for the glue to which they lead our attention. Part of our defence of the conclusion that it is semantically unfortunate that the cement of the universe gets called 'causation' involves identifying what we call conflicting centres of semantic gravity for 'causation'. Roughly, there is a folk centre, a philosophical centre, and a notional-world centre that has a proper place in science. It is the existence of this third centre that stops us from following Russell all the way to outright eliminativism about causation.

5.2 PHILOSOPHERS AND FOLK ON CAUSATION

To decide whether there are causes, or whether anyone seeks them, one needs to decide (i) what to count as a cause, and (ii) how to tell whether such things are real, or are sought. Russell partly settles (i) by reference to psychology. He also, as noted, regards psychological considerations as irrelevant to philosophy,[9] maintaining that the content of some cause-talk is anthropomorphic projection, failing to correspond to what is discovered by science. Russell does not claim, though, that the 'law of causality' itself arises from psychological considerations—he blames philosophers for the law. Philosophers might reply that in taking 'invariable succession' as causation they were following a long tradition, claiming to find the law of causation through psychological observation. This is particularly true of the history of the account Russell criticizes, namely, the Humean view that causes are event regularities or 'constant conjunctions'. Both Hume and his successors, including Kant (1933, 218) and Mill (1974, 326–7), derive the philosophical significance of constant conjunctions from purported observation of their importance to both everyday and scientific cognition. Naturalists, trying to shun appeals to intuitions and conceptual analysis, are particularly inclined to look for evidence in scientific psychology.[10]

---

[9] Russell (1913, 174) objects to one of the definitions of 'cause' in Baldwin's (1901) *Dictionary* that it is psychological, focusing on 'thought or perception' of a process where what is required is a definition of 'the process itself'.

[10] Melnyk (2003, 139–64) gives a Humean account of causation that does not rely on psychology. Melnyk's account of causation is parasitic on his account of explanation, and makes appeals to the existence of brute physical regularities that have no explanations. We agree that the PNC motivates belief in such regularities. If the universe is a block, these regularities will be standing structural properties rather than tensed relations. Melnyk's account seems subtle enough to allow for this too. But if we imagine a block version of Melnyk's view, it looks as if he would be forced to grant that all causation is formal and none is efficient; all that would then be left of *any* traditional notion of causation would be the occurrence of the word 'because' in explanations (see ibid. 141–2). Since we don't think metaphysics can be done as an exercise in semantics, we don't think this point sees off Melnyk's account. However, we do suggest that it would amount to

Some of what we know from the study of 'causal cognition' or 'causal learning' undermines the idea that our causal cognition is Humean. Contrary to early behaviourist dogma, associations are not all equally learnable. Rats find it much easier to learn (and show appropriate avoidance behaviour when cued for) an association between eating a certain food and nausea, and a loud flash and an electric shock, than either complementary pairing (Garcia and Koelling 1966). We might say that natural selection has projected a regularity in the world of rats, so that any given rat does not experience some types of phenomena as 'entirely loose and separate' (Hume 1975 [1777], 74). The projected regularities here, referred to by Garcia and Koelling as 'genetically encoded hypothes[es]' are not, furthermore, 'invariable uniformities of sequence'. Learning in the nausea case can take place even when the delay between stimulus and response is considerable (over an hour) and avoidance or aversion can also be learned following a *single* exposure.

The work of developmental psychologists already tells us a lot about the causal expectations of, for example, children around 3 months old, who are surprised when apparently cohesive objects seem spontaneously to fragment, to pass through one another, or not to exhibit a single trajectory through space and time (Spelke et al. 1995), or when apparently unsupported objects fail to fall (Baillargeon et al. 1995). Some headway is being made on the difficult question of what features of an object's behaviour and/or structure lead children to regard it as capable of self-motion (Gelman et al. 1995). Here too some events are not experienced as 'entirely loose and separate'.

It is also clear that we do not spontaneously think in terms of constant conjunctions, but rather in terms of networks of influence similar to directed causal graphs, or Bayes nets. Young children (between the ages of 3 and 5) can, *inter alia*, reason to unobserved causes, plan appropriate novel interventions to prevent a process they had only observed in operation (but never seen prevented), and make inferences about the 'direction' of causal dependence in cases of simultaneous change (see, for example, Gopnik et al. 2004). The evidence is thus inconsistent with a simple conditioning or associationist view where 'invariable uniformities of sequence' are what is learned.

The relevance of this to a naturalist should be clear: if 'our' concept of causation is an image of the kind of causal cognition that we actually engage in, then any feat of conceptual analysis starting from a model of 'Humean' constant conjunctions, or an appeal to alleged intuitions about such constant conjunctions, seems unlikely to be relevant to us, to science, or to metaphysics. To the extent that we have culturally universal intuitions about causation, this is a fact about our ethology and cognitive dispositions, rather than a fact about the general structure of the world. As we argued in 1.1 and

---

a backhanded concession that the substance of Russell's view—our topic here—might indeed be shown by physics to be correct.

1.2, intuitions are not evidence for their own content, in either science or metaphysics.

Empirical evidence in the specific case of intuitions about causation provides a particular exemplification of this general scepticism. Intuitive (in the sense of immediate, unforced) responses to causal questions are culturally varied. In particular it appears that adults in different cultures reason differently about causes. Chinese and American newspaper reports of very similar local multiple murders differed strikingly in the extent to which they referred to dispositional properties of the perpetrator (more common in America) and environmental ones (more common in China). Further, Chinese and American subjects in an experimental task involving counterfactual judgements about the murders had differing views about what sorts of changes (dispositional or situational) would have been likely to prevent the murder (Morris et al. 1995). Nor are our causal intuitions a reliable basis for basic scientific inference. It seems reasonably clear that people are relatively poor at judgements about conditional dependence, as measured in the Wason selection task (see Cosmides 1989); probability, where many will, for example, assert that the conjunction of two possibilities is more likely than one of the conjuncts alone (Tversky and Kahneman 1983), and in other areas, compared to communities of scientists. Our response in such cases is of course not to take our scientific practice as thereby impugned. Just as the scientific study of vision enables us to explain some visual illusions, so the study of causal cognition might be expected to show us that we were prone to 'causal illusions' (Gopnik et al. 2004). The proper task of philosophy and science in cases of causal illusion is to explain the answers people give to some causal questions, rather than to try to make the answers turn out to be correct.

The upshot of this section so far is twofold. First, human causal cognition is not Humean. One temptation to take regularity as definitive of causation is thereby removed. Second, our intuitions about causation are neither detailed enough nor stable enough for the entire folk concept as exemplified by this or that culture to be used as a basis for constructing the notional-world concept relevant to special sciences. To the limited extent that the folk concept of causation has a common core across cultural elaborations, its content is as follows. First, it construes causal relations as centred on some agent of change (animate or otherwise), thereby distinguishing agents and patients. Second, it postulates various transformative principles (often conceived as 'forces') proceeding *out from* an agent *to* the recipient of causal influence. Kim's (2005) endorsement of a 'generative' account of causation may illustrate the attractions of this feature of the core folk centre of gravity, as we suspect does Armstrong's (2004) contention that the philosophical theory of causation begins with our experience of 'biff' in acting in the world.[11] Third, the core folk notion of causation

---

[11] Norton (2003, 18–19), in trying to articulate the structure of the folk notion of causation, suggests that it has somewhat more structure than this. However, it seems to us that he reads

incorporates assumptions about time-asymmetry: causal influences flow from the past into the future. These three features seem empirically to exhaust the core folk concept.[12] The artificial intelligence researcher Patrick Hayes (1979) spent years trying to work out deductive models of the everyday concept of causation as the basis for designing a simulation of a person. These models began with the three features just identified, then added the knowledge people have of the mechanics of their own bodies to produce programs for distinctive manipulations of solid objects and liquids. Folk psychology would be conjoined with the core folk causal model to produce programs for causally influencing other agents.

It is open to a naturalist to allow a folk notion to fix what is to count as a cause for purposes of inquiry into anthropically constructed or centred domains and practices. However, the naturalist cannot allow anything but science to say whether there *are* such things as causes. With a watchful eye open for illicit appeals to folk notions, then, let us turn to consider science.

5.3 CAUSES IN SCIENCE

Russell maintains that 'advanced' sciences have no use for the 'law of causation' and don't seek instantiations of that law, even claiming that in such sciences the *word* 'cause' never occurs. Whether or not Russell is correct about the relevance of the 'law of causality' to science, many of the sciences at least claim, explicitly, to pursue and answer causal questions. By this we mean (for now) the weak claim that the output of many communities of scientists includes papers and other documents that refer to the 'causes' of various phenomena, or the 'mechanisms' that 'produce' them, or ways in which various effects can be 'inhibited' or 'prevented'.

Consulting a single issue of *Science* (volume 300—June 2003) we note that things referred to as being 'caused', 'determined', or being the 'result' of something within the first few pages of the issue (excluding letters, pieces on policy matters, and editorials) include 'broadening of lines' in spectroscopy, 'global extinctions', 'a gradual drop in IOM [Indian Ocean Monsoon] intensity over 5000 years', 'an electrical barrier', 'how genes are expressed in two closely related species of fruit fly', decrease in 'the reactivity of some state-to-state channels', 'efficient, spontaneous cell-cell fusion', 'the common cold', 'oceanic primary production', and 'pear blister canker and cadang-cadang disease of coconuts'. A search for

---

too much philosophical sophistication and restrictiveness into the folk concept. Norton himself inadvertently provides some evidence for this when he goes on to show how the various folk properties of causal relations that go beyond the ones we have identified are all subject to being waived in special applications.

[12] Recall our discussion of the containment metaphor in 1.1 where we also referred to 'causal oomph'.

articles in which the word 'cause' appeared in the online archives of *Science* between October 1995 and June 2003 returned a list of results containing 8,288 documents, averaging around 90 documents per month, in which the word 'cause' occurred. 'Effect' was more popular—10,456 documents for the same period, around 112 per month. 3,646 documents included the word 'influence', 8,805 the word 'response', 2,683 the word 'consequence'. (In case the reader is wondering, 'philosophy' turned up in 553 documents, much less often than 'cause', and approximately half as often as 'soil' with 1,007.)

So scientists talk about causes a great deal and, unless intolerable strain is to be placed on the notion of an 'advanced' science, or 'advanced science' is just a misleading way of denoting fundamental physics, Russell is simply wrong that the word cause 'never occurs'. This is a result of very modest significance. As already noted, most philosophers of science today agree with Russell that scientists don't seek laws of constant succession. What, then, are the things scientists call causes? We take it that what is being provided, or attempted, when scientists talk about causes is causal explanation. A well-established set of philosophical questions about explanation in general, and causal explanation in particular, arise here focused on the problem of what makes for a good (causal) explanation in a particular science, in some group of sciences, or in general. For present purposes we adopt, without argument, a broadly Kitcherian approach to these questions. As Kitcher has emphasized (for example, 1981), the explanatory relations accepted in any science are determined by the 'general argument patterns' (see 1.3) accepted in that science. For reasons noted above, and returned to below, we are much more concerned than Russell to take seriously the fact of the number and variety of sciences. We agree with Kincaid (1997) that scientists typically rely on special causal relations that are parochial to their individual disciplinary traditions, and that don't, at least on the face of things, reduce to or welcome analysis by reference to a single over-arching conception of the kind that philosophers typically seek.

It thus turns out that when contemporary scientists talk about causes what they are typically doing is documenting the very kind of relationship of multiple dependence that Russell takes to be distinct from putative 'causal' relationships. Russell's descriptive claim about what scientists say is incorrect, and he gives no compelling naturalist reason for saying that explanatory interdependencies shouldn't be called causes. Further, if we're correct that cognitive science tells us that our own cognition isn't Humean, then a key temptation (for a naturalist) to make event regularity definitive of causation disappears. Since in science causation doesn't appear to reduce to event regularity, we are left with no motivation for the Humean analysis. Might our verdict then be that scientists instead join the folk in using causal ideas to indicate hypotheses about 'biff' and 'flow'?

Can we seriously attribute to scientists such hazy and immature metaphysical generalizations as refer to 'flow' and 'biff'? If the resolution we are canvassing at this point were intended as anything more than a way-station in the argument, then as naturalists we would have to give this question careful attention. Since we don't ultimately attribute the aforementioned hazy metaphysical conviction to scientists, however, we can just suggest how someone might most plausibly try to do so. First, one would maintain that scientists have no distinctive *general* idea of causation of their own; they just inherit folk custom here, and then tighten application in any given instance (so as to make causal claims suitable for testing and measurement) when they hypothesize or aim to discover specific *mechanisms*. It is a regularly encountered thesis in the methodological literatures of the special sciences that explanations have not been provided until a mechanism, of a kind distinctive to the special science in question, is specified. (See, for example, Elster 1989 and Kincaid 2004). 'Mechanism', as this idea is used outside of mechanics, typically seems to mean nothing more definite than 'stable, quantitatively parameterized and measurable transmitter of influence'—which in turn looks as if it could be rendered without loss as 'biff-flow channel in a specified model'.[13] The substantive point here is that 'influence', as restricted to the sort of thing one needs a mechanism to transmit, is plausibly equivalent to the folk concept that Armstrong calls 'biff'.

These reflections might seem at first glance to furnish a basis for resolving all questions about what a contemporary naturalist ought to think about Russell's thesis. Neither the folk nor scientists conceptualize causation in a Humean fashion, in part because both consider more complicated interdependencies than bare regularities, and both can (more arguably) be taken as generally supposing, when attributing causal influence from some x to some y, a 'flow of biff' from causal agent x to patient y. If we were to accept all this, then we could say that we let folk custom fix what general sort of relation counts as causal, and then assign to science the job of telling us in detail what instantiates the set or network of actual causal relations.[14] Then our verdict on Russell's thesis would be that Russell conceded too much semantic priority to Humeans after all, and was thereby blinded to the compatibility of everyday and scientific ways of thinking.

---

[13] Consider, for example, how little additional structure the following examples of causal mechanisms given by Kincaid (1997) have in common: strategies of organizations produced by the mechanism of differential birth and survival in different environments (sociology) (19); rational expectations as mechanisms explaining monetary policy failures (economics) (140); signal sequences attached to proteins as protein-sorting mechanisms in cells (cellular biology) (53); protein structures as mechanisms determinant of binding strengths in antibodies (biochemistry/physical chemistry) (60); distributed rule following by networked individual people as the mechanism that steers an aircraft carrier (cognitive anthropology) (124); homeotic genes as determinants of development paths from imaginal disks to specific organs in organisms (developmental biology/genetics) (61). In each special science there are background theories about what kind of processes count as mechanisms.

[14] Note that the fit between the folk notion and the output of science need not be perfect on this account—the folk can be surprised about what causes what, and also be called upon to make various modest revisions to their view of what causes are without being subject to the wholesale violence of Russell's proposal that there are no causes at all.

This sort of resolution would be better than either simple eliminativism about causation, or an attempt to save causation by defence of Humeanism. It is very close to a proposal of Norton (2003), whose general view is in turn the closest one to ours that we find in the literature. However, we think it is not quite adequate. For one thing, it is silent on any connection between causal relations and ontological unity. Norton explicitly denies such a connection, for reasons similar to Kincaid. Certainly, it is a plausible view to hold that causation has nothing to do with the glue the unificationist seeks, and that Kim and others are likewise confused in regarding causal multiplicity as being a problem for folk metaphysics. But unless one thinks that fundamental physics is the only metaphysically serious science—as Norton's failure in his paper to consider any cases from special sciences as possible counterexamples to Russell's thesis suggests that he might—then leaving the problem at this point will be fully satisfying only to the advocate of metaphysical disunity.

As we have urged, Russellian eliminativism about causes is a plausible—indeed, we think strongly PNC-motivated—position in *fundamental physics*. No resolution of the problem of unity in the face of multidisciplinarity, with respect to causation or any other putative glue, can by itself resolve the tension between Russell's thesis *as applied to fundamental physics* and causal process theories. An advocate of metaphysical unification could just deny that causal process theories have got anything right at all, and set off in search of alternative glue. But this would imply an error theory of the underlying metaphysics of special sciences—Kincaid (2004), for example, would come out as flatly wrong—*and that* in turn leads back to the position of Dennett's we rejected in 4.2, according to which fundamental physics studies the real *illata* and special sciences study ontologically second-rate *abstracta*. This would be reductionism (though not of Kim's kind), and we fear it is where Norton's position leads.

Imagine a philosopher who self-consciously chose this route (unlike Dennett, who falls into it by mistake), concluding that special sciences (but not fundamental physics) are pragmatic pursuits made necessary by the conjunction of human epistemic limitations and parochial concerns. For such a philosopher it might not be surprising if it turned out that minimal conceptual revision was necessary for understanding the role of causal generalization and explanation in special sciences. But it then would be surprising if the homely model of causation were vindicated in application to fundamental physics, since this would amount to the discovery that our natural conceptual apparatus had been well evolved for domains of no consequence to our fitness. (As we argued in 1.1 and 1.2, this problem should trouble the neo-scholastics.) The more ambitious unificationist hope, that special sciences and physics can justifiably be interpreted as pursuing an objective and coherent description of reality in tandem, should lead us to look for comprehensive revisionary principles first, then ask how these can accommodate the practices we find in special sciences, and what relations these have to patterns of folk usage.[15]

We can make the above agenda more explicit by framing it in terms of the theses with which we have been contrasting Russell's, viz. causal process theories and Kim-style reductionism. The point of process theories, as discussed earlier, is to distinguish real causal processes from merely apparent ones. Salmon, like Russell and like us, holds that, on questions about objective reality, not only should science trump anthropocentric perspectives (that is, we should be naturalists), but that within science fundamental physics should trump the special sciences whenever they conflict. Thus for Salmon the paradigm instances of real causal processes are to be sought in fundamental physics, which he takes to be microphysics. Such commitments fuel the recent controversies based on worries about causal exclusion, with the added distraction that the promoters of the worries ignore actual physics in favour of an imaginary physics that hides the problem.

Kitcher's (1989) early critique of Salmon centred on just this point, arguing that our identification of causal interactions in most, probably all, macro-processes depends more on our knowledge of the general causal structure of the macro-world (as uncovered by sciences other than fundamental physics) than it does on the sorts of micro-process paramount for Salmon. However, for Kitcher such considerations only give reason for not attempting to extract direct methodological guidelines for conducting the business of the macro-sciences from process accounts, and do not seriously undermine the value of Salmon's approach and its descendants when they are understood as saying something about the kinds of real structures in the world that science aims, in the limit, to discover. That is, in Kitcher's view a sound point about explanatory practice in science need not by itself establish anything about fundamental ontology. This is why Kitcher is able to describe his approach and Salmon's as complementary, characterizing his own work as analysing 'top-down' explanation, where phenomena are explained by having their roles in wider ensembles of regularities fixed. He contrasts this with 'bottom-up' explanation, the sort he says is analysed by Salmon—who in turn explicitly endorsed Kitcher's suggested complementarity (Salmon 1990). The problem with this rapprochement is that it sacrifices the ambition for unification of fundamental physics and special sciences, if it is agreed that special sciences explain by reference to causal processes but fundamental physics does not. This is slightly ironic in light of the fact that Kitcher's overarching aim is to defend the idea that the scientific value of explanation consists in the unifications it furnishes. The Kitcherian rapprochement, at this point in our dialectic, supports explanation as unification within disciplines, but frustrates general unification—the very point of metaphysics, according to us.

---

[15] We didn't worry about an analogous issue in the case of cohesion because, as argued, cohesion isn't a folk concept.

We have now argued our way into a tight corner. Getting out requires us to seek an account that (i) leaves scientists free to talk about causal processes without risk of embarrassment in the company of metaphysicians, but (ii) follows Russell in denying intuitions (folk or Humean) about causation *any* role in informing the metaphysical foundations of physics, or science more generally. The view we outline joins Russell, Redhead, and Norton in leaving the folk concept of causation intact (as our view of the concept of things leaves that concept intact, rather than claiming like Eddington that it applies to electrons but not to tables). We appeal to two considerations, neither of them derived from natural or folk intuitions, to explain why the core folk concept of causation need not conflict with naturalistic philosophy of science or naturalistic metaphysics. One consideration is the very thin-ness of the core folk conception discussed in 4.7. The folk are committed only to the existence of processes, but not to the metaphysical idea that causal processes are glue. The second consideration will be a set of reasons for thinking that if Russell, Redhead, and Norton are right about fundamental physics, then the exact sense in which they might be right—given the actual state of contemporary physics—suggests physical and metaphysical explanations for the existence of processes, while implying that causal processes are not metaphysical glue. Our view grants Salmon and other process theorists a good deal—in particular, it accords with their belief in objective causal relations, and with their idea that those relations should be explicated in terms of processes. In requiring of physical theory that it explain why special sciences study causal processes as real patterns it serves unification. However, in elaborating on how causation might play no role in fundamental physics, it accommodates Kitcher's objection to Salmon and locates what is right in Russell's thesis once his primitive reductionism is set aside.

5.4 LETTING SCIENCE HOLD TRUMPS

As we pointed out above, one premise of Russell's, deriving from his naturalism in application to the state of empirical science when he wrote, is a direct and simple form of reductionism about the relations among the sciences. When he champions letting science hold trumps on conceptual revision, he supposes that it is specifically fundamental physics that has the cards. He argues that because the functional generalizations of physical theory involve no temporal asymmetries, whereas causation in the anthropomorphic sense depends on such asymmetry, the universe itself cannot have causal relations as a feature, except in the sense that the universe includes cognitive systems and cognitive systems anticipate by constructing causal relations. He thus assumes what current worriers about causal exclusion do, namely, that whatever general patterns of relations are identified by any sciences must ultimately be (at least in principle) identifiable and explicable by fundamental physics.[16]

When Russell wrote his essay this was still a reasonable bet. However, we have been arguing throughout this book that the bet has been failing almost since the moment Russell joined others in putting his chips on it—as noted above, de Broglie's work was just around the corner. The weak version of physicalism supported by current scientific practice, the PPC, is too weak to license Russell's inference from eliminativism about causation in fundamental physics to eliminativism about causation in special sciences. If physical theory abjures reference to causal relations but biology or economics invokes them, this generates no contradiction.

Although the PPC doesn't pack the whole punch of mereological reductionism, and carry us all the way to Russell's eliminativism about causation, the PNC does pack enough punch to allow us to take Russell's reasoning as a template, requiring us only to complicate it slightly. By the PNC, if causation does not figure in fundamental physics, then the class of causal relations should not be supposed to pick out an extra-representational real pattern in metaphysics. This is compatible with the possibility that metaphysics and physics can explain why a notional-world concept of causation is necessary for organizing special sciences that successfully identify real patterns.

We want to see where reasoning that resembles Russell's leads if we replace his reductionist version of the priority of physics with the weaker PPC. Consider the set of fundamental physical theories currently being taken seriously by physicists—that is, those used or investigated in funded research programmes that aim to unify physics and so shade into metaphysics (the serious candidates for accounts of quantum gravity reviewed in 3.7.2). Suppose, following the arguments of Chapter 4, that none of the generalizations in these theories invoke the kinds of causal asymmetries characteristic of the flow (or process) concept of causation, while some or all special-science causal principles do presuppose such asymmetries (as Kincaid persuasively argues). Then three possibilities are open with respect to the status of these asymmetries in metaphysics.

One possibility is that fundamental physics ultimately endorses a 'background-free' theory of the sort promoted by Smolin (2000). (Recall from 3.7.2 that background freedom means that the theory has no, or almost no, structure that is independent of dynamics.) If this is the future of physics, then the

---

[16] Neither Russell nor the causal exclusionists expect physics to step in and take over the explanatory projects of the special sciences. Kim (1998), unlike Russell, has a sophisticated account of why this won't happen that isn't just pragmatic. That is, Kim argues that special sciences study the distributions and effects of properties that are 'micro-based' but not themselves microproperties; and these latter are the direct targets of physical generalizations. However, because for Kim special-science properties supervene without cross-classification on physical ones, physics must still be able to identify and explain, at least in principle, all general relations that hold among groups of these micro-based properties.

search for metaphysical glue is forlorn, and the substance of radical disunity views might be satisfied. Smolin favours Loop Quantum Gravity (LQG) as the candidate unifying theory for GR and QM in part because he thinks it is (largely) background free.[17] Notice, however, that even here both Russell and the causal process theorists could feel substantially vindicated. The eliminativist would not be able to rely on the claim that causation is based on asymmetry and fundamental physics insists on universal symmetries. However, the core folk causal notion is committed to more than just asymmetry; it is anchored in the anthropomorphic experience of flow of influence. In the limiting case of background freedom, nothing fundamental will be available to stand proxy for the universal direction of causal influence, and so the anthropocentric experience fails to find any analogical grip there at all. At the same time, process theories might be necessary for explaining and predicting the real patterns in our part of the universe, in which the Second Law holds sway. It would be the responsibility of LQG to explain the local asymmetries, which would in turn explain the utility of time-asymmetric processes as locators.

However, a completely background-free physics is in any case just a limiting possibility, which most physicists (including Smolin) regard as unlikely. (We indicated the grounds for such doubt in 3.7.2.) We can sort other perspectives in quantum gravity by reference to the distance they stand away from this limit. Shifted just over from it are possible theories that give only geometrical properties of the universe, which neither constrain nor are consequences of any local processes. Here talk of 'glue' still seems inappropriate, and the following would all look like motivated speculations for the metaphysician: the universe is minimally unified, there is no causation at the level of fundamental physics, process theories might still describe every real pattern we can measure except some features of the large-scale structures of spacetime, but if they do this cannot be explained directly by fundamental physics, though they must be compatible with it.

A more interesting possibility (with respect to the dialectic of this chapter), which is compatible with both the PNC and the Russell–Redhead–Norton thesis about physics, is that fundamental physical theory helps us to say something about what all of the parochial special-science models of causal processes have in common, though these processes make no first-order appearance in fundamental physical theory itself. If this possibility obtains, the metaphysician is justified in studying physics in search of universal glue, but causation cannot itself be that glue. The triumph of superstring/M theory would support this metaphysical picture. Investment of energy in superstring/M theory is usually justified by the fact that it takes seriously the alleged induction over the history of physics that respect for symmetries is the key principle of any would-be 'fundamental'

---

[17] We worry that this may be an instance of a physicist committing a sin more typical of philosophers: letting metaphysics drive speculations in physics, instead of the other way around.

physical theory (see 3.7.2). As we have argued, the folk concept of causation may not be committed to very much, but it insists on asymmetry between causes and their effects. Thus if M theory turned out to yield the content of fundamental physics, then causation could not be universal glue. ('Stringhood' might be the universal glue.)

Recall that M theory and LQG are not strictly exclusive of one another. There are possible extensions of M theory—albeit not yet mathematically formulated—which would explain why LQG holds (Greene 1999; Smolin 2000). This would be a particularly inviting theoretical environment for reconciling Russell and the process theorists. Though LQG can be interpreted in the context of a block universe, it is usually presented in tensed terms. Thus it has resources for potentially providing a correct description of the historical development of the region of the universe in which we live, or even of the whole of what physically exists, that would be given in terms of processes. M theory might then offer the unifying structural account at one higher level of abstraction. At this level Russell's thesis would be confirmed.

We claim that on all these possible futures, there might be real processes—chemical ones and evolutionary ones and macroeconomic ones (etc.)—but that there is nothing they would have in common beyond satisfying the thin folk conception of time-asymmetric influence-flow described earlier. We are thus in basic agreement with the claim of Kincaid, cited earlier, that many of the mechanisms for transmitting influence found across the sciences have nothing more general in common than that they transmit influence. On some scenarios now seriously entertained (background freedom), fundamental physics would be merely compatible with asymmetric influence-flow in some regions of reality. On another account—M theory—there would be no fundamental asymmetries, hence nothing corresponding to causal processes at the fundamental level; yet fundamental physics would explain why such asymmetries hold of real patterns measurable in special-case regions where values of some of the fundamental eleven dimensions of strings take vanishingly small values.

It should now be clear in what sense our view accords with Russell's. Causation, we claim, is, like cohesion, a notional-world concept. It is a useful device, at least for us, for locating some real patterns, and fundamental physics might play a role in explaining why it is. Or, if fundamental physics does not directly do this, then the PPC at least requires that such time-asymmetry as appears to be characteristic of our region of the universe be compatible with fundamental physics; and the ubiquity of such time-asymmetry would in turn explain the value of the 'causation' locator in special sciences. However, we think the current state of fundamental physics supports Russell's, Redhead's, and Norton's expectations that it will not directly generalize over causal processes.

How much does this allow us to concede to causal process theories? Salmon initially sought to provide an account of causation as universal glue. Taken strictly, this is incompatible with Russell's thesis. However, faced with the problem that quantum theory requires uncaused parameter values, Salmon (1990) ultimately expressed stark pessimism about the idea of causation as universal glue. Note that this reason, while different from those we have pursued here, is a PNC-compatible motivation. What Salmon considered more important to the enterprise of process theorists was the attempt to distinguish cases of 'real' causation from pseudo-processes; and process theorists still think that this enterprise is getting somewhere. Now, this may also be thought to be incompatible with Russell's thesis. After all, if causation isn't an extra-representational real pattern, then how can there be a metaphysical distinction to be drawn between real and pseudo processes?

The answer to this question rests on reference to Russell's mistaken belief in ontological reductionism. For a non-reductionist, the failure of causal concepts to appear in generalizations of fundamental physics doesn't imply that these concepts simply denote fictions.[18] As discussed in 4.5, all our evidence tells us that we live in a region of the universe—which might or might not be coincident with the whole universe—in which the degrees of freedom of every system we approximately isolate for measurement and projection is restricted by various asymmetries. This region is highly isotropic, and supports robust (that is, projectible across all counterfactual spaces within the boundaries of the region) distinctions between sources and recipients of information. That is sufficient to establish the scientific utility of directional flow. This in turn implies that many of what we refer to as causal processes are real patterns, even if this does not mean that they exemplify any extra-representational real pattern of general 'causation'. We said earlier that we do not attribute to scientists a metaphysically serious commitment to hazy folk intuitions about biff and flow. We can now say what we meant by that. Taking directed, irreversible transmission of influence, plus conservation principles, for granted is rational when you know you're taking all your measurements from regions governed by the Second Law. This practice conforms so closely to what is enjoined by a folk metaphysical endorsement of causation that no serious risk of misunderstanding arises if the scientist helps herself to the culturally inherited folk idea of causation. But she need not thereby endorse folk metaphysics, any more than she must endorse folk physics. It is either to fundamental physics directly, or to some domain less general than fundamental physics but more general than our domain of immediate measurement (for example, LQG, in the scenario where that is

---

[18] Norton (2003, 3) says something that sounds similar: 'I will characterize causal notions as belonging to a kind of folk science, a crude and poorly grounded imitation of more developed sciences. While this folk science is something less than our best science, I by no means intend to portray it as a pure fiction.' As we will go on to do, Norton characterizes this not-altogether-fictional status in terms of 'heuristics'. However, on our view special sciences, where causal relations are ineliminably quantified over, are not 'crude and poorly grounded imitations' of physics; they track real patterns that fundamental physics does not and cannot. Therefore, the sense in which we regard causal relations as 'heuristics' separates them more sharply from fictions than does Norton's sense of the term.

explained by M theory), to which one must turn for an explanation of how regions governed by the Second Law can come to be and remain stable (at least temporarily).

Is this enough to show that there are real processes, and so also cases where we might misattribute process, that is, pseudo-processes? Or does it merely show that it is useful for us limited beings, stuck in a restricted region of reality governed by special asymmetries, to book-keep our experience in terms of the idea of causal process? This question amounts to asking once again how it is possible to be a naturalistic realist (in denying anthropocentric instrumentalism) and a non-reductionist (in denying that fundamental physics exhausts or strictly implies everything that is true) at the same time. That general question is of course the subject of this book as a whole.

ITSR explains why parochial causal concepts are robust in special sciences, but also why nothing stronger than the thin notion of flow or process unifies them. First, special sciences, by definition, are not constructed from the vantage point of the limit. By this we mean not only that they aren't constructed from the physical limit; they also aren't constructed from the limits achievable at their own scope boundaries. They are developed in, and are accounts of parts of, a notional world of informational asymmetries. We don't know, until fundamental physics tells us, whether in this respect the notional world agrees with the correct description of the actual world available from the limit. So causation, as it figures in science, is a notional-world construct. However, as we argued in 4.5, this notional world provides a useful heuristic for locating some real patterns. It is because special sciences are concerned with relatively isolatable regions of the universe which involve cross-classification from the perspective of less-than-fundamental but nevertheless limiting domains for them, that the causal relationships on which they variously focus will appeal to different aspects of their information-carrying potential. This is equivalent to Kitcher's and Kincaid's point that explanatory argument patterns distinguish special sciences, and are reflected in parochial restrictions on what kinds of relations count as causal for each of them.

This perspective helps us make sense of the relationship of folk causal concepts to scientific applications. RR makes it a necessary condition on a pattern's reality that it not be informationally redundant in the complete account of the universe. When actually formulating scientific theories short of this limit, we cannot know that our posited patterns would satisfy this criterion, but we care whether they do, and so we make our existence claims at least implicitly provisional. Furthermore, as science progresses we adjust our ontology in accordance with our concern for ontological parsimony.[19] However, practical folk have no systematic reason to be interested in this constraint. Nor could natural selection attend

---

[19] As naturalists, we do not claim that greater relative parsimony provides any reason to think a scientific theory relatively more likely to be true. The point is simply that, as emphasized in

to it when it designed the native anticipatory apparatus used by the practical folk. In coping with problems of scarcity, tracking the trajectories through local space and time of bundles of rewards is almost everything. Attention to wider informational dynamics in which processes are embedded typically delivers few if any additional payoffs, and may get in the way of payoff maximization because of computational costs.[20] Therefore modelling causal relations as sequences of collisions of objects in time is a sensible heuristic. Furthermore, for social animals like people, or mobile asocial animals that prey on and/or are preyed on by other mobile animals, it might even make sense to model all causal relations on volition, with rocks amounting to limiting cases of agents with minimal sophistication of purpose and adaptability. Such models of causation won't generalize when one tries to do science, that is, when one's scope of concern widens beyond what nature could directly design creatures to worry about. However, special extensions of the thin concept of flow causation, such as animism, can then be dropped, leaving us with the kind of stripped-down notion of process sought by Salmon and his successors. What these theorists give us in so far as their analytical efforts are successful isn't *universal* glue, but it is indeed a binding agent that holds special sciences together in parts of reality where they have a point.

Let us summarize the argument of the past few sections. We agree with Russell that scientists generally don't seek 'invariable uniformities of sequence'. Because we're persuaded that cognitive science shows that everyday people don't anticipate in terms of such uniformities, we're not convinced that this fact is important. Second, we asked, if they don't, does it follow that they don't seek what are legitimately called causes? We worked up some concern over a seeming tension lying within the basis for Russell's answer. On the one hand, his crucial premise for justifying the answer 'no' is that physicists don't mention causes in their fundamental generalizations. But since special scientists carry on about causes all the time, Russell's denial entails the conclusion, unacceptable for a naturalist, that most scientists are ontologically confused in ways philosophers can diagnose. We eventually found the solution to this tension just where a naturalist should want it, in correcting empirical assumptions about physics that weren't borne out by the subsequent development of science. There is no justification for the neo-scholastic projection of causation all the way down to fundamental physics and metaphysics. Nor, however, does science presently motivate us to conclude that special sciences don't measure real patterns that must be modelled as processes.

Chapter 4, it does not help to unify science, and thus makes no contribution to metaphysics, to say that objects generalized over only by redundant theories exist.

[20] This statement needs some qualification. Some folk engage in practical activities of a kind that evolution could not anticipate, like buying and selling assets in futures markets. When you do that sort of thing, it is wise to book-keep your structures and relations more like a scientist.

5.5 LAWS IN FUNDAMENTAL PHYSICS AND THE SPECIAL SCIENCES

We now return to the issue left at the end of Chapter 4. There, we mentioned the dispute between Earman and Roberts (1999) and Kincaid (1996, 2004) over whether fundamental physics and special sciences differ from one another with respect to their capacities to discover exceptionless laws. Our aim now is to show how the view of the relationship between fundamental physics and special sciences defended in this and the preceding chapter, and in particular our view about the place of the concept of causation in this relationship, allows us to recover the central insights of both parties to the debate. This continues our general strategy of defending our position by appeal to its achievements of consilience.

Of the two views in tension, Kincaid's belief that special sciences are just as much in the business of discovering laws as fundamental physics is the more fashionable among current philosophers. Part of the motivation that leads them to favour this opinion is naturalistic in just the sense captured by the PNC. Saying that special sciences do not or cannot achieve something—discovering exceptionless laws—that fundamental physics sometimes achieves seems at first glance to impugn special sciences and find them to be 'metaphysically second rate' on the basis of a philosophical judgement. But this is just what the PNC forbids. We have pronounced against demands to implicitly (or explicitly) revise special sciences because they fail to respect traditional intuitions about the causal closure of the physical. Thus it might appear that we are compelled to join Kincaid, and also Fodor (1991), Lange (1993), and Pietroski and Rey (1995) in maintaining that the philosopher's idea of lawhood should be based just as firmly on what special scientists take to be 'their' laws as on exemplars from fundamental physics. If one endorses this stance, and also agrees with these authors (including Kincaid 1996 but not Kincaid 2004—see below) that special sciences do not discover exceptionless laws, but instead discover only laws that are true subject to *ceteris paribus* conditions or 'provisos' ('*ceteris paribus* laws'), then one must reject as a badly motivated philosophical dogma the old doctrine that being exceptionless is among the necessary properties of a law.

Motivated by his strong naturalism, Kincaid works with a notion of lawhood that is highly deflationary by historical standards: 'A law', he says, 'picks out a causal force or factor' (Kincaid 2004, 179). By this he does not mean that picking out a causal factor is necessary for something's being a law. His naturalistic scruples are so strong that he doubts it makes sense to seek general necessary (or sufficient) conditions on lawhood, because this would have to rest partly on a priori conceptual analysis. Rather, what he maintains is roughly the following. All scientists—fundamental physicists and practitioners of special sciences alike—seek generalizations that non-redundantly predict and explain phenomena (which requires some degree of support for counterfactuals; see Kincaid 2004, *ibid.* 174, and discussion below). The main, though not exclusive, way they do this is by identifying causal factors that govern ranges of situations. Whenever a science succeeds in doing this, it has *ipso facto* discovered a law. Or, at least, so Kincaid asserts. But even if we agree with him that special sciences identify causal factors, why should we assimilate this to the discovery of laws?

One premise identified by Earman and Roberts as a ticket to this inference, but which is abandoned by Kincaid (2004), is the claim that fundamental physics doesn't discover exceptionless laws *either*; that, as Earman and Roberts put it, 'it is provisos all the way down to fundamental physics' (Earman and Roberts 1999, 440; this is the view defended by Cartwright 1983). If this were right then the general conclusion of Kincaid, Lange, etc. would indeed be irresistible for a naturalist. The idea of an exceptionless law would then have no PNC-compatible interest. Were that so then we might as well grant Kincaid his deflationary interpretation of lawhood, because there would be nothing else of well-motivated significance that 'discovering a law' could amount to. (Of course, some minimal conceptual constraints are still lurking in the background here. If identifications of causal factors supported no counterfactuals, so were just summaries instead of generalizations, it would be perverse to refer to them as laws, since all connection between the traditional concept and the deflated one would have dropped out. We will return to Kincaid's interesting remarks about the modal force of what he calls 'laws' shortly.)

This version of the argument is the one of immediate interest to us, because it turns directly on claims about the relationship between fundamental physics and special sciences. It is also the version contested by Earman and Roberts (1999). Later, we will consider the modified position defended in Kincaid (2004). At that point we will argue that he backs into a view that is implicitly closer to ours than his rhetoric (which remains largely consistent with his earlier position) suggests. This will complete our argument for the consilience of ITSR.

Let us first summarize the 'generic' Kincaid position (that is, the part of the position that is common to its 1996 and 2004 articulations). We of course agree with him that it is not among the rights of the philosopher to pronounce some institutionally *bona fide* sciences methodologically or metaphysically defective. We also agree, for the reasons explored in 4.5 and 4.6, that identification of types of causal process is the principal sort of achievement at which special sciences aim, and at which they regularly succeed. Furthermore, just as Kincaid says, it is by means of such achievements that they non-redundantly predict and explain phenomena—that, in our terms, they track real patterns. (To cite the leading example discussed in Kincaid (2004), economists over the years have documented, by careful measurement of found variations, how changes in supply and demand relationships cause changes in prices, under increasingly rich and detailed specifications of kinds of circumstances.) We agree with him that discovery of causal factors does not imply prediction of exceptionless regularities, since no causal factor identified by a special science empirically manifests itself in absolutely all circumstances. Indeed, we join Kincaid (2004, 171–2) in agreeing with Cartwright (1983) that exceptionless *causal* regularities are not even predicted by any part of physics.[21]

However, we think that Earman and Roberts are correct in their claim that fundamental physics discovers something of a kind that special sciences don't; and we call this kind of something a universal real pattern. We argued in the preceding sections of this chapter for Russell's and Redhead's thesis that these real patterns are not causal. They are instead structural, in the sense articulated in Chapters 2 and 3. In light of our agreement with Kincaid about what special sciences do, this means there are, according to us, two respects in which the achievements of fundamental physics differ from the achievements of special sciences: the former are exceptionless, and they do not consist in the identification of causal factors.

We should reiterate, in this context, what we mean in saying that the generalizations of fundamental physics are exceptionless. We do not mean that they potentially predict every event or process. Rather, we mean that any measurement taken anywhere in the universe is a potential counterexample to them. This is not true of the generalizations of any special science. At one level these two claims are tautological: vulnerability to counterexamples from anywhere is the basis on which we have defined 'fundamental' physics, and it is restriction of the scope of its generalizations that defines a science as 'special'. However, these definitions are not based on a priori conceptual reflection. They are part of the content of the PPC, which is an institutional norm justified by appeal to the history of science.

We thus directly disagree with philosophers, including Kincaid (1996), who believe in 'provisos all the way down'. Earman and Roberts (1999) offer what we regard as a compelling argument for our side of this disagreement. To a large extent, the rival opinions must obviously be based on interpretations of the actual history of physics. We are not aware of a convincing history of recent physics that succeeds in undermining Earman and Roberts's following remarks:

Typical theories from fundamental physics are such that *if* they were true, there would be precise proviso-free laws. For example, Einstein's gravitational field law asserts—without equivocation, qualification, proviso, *ceteris paribus* clause—that the Ricci curvature tensor of spacetime is proportional to the total stress-energy tensor for matter-energy; the relativistic version of Maxwell's laws of electromagnetism for charge-free flat spacetime asserts—without qualification or proviso—that the curl of the E field is proportional to

---

[21] Cartwright, Kincaid, and we endorse the same argument for this conclusion where non-fundamental physics is concerned. Whatever holds of special sciences in general holds of non-fundamental physics. Cartwright, unlike us, does not believe there is any fundamental physics (in our sense). So we need an additional basis to Cartwright's for extending her claim beyond the special sciences. We provided that basis in 4.5 and 4.6 when we argued that fundamental physics does not describe causal regularities.

the partial time derivative of the **B** field, etc. We also claim that the history of physics and the current practice of physics reveal that it is the goal of physicists to find such strict, proviso free laws ... [W]e believe that a fair reading of [that history] shows that when exceptions are found to the candidates for fundamental physical laws, and when the theorists become convinced that the exceptions cannot be accommodated by explicitly formulated conditions in the language of the theory, the search is on for new candidates. (1999, 446)

To resolve this standoff over the interpretation of science satisfactorily, Earman and Roberts owe some explanation of the fact that important recent philosophers have taken the opposite view. They find this explanation in a misinterpretation (explicitly advanced by Giere 1988a, Fodor 1991, and Lange 1993) of an insight due to Hempel (1988). The insight in question is that a typical theory T in an advanced science will not have any logically contingent consequences that can be stated without use of any theoretical terms introduced with T. To state a testable consequence of T that uses only antecedently understood vocabulary, one will typically have to add a 'proviso' that restricts the domain of the prediction by reference to the new vocabulary of T. In the context of contemporary philosophy of science, putting everything in these hypothetical-deductive terms may be thought quaint. Earman and Roberts therefore reformulate Hempel's insight (restricted now to theories of fundamental physics) in the terms of the semantic view of theories as follows:

For a typical theory T of fundamental physics, there are no logically contingent conditions on the empirical substructures of the models that hold across all models; but there are locally contingent conditions on empirical substructures that hold across all models in which some proviso P is true, where P places constraints on features of models other than their empirical substructures. (Earman and Roberts 1999, 443)

Hempel's main example is drawn from classical celestial mechanics. Application of Newtonian mechanics to predict celestial motions presupposes a proviso to the effect that no forces besides those of mutual gravitational attraction act on the planets. The proviso is not a proper part of the theory itself; but it cannot be stated without use of a concept introduced by the theory, viz. 'forces' (in Newton's sense of them).

Earman and Roberts argue that aspects of this example have caused (at least) Giere, Fodor, and Lange to confuse Hempel's provisos with the provisos incorporated into *ceteris paribus* theories:

We fully endorse Hempel's insight. But his example and its mode of presentation are unfortunate ... [I]t uses an idealization (no forces other than gravitational forces) and/or an approximation (the total resultant forces on the planets are given to good approximation by the gravitational force component). Approximations and idealizations are widely used in physics, and their usage raises a host of important methodological issues. But Hempel's key point is independent of these issues. In the case at hand, it is in principle possible to do without any idealization or approximation: there is nothing to prevent the introduction of a specific postulate into the theory which specifies the kinds of forces that occur in nature, and there is nothing in principle that prevents the exact specification of the values of each of these forces acting on the planets of the solar system ... Even so, Hempel's key point stands: the said specification requires essential use of the [theory's] vocabulary. Hence, with or without idealizations and approximations, the theory by itself, without conditions of application stated in the [theory's] vocabulary, cannot be expected to yield non-trivial predictions stated purely in the [pre-theoretical] vocabulary. (Earman and Roberts 1999, 445)

Failure to notice that idealization and approximation are incidental features of Hempel's example have, Earman and Roberts show, led the philosophers mentioned above to misread him as having demonstrated that all theories, including those of fundamental physics, require *ceteris paribus* provisos to be stated truly. But provisos in Hempel's sense—the provisos required for operationalizing theories—are not *ceteris paribus* provisos. Thus Hempel's insight does not imply that theories in fundamental physics need be *ceteris paribus* theories. Laws in fundamental physics need not be true only *ceteris paribus*.

Hempel's insight nevertheless does have implications for theories in special sciences (including theories in non-fundamental branches of physics, such as statistical mechanics and, perhaps, thermodynamics). As Earman and Roberts put the point:

Phenomenological physics and the special sciences take as their subject matter entities, properties, and processes that can be observed independently of any particular theory of fundamental physics. Thus the pronouncements of these sciences will impose conditions only on the empirical substructures of the models of any theory of fundamental physics ... It then follows from Hempel's insight ... that any generalization that these sciences discover will not be true across all models of any of our fundamental physical theories. (1999, 447)

The consequent here is hardly surprising, and we don't need Hempel's insight to arrive at it. The point of noting that it follows from Hempel's insight is to draw attention to an irony: an argument that has been used to collapse an important part of the distinction between fundamental physics and special sciences—that the former aims at universal laws and the latter only at scope-restricted, physically contingent generalizations over pre-demarcated samples of evidence—in fact reinforces this very distinction.

As noted in 4.5, Earman and Roberts do not conclude from this that special sciences are in any way impugned by emphasizing that they do not discover universal laws. In exploring the alternative sources of epistemic value in special sciences, Earman and Roberts consider one of Kincaid's (1996) examples of a kind of proposition that much special science activity aims at confirming: 'In population H, P is positively statistically correlated with S across all sub-populations that are homogenous with respect to the variables $V_1, \ldots, V_2$' (ibid. 467). If we insist on treating this as a law, it must be a *ceteris paribus* law. But we need not treat it as *any* kind of law to agree that it has explanatory value.

We typically use propositions of this sort to explain individual events or sets of events. As Earman and Roberts say 'there seems to be no compelling reason to suppose that in order to shed light on the causes of individual events, it is necessary to cite any general laws. So [this sort of proposition] could be useful for providing causal explanations, even if it doesn't state or imply a law' (ibid. 468).

Earman and Roberts conclude that while special sciences can do first-rate work in furnishing causal explanations, where they do resort to generalizations that incorporate *ceteris paribus* provisos, this must in every case show that there is more science to be done somewhere; such a generalization cannot be an ideal limit-point of knowledge because a *ceteris paribus* clause is a locator for a specific area of ignorance. Since part of the traditional concept of a 'law' is that it represents an ideal knowledge achievement, and since there really are 'laws' in this sense—in fundamental physics—Earman and Roberts defend a semantic preference for restricting the concept and denying that there are *ceteris paribus* laws. Their motivation is pragmatic: calling *ceteris paribus* generalizations 'laws' implies that our scientific work is finished along dimensions where it really isn't, and where we're in a position to know it isn't.

Saying that special sciences furnish causal explanations, while fundamental physics aims at discovering universal laws, comes very close to the view we articulated in 5.4. Special sciences, according to us, use causal ideas as heuristics for locating real patterns. Fundamental physics is in the business of describing the structural properties of the whole universe. These properties are not causal relations. Special sciences and fundamental physics are thus mainly different from each other in a way we find it very natural to express by saying: fundamental physics aims at laws, whereas special sciences identify causal factors.

In the context of this dialectic, Kincaid (2004) offers an interesting new move. Its novelty is slightly obscure in his paper, because the rhetoric there implies that he is just reiterating and defending the position taken in Kincaid (1996). However, he is not. He still aims to eliminate or at least de-emphasize the boundary (with respect to the status of laws) between fundamental physics and special sciences. But now he does so not by arguing that laws in fundamental physics are *ceteris paribus* laws just like those in special sciences. He argues instead that laws in special sciences—and not just claims about statistical correlations, but claims with the sorts of logical forms that have usually led them to be treated as putative covering laws—are not *ceteris paribus* after all. We can clearly see his general strategy for defending this idea, despite the fact that his only example of how *ceteris paribus* provisos are to be eliminated from consideration is unfortunately drawn from classical physics instead of a special science. Here is what he says:

The problem of analyzing ceteris paribus laws can be ignored, because we need not think of laws in the social sciences as qualified ceteris paribus in the first place... A law picks out a causal force or factor. The law of universal gravitation, for example, asserts that there exists the force of gravity. It does not describe a regularity, for gravity is not the only force and the law of gravity is silent on how other forces might combine with it. Since it does not claim to cite a regularity without exceptions, there is no reason to think of the law as qualified ceteris paribus. The law is true if there is indeed a gravitational force. Laws in the social sciences work in the same way: they claim to identify causal factors and make no commitment by themselves to what other causal factors there might be and how they might combine. (Kincaid 2004, 179–80)

Obviously, we do not agree that laws in fundamental physics pick out causal factors; so even if we were persuaded by Kincaid that picking out causal factors in special sciences should be thought of as identifying laws, we'd be left with two kinds of laws, one kind for fundamental physics and one kind for special sciences. But let us set this aside. What mainly interests us about Kincaid's 2004 position is how radical the deflation of the idea of a law has become. Kincaid's laws are now not even generalizations about causes, of the kinds that provoke our Russellian worries; they are just identifications of 'explainers'. Scientists now discover laws merely by—to put it in the terms of our view—locating real patterns.

This is a highly eccentric meaning to suggest for 'law'. (On Kincaid's view, there are clearly far more laws than people have thought.) Indeed, it is so much so that the possibility looms that Kincaid's disagreement with Earman and Roberts has turned semantic.[22] But an interesting motivation for Kincaid's radicalism lurks elsewhere in his paper.

Among the objections to his 1996 view that Kincaid (2004) considers is that 'claims citing causal factors may be picking out accidental truths that do not support counterfactuals and cannot predict unobserved phenomena' (173). As part of his answer to this objection he argues against the idea of a sharp distinction between the necessary and the contingent in the first place. 'All causal claims hold to some extent by necessity,' he says; 'or, in other words, being an accidental generalization is a relative matter' (ibid.). This will be a startling remark for many philosophers. Its motivations are clearly naturalistic. Let us follow the reasoning a bit further:

Think ... about whether laws are necessary. Let's assume that a relation is necessary when it holds across different possible arrangements of things. Then the claim that fitness is a causal factor in inheritance is also necessary, to a degree. In any world in which there is differential survival and inheritance of traits, fitness is a causal factor—even if different species exist or even if those organisms have a different physical basis for inheritance than DNA/RNA. Yet the laws of evolution by natural selection are accidental in that they hold only of systems with the right inheritance and competitive characteristics. They need not describe other life forms that do not meet these criteria. Even the claim that 'all the coins in my pocket are copper' looks less accidental if there is some causal mechanism that excludes non-copper coins—maybe my pocket has holes in it and only pennies do not

---

[22] Earman and Roberts grant that special sciences pick out causal factors. They would just not agree to call doing that discovery of laws. The substantive issue between them and Kincaid seems to disappear if he agrees that laws in fundamental physics are not *ceteris paribus* and that (as follows from this) it is not 'provisos all the way down'.

fall out. And the basic laws of physics are accidental in that the values of the fundamental constants are apparently the result of chance events in the big bang that need not have happened. (ibid. 174)

Again, put aside the language in this passage that folds in Kincaid's view that all laws describe causes. Then the general point is one that a naturalist should uphold. Naturalism is, among other things, the metaphysical hypothesis that the structure of the objective world is not constrained by any reasons or standards of reasonableness. The hypothesis is motivated by the fact that science conducted in accord with it has discovered far more than inquiries that ignore it. Furthermore, if the universe is limned by a singularity, as physics suggests it is, then the explanation of the fact of the universe's existence cannot be speculated upon in a PNC-compatible way; speculation here is empty. It then indeed follows that necessity is relative. If there are fundamental physical facts—if the world is not dappled—then at least some of these facts, those that are not explained by some of the others, are brute contingencies. The point here is not that we must accept brute contingency in physics; QM forces us to recognize that possibility anyway. Rather, the point is that it is inconsistent with naturalism to suppose that any non-logical or non-mathematical fact is ultimately necessary.

The need for the adjective 'ultimately' marks recognition of the fact we can indeed make sense of *relative* necessity, as Kincaid says. Physicists are presently making some progress working on theories that are fundamental, in the sense we have defined. If there are structural facts about the whole universe, and these facts constrain all the facts about all particular regions of the universe—the conjecture institutionalized by the PPC—then the only necessity in nature is furnished by these constraints. The constraints—that is, the structures themselves—are real patterns.

Thus we agree with Kincaid that discovering laws is, in the end, just a matter of correctly describing real patterns. We don't agree, for reasons given in the previous section, that all of these real patterns—in particular, those identified by fundamental physics—are causal patterns. Furthermore, we think that the fact that information about the real patterns identified by fundamental physics is available from every measurement point in the universe renders fundamental physics sharply distinctive among the sciences; thus we agree with Earman and Roberts that fundamental physics discovers real patterns that are of a higher order of relative necessity than what is discovered by the special sciences. *From the point of view of those engaged in special science activity, fundamental physics gives the modal structure of the world.* For philosophers who insist that talk of 'higher and lower orders of necessity' just makes a nonsense of the idea of necessity altogether, the italicized claim operationalizes what we mean by such talk.

Thus we have a semantic preference for applying the idea of 'law' coextensively with Earman and Roberts. Yet we are also in substantial agreement with Kincaid (2004). It makes little difference to his view of special science generalizations whether fundamental physics has unique aspects. His first claim that truly matters is that special sciences are mainly in the business of picking out causal relations. We refine this when we say, as we did earlier in the chapter, that special sciences are mainly in the business of locating and measuring real patterns in specific regions of the universe using a (thin) concept of causation as the primary heuristic. Our work in the previous sections of this chapter explains, in a world where Russell and Redhead are right about physics, the basic fact Kincaid emphasizes: special sciences seem to be all about causes. Indeed, this is so much the case that it is easy to understand how Kincaid finds the identification of discovery of laws with discovery of causal factors plausible; all one needs to succumb to this temptation is a bit more seduction by disunity hypotheses than we think the scientific evidence warrants.

The core (2004) idea of Kincaid's we applaud is that discovery of laws consists simply in *finding certain structures*, which we gloss as confirmation of real patterns. According to us, a law is simply a real pattern, described by a structural claim, that is hostage to disconfirmation by any measurement taken anywhere in the universe. Because we think fundamental physics describes some such real patterns, we believe there are universal laws. We do not believe they are about causal factors. But then we believe that special sciences are precisely in the business of describing real patterns of less-than-universal scope. All such patterns we can locate will be governed by the asymmetries necessary for application of causal heuristics. These heuristics have proven extremely powerful in the history of special sciences, and dominate its practice. This can lead someone who is committed to unification of the sciences to try to get too much metaphysical work out of the idea of causation; that is, to model fundamental physics on the special sciences and then promote causation as universal glue.

According to us, those who try to recruit causation to do more metaphysical work than it can handle typically make two mistakes. We have just discussed one of them. The second is to read too much special philosophical structure into the concept they over-extend. Then the result is a picture that is not only mistaken in taking causation to be universal glue, but furthermore takes the causation in question to consist either in universal regularities of sequence or in microbangings. In these days of metaphysical realism the latter interpretation is far more common. However, we argued in 5.2 that all that is generally important in the idea of causation is information flow along asymmetric gradients. We fall into basic metaphysical error when we over-interpret the evidence and think we have discovered that the world is held together by microbangings. Indeed, we go wrong as soon as we think there are any microbangings, because that question is for fundamental physics to settle, and it now speaks against them. Once we eliminate microbangings from our picture of fundamental physics, there can surely be no temptation to read them into the causal processes studied by special sciences, where regularities are all statistical. For example, to suppose that the relationship between an increase in the money supply and a rise in prices consisted in microbangings would require belief in mereological reductionism *and* belief that fundamental physical causal connections are microbangings. This example from macroeconomics draws on the extreme case where no intuitions not driven by these strong metaphysical hunches will be pulled in the opposite direction. But then we need simply observe that phenomenological physics and chemistry are just like macroeconomics in the relevant respect. From the point of view of fundamental physics, what they all generalize over are strong statistical attractors of the kind physicists call 'universalities' (Batterman 2002). Thus even mereological reductionism is not enough to save microbangings once they are expunged from fundamental physics.

In this chapter we have argued that things are pretty much as they seem on the surface in special sciences. When inquiry is going well, special scientists successfully track real patterns. They do so by book-keeping them as individuals interacting in causal processes. Because fundamental physics merely constrains special sciences, rather than providing their mereological reduction base, there is no PNC-compatible motivation for seeking counterparts to the individuals and processes in fundamental physics. In that case, there is no motivation for including them as elements of our metaphysics; our metaphysics need merely be compatible with the special-science methodology that makes use of the book-keeping heuristics. The metaphysics of ITSR is so compatible. It doesn't imply that the universe is asymmetrical in a way that would explain the utility of the heuristics, but it explains why sciences that gather measurements from specific perspectives would use individuals and causal processes as locators. Such weak, but non-trivial, unification is the metaphysic that the empirical evidence currently justifies. Nothing stronger has any naturalistically acceptable justification at all.

5.6 REAL PATTERNS, TYPES, AND NATURAL KINDS

In much philosophical discussion the concepts of '(natural) kind' and 'law' are yoked together, on the general supposition that natural kinds are what laws are true of. In the present section we turn to kinds. This requires more than simple extrapolation because of the presence of an influential argument for metaphysical realism concerning natural kinds. Indeed, this argument, due to Putnam (1975b), was probably the most influential one for metaphysical realism in general in the immediate period of the decline of logical empiricism. It begins with the claim that if intensional meanings of referential terms determine their extensions, then significant theory changes imply changes of reference. This is embarrassing when we try to say sensible things about the history of science. For example, we want to say that until recently zoologists falsely believed that giant pandas weren't bears but were more closely related to red pandas. But the place of a kind of organism on the phylogenetic tree is its most important property where classification is concerned. Therefore, if the term 'giant panda' was intended twenty years ago to mean 'one of two kinds of panda', and intension determines extension, then zoologists twenty years ago could not have been referring to, and expressing false beliefs about, one of several kinds of bear. But this is silly. Of course zoologists of twenty years ago referred to the same group of animals by 'giant panda' as do zoologists now. Both groups refer to '*those* animals, whatever they are like', where 'whatever they are like' is elliptical for 'whatever properties are the basis for there being true generalizations about them'. Twenty years ago, zoologists thought that true generalizations about giant pandas were functions of their being a particular kind of panda; now they know that these generalizations instead depend on their being a particular kind of bear. This is so obviously the better way to think about the history of science that Putnam concludes we must reject the idea that intension determines extension in the case of scientific kind terms. Instead, we should suppose that such terms have whatever extensions they do independently of our beliefs about them, and then try our damnedest to get our beliefs and our language in tune with these independent facts.

On this sort of account, explanation, generalization, and natural kinds form an interlocking analytical circle. As Hempel stressed, and as continues to hold good whether one is closer to Salmon or Kitcher on the rest of the story, to explain something requires at least showing that it falls under the scope of some established generalization. That there can be true generalizations, claims good today of this case and good tomorrow of that one, is a function of the fact that the world is relatively stable. Saying that there are 'natural kinds' in the world is, on first pass, merely one way of expressing this fact. 'Relatively stable' implies that some background stays the same while some foreground shifts. Then natural kinds are just the clusters of properties that stay clustered while we track some more transient ones, and which we use to keep our referential grid constant while we do so.

Thus, as Laporte (2004) argues with the help of many examples—in a nice demonstration of mostly PNC-consistent metaphysics—'a natural kind is a kind with explanatory value' (19). Laporte thinks that this insight invites us to relax about the extent to which types absolutely are or aren't natural kinds; naturalness, as he says, is a context-sensitive property. As all biologists know, boundaries in nature are both fluid and fuzzy, and on evolutionary time scales beyond a certain magnitude it ceases to make sense to treat bears and pandas as different kinds of things. In fact, one wonders at this point, why not relax to the point of letting go? Could anything be regarded as a kind at all if the kind in question had no explanatory value in any context?

Because, on the one hand, it looks as if the answer to this question is 'no', but, on the other hand, philosophers are convinced that some kinds are natural while others are not, they typically feel a need to go looking for a basis on which to distinguish between 'real' and merely artefactual kinds. This basis is commonly located, by Laporte as by others, in the idea that some kinds are sorted by reference to necessities and others by reference to mere accidents. Here is the relevant intuition: my computer and my copy of Laporte's book might both be instances of the kind 'paperweight' but this is an accident. There are nearby possible worlds—ones in which my copy of Laporte is a virtual download, for example—in which it is still a copy of Laporte but it isn't a paperweight; but there are no possible worlds in which giant pandas are still giant pandas but they aren't bears. This is supposed to explain our 'intuition' that giant pandas, and bears, are natural kinds but paperweights aren't.

Dupré (2004) strikes a blow on behalf of PNC-compatibility in metaphysics when he objects (specifically to Laporte) that he has never heard any biologist talk about possible worlds or rigid designation, and doubts that many could be easily persuaded to do so. We approve of the spirit of this complaint, but it is somewhat unfair to Laporte. If Laporte's claim were that names of natural kinds must rigidly designate because of facts about reference, then indeed this would accord with the beliefs of few scientists. Naturalists insist that there are no such facts, and that if there somehow were such facts—say, there's a language organ and it explodes if people try to let names change reference across possible worlds—this would be a good reason for scientists to avoid natural language wherever possible. But we think it is clear that in Laporte's case he is just using the philosophical jargon to characterize externally what he takes to be a fact about scientific practice. Scientists treat certain properties as if they were essential to a kind's being the kind that it is. Laporte shows by his many examples that they are not very consistent with respect to principles on which properties are promoted to essences. Sometimes the honoured properties are properties philosophers would traditionally call 'intrinsic', and sometimes they are functional.

Nevertheless, Dupré is right that Laporte, unlike him, thinks that (some) classifications have metaphysical import. This is because Laporte thinks that scientists' judgements about essential properties broadly reflect, sometimes sloppily, their judgements about modal regularities. We agree. However, we concur with Dupré (but not for his reasons) that describing this feature of science in terms of rigid designation of natural-kind terms is very misleading.[23]

Philosophers sometimes invoke natural kinds as if their existence would explain the possibility of scientific explanation (in something like the way that the existence of photons explains the possibility of optics). This is characteristically neo-scholastic. That anything at all can be explained, and that properties stay clustered together under various sorts of shifts in circumstance, are two expressions of one fact: that the world is relatively stable. Philosophers who invoke possible worlds to explain intuitions about especially *natural* kinds should most charitably

---

[23] Note that we do not say 'false'. We do not think that the proposition '"Giant pandas" pick out bears in all possible worlds' is the sort of claim that could be true or false of the world *simpliciter* (as it is intended to be by most philosophers who say this sort of thing), though it could be true or false if 'possible worlds' were given some special restricted sense. All could be well with the locution, including even truth, if the restriction in question were motivated PNC-compatibly. But what we argue below is that there are no such motivations.

be interpreted as seeking to explain what it is about the stable bits of the world that distinguishes those bits, in general, from the less stable bits (especially the bits we can readily rearrange to suit our transient purposes). Are these philosophers asking a PNC-compatible metaphysical question?

The answer to this turns entirely on how we interpret 'in general'. In Chapter 4 we discussed the views of some systems theorists, both physicists and biologists, who think that natural selection, and therefore life, depends on the stability of pools of information made possible by thermodynamic asymmetry. When these scientists wonder about 'general stability' they are certainly engaged in well motivated inquiry (by the lights of the PPC). Note that in order for them to pursue this inquiry, they must consider, for contrastive purposes, possible worlds in which the Second Law doesn't hold. As we saw, there isn't at the moment a consensus on whether this world is in fact physically possible; if there were, then it would be a closed, rather than an open, question as to whether thermodynamics is part of fundamental physics.

Suppose first that the question ultimately comes to be settled in favour of the conclusion that thermodynamics is fundamental. This will mean that we'll come to see that we can't consistently model a world, or parts of a world, absent the Second Law. In that case whatever aspects of relative stability in our part of the world are explained by thermodynamics will thereby be as fully explained as they can be, while the laws of thermodynamics themselves will be (one or more, depending on how many basic laws are needed) inexplicable brute facts. On the other, suppose that thermodynamics is found not to be fundamental. This implies that fundamental physics will explain why its generalizations hold in some parts, including our part, of the universe. Notice that on the second outcome more stability properties get explained than on the first outcome. This reflects the insight of Friedman (1974), clarified by Kitcher, that the extent of explanation is inversely related to the proportion of facts that must be taken as brute.

In this book we have defended the claim that scientists treat the space for modelling as modally constrained; the constructive empiricist makes a mistake in denying this. What we mean by 'modal constraints' are those generalizations about structure that apply universally across the domain. Commitment by scientists to unity is embodied in their respect for the PPC, since it reflects the so-far successful bet that one part of science constrains all domains and that part is entirely within the preserve of physics. This subpart of physics is in turn just that part that physicists will conclude cannot be itself explained. Thus: how much of the stability of the world is explicable is itself a matter for empirical science, not philosophy, to determine. We could only have a full determination to the extent that we had a completed physics. But because current physics allows that thermodynamics might not be fundamental, we can see, in outline, how explanations of it would go if it isn't. (Each candidate framework in fundamental physics, as reviewed in 3.7.2, offers an alternative sketch.)

This picture offers quite a lot of scope for asking and trying to answer sensible questions about general conditions for stability, including recursively conditional stability. (For example: if system S is at equilibrium then this restricts dynamics of sub-systems $s_1, \ldots, s_2$ with their own local equilibrium properties, and so on.) But here is an interpretation of the philosopher's question that is not sensible: how or why is there any stability in the universe at all? By this point in the book, we have available a number of things we can say by way of rejecting this question, each of which has been given a clear meaning. Taking this question seriously violates the PNC. It ventures not just beyond actual innocent phenomena, but beyond all possible phenomena. It asks us to compare the actual world with one in which there are no real patterns, but this activity has no constraints and therefore no epistemic point. We invite philosophers to ask themselves: how much of the discussion of natural kinds is trying to ask this pointless question?

Natural-kind talk is perfect for disguising the pointlessness of the question, because of the way in which it appeals to the old intuitions that in some sense 'before' there are structures and theories and models there are self-subsistent *things*. These intuitions, as we have discussed, are incompatible with the objective existence of paperweights. But they have us imagine that giant pandas could (in some extra-physical sense) be a type regardless of the context, just as long as the properties essential to their kind were glued together in the appropriate way. This is beguiling because, after all, some cohesive objects in our notional world are so effective at resisting entropy that we can transport them to radically new environments in space and time and yet relocate them. But for a naturalist it is beguiling *nonsense*. Nothing in contemporary science motivates the picture. (Take giant pandas to Saturn, or 6000 Mya backward in their light-cone. It's easy *to think about*, isn't it? But organisms are unusually strongly cohesive real patterns, unlike many real patterns studied by scientists. Now imagine taking the market in airline risk derivatives to Saturn or 6000 Mya backward in its light cone. That was a bit harder even to imagine, wasn't it?)

We said above that we agree with Laporte that scientists' judgements about essential properties broadly reflect, sometimes sloppily, their judgements about modal regularities. How shall we flesh out this agreement without reference to natural kinds as kinds of *things* constituted out of essential properties?

We contend that everything a naturalist could legitimately want from the concept of a natural kind can be had simply by reference to real patterns. Real patterns are of course modally significant: projectible by definition, and represented by structures. The obvious way in which a would-be defender of natural kinds would respond to their proposed replacement by real patterns would be to try to show that the idea of a natural kind allows us to mark off an important difference *within* the real patterns. This, one might think, is trivially easy for her to do. We have said, after all, that everything that exists is a real pattern, and that Napoleon and the table in the corner are among the real patterns. But these are among the paradigm cases of what natural kinds are not supposed to be. So whatever it is that makes giant pandas a natural kind and the table in the corner not a natural kind is lost sight of if our ontological typology has only real patterns in view.

This objection establishes only that in translating Laporte's good sense into our terms, we have to say somewhat more than just that successful scientific theories are about real patterns; it does not show that we need to find criteria for marking off a special subclass of the real patterns such that they can be surrogate natural kinds for all purposes to which philosophers have tried to put natural kinds. (Laporte, given his acknowledgement of the context-relativity of natural-kind-hood, must agree.) So let us duly say some more.

For any given real pattern, there will be a specific range of locators from which measurements carry information about it. The universal real patterns, those that are studied by fundamental physics, are measurable at all locators. The real patterns of solid state physics are measurable from all locators within a certain temperature band. The real patterns of biology are measurable from many points near the surface of the earth and after 3500 Mya, but no others (as far as we now know). The table in the corner is measurable only in the tiny locator sets along its historical spacetime trajectory and the spacetime trajectories of the people who register it in a way sufficient to pick it out from other very similar tables, and for as far along their trajectories as these registrations are recoverable. Let us say that to the extent that a real pattern is fully measurable from many locators it has high indexical redundancy. This is not to be confused with *informational* redundancy (which undermines pattern reality) as invoked in the definition of a real pattern. The idea is that for patterns with high indexical redundancy, most measurements carry no new information about them to most measurers. For example, most people could get new information about my left shoe by any measurements they take of it, while people earn Nobel prizes for finding ways to take measurements that bring anyone new information about the real patterns of fundamental physics. To be objective, indexical redundancy should be defined modally, that is, in terms of physically possible perspectives. Let us say that a real pattern's indexical redundancy is an inverse function of the size of its 'measurement basin' in locator space (however many dimensions science turns out to require to fix that) multiplied by the proportion $x: 0 > x > 1$ of physically possible perspectives from which information about the real pattern not available from other perspectives is detectable.

As a first pass, one is tempted to suggest that what philosophers have typically meant by 'natural kinds' are real patterns with high indexical redundancy. Notice that in general there are two ways in which a real pattern's indexical redundancy will tend to be reduced. One is for it to be measurable from only small regions of the universe. A second is for it to be extremely complex, since then no matter how 'small' it is, there will be many physically possible perspectives from which non-redundant information about it is obtainable. This corresponds to the two main features of types of things that incline philosophers not to regard them as natural kinds.

A complication here is that paradigm instances of natural kinds in many philosophical discussions have been biological taxa. As real patterns go, these have quite low indexical redundancy, since, on the scales at which we consider the universe, most of them flicker briefly into and out of existence in a minuscule bit of space. However, we can turn this complication to our advantage. Ever since the suggestion was made in a classic paper by David Hull (1976), it has become increasingly common for philosophers of biology—and even, in one of the rare cases of philosophy influencing science and thus obtaining ratification of PNC-compatibility, for biologists—to regard biological taxa as individuals. The basic motivation for this is that taxa are historically unique. In the debate around this proposal, what is being contrasted with individuals are universals; and of course the individual/universal distinction is not the same as the natural kind/artefactual kind distinction. However, it seems that if a philosopher accepted that biological taxa are individuals but still wanted to promote them as exemplary natural kinds, this philosopher would then have to be willing to admit Napoleon as a natural kind. This would be fine with us—'natural kind' isn't our notion anyway, and why should not 'Napoleon' be a type, of which 'Napoleon in 1805' and 'Napoleon in 1813' are instances? (Indeed, these can themselves be, recursively, types. There are true generalizations, of abiding interest to historians, about the 1805 edition of Napoleon that do not apply to the 1813 edition and vice versa.) To the extent that a philosopher is willing to admit individuals as natural kinds, this philosopher is choosing to emphasize the objectivity of the nature of the kind, rather than its tendency to feature in generalizations, and hence in explanations, as emphasized by Laporte. This hypothetical philosopher seems to have no need of any surrogate for natural-kind-hood more specific than that of a real pattern in general.

Thus we conclude our discussion of natural kinds as follows. To the extent that philosophers who believe in self-subsistent individuals are persuaded by Laporte's analysis of natural kinds, they should grant that we can get the work they want out of the natural kind/artefactual kind distinction by invoking real patterns with high indexical redundancy. To the extent that they resist Laporte's analysis because of its tolerance for contextualization of natural-kind-hood, when what attracts them to the concept is its focus on objectivity, they should be open to just replacing reference to natural kinds with reference to real patterns.

Twenty years ago, zoologists tracked some real patterns from a certain set of locators, and as a result obtained some generalizations about a group of organisms they called 'giant pandas'. They also then had some generalizations about two other groups of organisms, red pandas and bears. We have since learned that these zoologists had two general groups of false beliefs going with their true ones. First, they thought that the conjunction of the two real patterns named by 'giant panda' and 'red panda' was itself a real pattern. Second, they thought that when they measured the real pattern named by 'giant panda' they were not measuring the real pattern named by 'bear' and vice versa. Learning that both of these beliefs are false is equivalent (if we assume we know that red pandas aren't bears) to learning that either giant pandas are bears or bears are giant pandas. They never imagined that everything projectible of giant pandas might be projectible of bears; but now they know that everything projectible of bears is projectible of giant pandas. So giant pandas turned out to be bears, not (giant pandas and red pandas = pandas).

We share the standard scientific realist's commitment to the view that scientific inferences are intended with modal force, and that this is crucial to the possibility of scientific explanation. We also share her opinion, contra that of the constructive empiricist, that explanation and unification are goals that normatively regulate the practice of science. We have now shown how to maintain these aspects of realism without violating the PNC by conjuring up real essences or magical 'referential cables' that transcend empirical verification. 'Natural kind' is a more elegant phrase than 'real pattern of high indexical redundancy'; but then folk metaphysics generally makes for better poetry than scientific metaphysics.

6

Conclusion—Philosophy Enough

Don Ross and James Ladyman

We close by returning to our opening themes and adding some further reflections on them that would have been premature in advance of the arguments presented over the preceding chapters, and by putting our views in broader context.

6.1 WHY ISN'T THIS DENNETT? WHY ISN'T IT KANT?

In 4.2 we criticized Dennett for dividing reality into first-class illata and second-class abstracta. Now we have argued that fundamental physics directly studies extra-representational real patterns, while special sciences study real patterns located by means of the notional-world concepts of cohesion and causation. How is this not just Dennett's distinction after all, with the addition of a special analysis of what abstracta are?

Dennett's abstracta are approximate descriptions of the illata, where the approximations in question usefully serve human purposes. By contrast, special-science real patterns are not, according to us, approximations of (fundamental) physical real patterns. The PPC is the only physicalist principle motivated by actual science, and it is too weak to support any form of reduction of special-science real patterns to physical real patterns. (Instances of Nagelian reduction occur, but this has nothing to do with physicalism.) The basis for our confidence that the special sciences often successfully track real patterns is the no-miracles argument as defended in Chapter 2, not any claim about how the special sciences inherit ontological seriousness from their relationship to physics.

We indeed argued that special-science kinds do not exist in the way that folk metaphysics takes them to do, that is, as self-subsistent individuals that constitute history by banging into each other. The basis for this denial is the failure of fundamental physics to describe a world unified by the organizing principles of the special sciences, viz., cohesion and causation. But the fact that special sciences are organized by notional-world principles doesn't make the subject matter of special sciences, the real patterns they study, into notional-world subject matter. Special sciences study local reality at scales where distinctive, irreducibly dynamical, real patterns are manifest. We do not know what the general *positive* relationship between the real patterns of fundamental physics and the real patterns of special sciences is, though we sometimes empirically discover particular such relations. Indeed, we do not even know whether there is any single general positive relation. This is the bite in our claim that the history of science warrants no metaphysical thesis of physicalism, but merely the much weaker institutional norm of the PPC. The PPC is *just* strong enough to suggest the unity of science, because the best metaphysical explanation for the PPC is that all locators tracked by special sciences bear information about the real patterns studied by fundamental physics, or equivalently, the modal structure studied by fundamental physics constrains the modal structures studied by the special sciences. (But modal relationships among locators described by special sciences generally fail to be detectable when these locators are redescribed by physics.)

This is why our view isn't Dennett's. But now it might be objected that, in explaining this, we have turned our position into Kant's, though with new terminology. How is it that special-science objects of generalization don't stand to the fundamental physical structures as phenomena stand to noumena in Kant's system? After all, it is the essence of our view that we resist substantivalizing the fundamental physical structures. We *say* these structures describe real patterns, but since we can only represent the real patterns in question in terms of mathematical relationships, in what sense are these real patterns 'real' other that in which, according to Kant, noumena are real? This will be an especially tempting line for someone who derives their interpretation of Kant from the work of Michael Friedman (1992), which emphasizes the realist and naturalist, as opposed to idealist and transcendentalist, aspects of Kant's thought. (Given the extent to which we revive the spirit of logical positivism it is no surprise, on Friedman's (1999a) reading of the positivists as pursuing a neo-Kantian agenda, that such a similarity is apparent. The positivists thought that formal logic enabled them to drop Kant's transcendentalism in favour of conventionalism without thereby abandoning his commitment to science's claim to epistemic authority.)

Then, to make it even easier to read us as Kantians, we claim that people *need* to organize the local domain of reality by means of the notional-world book-keeping principles of cohesion and causation. Are these then not just new candidates for 'principles of the understanding' that structure phenomena? Finally, we say we don't know, in general, how special-science real patterns are related to physical real patterns, or even if there is any such general relation. Doesn't this just echo Kant's insistence that trying to account for phenomena by reference to noumena is empty metaphysics? Even our version of verificationism, it might seem, is just Kant's version.

However, our differences from Kant are profound. Unlike Kant, we insist that science can discover fundamental structures of reality that are in no way constructions of our own cognitive dispositions. For this reason, our current best theory of the fundamental real patterns at any time is open to modification; the structures are determined empirically, not a priori.

It is true that, like Kant, we distinguish the propositions to be taken seriously in science and metaphysics by reference to human constructions. But for us these are literally constructions: our scientific institutions. As collective constructions, the institutional filters of science need not mirror or just be extensions of individual cognitive capacities and organizing heuristics. They have shown themselves to have a truth-tracking power—partly thanks to mathematics—that bootstraps the process of scientific learning beyond the capacities of individual minds. Furthermore, the limitations on the sorts of heuristics we must use for doing special sciences arise not from limitations of the kinds of minds we have, but from the irreducibly asymmetric nature of the local spacetime in which all our measurements must be taken. We are persuaded by Humphreys (2004) that a point is rapidly coming, if it is not upon us already, at which most of our science will necessarily be done by our artefacts because the required computations are beyond us, and we think there may well be a time after which people will no longer be capable of understanding what their artefacts tell them. The artefacts will, according to us, be studying real patterns. Yet even then, we expect, the artefacts will need to manipulate object-oriented frameworks because of the nature of the local part of the universe.

We grant that *some* elements of the prevailing world-view, namely cohesion and causation, are notional-world ideas, and in so being resemble abstracta and noumena. Of course it would be strange if, according to us, Dennett and Kant arrived at their views of these things on no basis whatsoever. But cohesion and causation are precisely *not* scientific ideas: they are metaphysical ones. Science, including special sciences as full, first-class, citizens, tells us to banish them from our first-order ontology. They do not then lead a whole parade of special-science objects of attention into metaphysical purgatory. Prices, neurons, peptides, gold, and Napoleon are all real patterns, existing in the same unqualified sense as quarks, bosons, and the weak force. We use concepts of cohesion and causation to keep track of the former but not the latter. This has no more direct ontological significance than does the fact that people who study peptides and bosons often wear white coats, while people who study prices generally don't.

6.2 A REDUCTIO

As defenders of the claim that naturalism has radical consequences which few putatively naturalistic metaphysicians have taken sufficiently seriously, we note the existence of an unlikely ally in Michael C. Rea. Rea is an avowed foe of naturalism from the most traditional perspective of all. His general view will be gleaned from a statement that closes his (2002): '[T]hough ultimately I reject it, I think that in fact naturalism is the most viable research program apart from a brand of supernaturalism that warrants belief in a suitably developed version of traditional theism' (226). Rea's strategy for arguing against naturalism is to show that it has consequences he thinks few philosophers would be disposed to accept. We agree with him that it has these consequences and that few philosophers are disposed to accept them. But the conclusion we draw from this, unlike Rea, is that there is something wrong with the thinking of the philosophers, not that there is something wrong with naturalism.

According to Rea, naturalism is a research programme rather than a substantive metaphysical thesis in itself (2002, 54–72). We agree; the content of our naturalism is given by the PNC, and the PNC is a claim about how to do metaphysics, not a first-order claim about what there is. Rea's formulation of the naturalistic research programme in philosophy is as follows: '[N]aturalism is a research programme which treats the methods of science, and those methods alone, as basic sources of evidence' (67). If 'methods' here were replaced by 'beliefs that survive the institutional filters of science' then Rea's description would apply to our naturalism. Though this difference is not trivial, we think that our naturalism is of the right tenor to be an instance of the sort Rea aims to portray and combat.

Rea's basic strategy is to show that naturalism leads to denial of 'realism about material objects (RMO)' (77), where 'material object' is defined by reference to 'non-trivial modal properties' (81). These are properties, such as possibilities and necessities with respect to persistence, that figure in traditional non-naturalistic metaphysics. They are indicated as the sorts of properties one must investigate to solve 'the problem of material constitution', which is the problem arising from 'scenarios in which it appears that an object a and an object b share all of the same parts and yet have different modal properties' (79). There is a question about 'how we could possibly learn (or acquire justified beliefs about) the truth or falsity of some of the crucial premises in puzzles that raise the problem' (80). Rea's example here is the 'puzzle' about how a person could be identical to a collection of particles and be the same person at an earlier and later time while being different collections of particles at those times. 'Philosophers of every persuasion', Rea announces, 'have devoted attention to the problem of material constitution, and virtually nobody responds to the puzzles that raise it by saying that we cannot have justified beliefs about the modal properties associated with various arrangements of matter' (80).

That is of course exactly what we say. The claim that a person is identical to a collection of particles has no basis in any science. The concept of 'material object' at work in Rea's discussion has no counterpart in real physics. There is therefore no reason to believe there are any such things. And the particular kinds of modal properties the philosophers are sure we have justified beliefs about are properties unknown to science.

Rea knows all of this. In fact, his knowing it is fundamental to his argument. He takes lengthy and careful pains to show that one cannot be a sceptic about justified belief in modal properties associated with arrangements of matter ('MP-beliefs') and yet be said truly to believe in 'material objects'. We of course claim a genuine naturalist *must* be a sceptic about MP-beliefs. Therefore, we agree with Rea that naturalists should not believe in 'material objects'. The 'material objects' in question are not what physics (or any other science) studies; they are pure philosophical inventions. Rea thinks it should shock naturalists to find that they shouldn't believe in such objects. We agree that many will be shocked, and take this as evidence that their naturalism comes up short of the real thing.

Like most philosophers, Rea combines some accurate beliefs about science with some quaint ones. At one point he says that a

> common strategy for defending realism about various kinds of problematic objects is to argue that belief in or talk about such objects is *indispensable* for science ... No doubt it *would* be impossible to engage in scientific theorizing as we know it without presupposing the existence of material objects; and as we have seen, belief in material objects carries a commitment to modal properties. Thus one might argue that our particular MP-beliefs are justified on the grounds that we have to form some MP-beliefs or other, and the ones we in fact form have allowed us to formulate a wide variety of very successful scientific theories. (129)

Given what is meant here by 'material objects'—extended individual things that 'take up' space and 'persist through' time thanks to 'intrinsic' modal properties that ground their identity—we think it would be nothing short of preposterous to suppose that belief in them has been indispensable to the construction of successful physical theories. But then Rea nevertheless rejects the reasoning he has canvassed, which might have rescued 'material objects' for naturalists, on the basis of a sensible view of science:

> '[E]ven if we agree (as seems reasonable) that theories that quantify over familiar material objects like cells or atoms are in some sense committed to the existence of such things, there is no reason to believe that simply by quantifying over such things a theory is committed to any philosophical thesis about what those things are' (ibid.). Indeed.

Because Rea's book advances its case by way of a complex argument about putative philosophical facts that we deny could be subject matter for justified belief one way or the other, we cannot express an opinion on the soundness of his conclusion that either naturalists must adopt the radical view we defend or believe that the world was designed by a traditional sort of deity. But if his conclusion is *true*, then just a few obvious logical steps would take us from it to our main theses by an unexpected alternative route. We welcome this surprising additional consilience.

6.3 NEO-POSITIVISM

The most general claims of this book can be summarized as follows. Taking naturalism seriously in metaphysics is equivalent to adopting a verificationist attitude towards both science and metaphysics. On the basis of this we arrived at the scientistic stance, where this is our dialectical combination of realism and empiricism. This in turn, when applied to the current near-consensus in science as a body of input beliefs, yields the details of our information-theoretical structural realism as a body of output (metaphysical) beliefs. The main part of this book has been devoted to arguments for ITSR. In coming back to the abstract philosophical topics with which we opened, we re-focus attention on the scientistic stance.

We take it that the consistency of our verificationism (which, recall, is verificationism about epistemic value, not meaning) will only be doubted by a philosopher who supposes that no verificationist can have any truck with objective modality. To this our response is to remind readers of the deflationary nature of our commitment to modal structure. We claim that science has provided evidence that some structural properties are properties of the whole universe. The evidence for all generalizations of this sort is testable. We then claim that the rational goal of seeking a unified account of reality justifies us in framing new theories on the assumption that some properties of the whole measured universe are properties of the whole measur*able* universe. (Which such properties this applies to is a function of the details of physical theory. Examination of these details, we have argued, shows that quantum properties are such properties, while the jury remains out on thermodynamic properties.) Since the scope of the measured universe is constantly expanding with the progress of science, the stronger claim about the measurable universe is being constantly checked with respect to confirmation and disconfirmation. We assert no claims intended with ultimate modal force. Thus, for example, if we say that Heisenberg's Uncertainty Principle is a law of nature, we do not mean that it is true in every logically or semantically possible world; despite its modal inflections, science does not aim to describe such worlds. We mean only that the Uncertainty Principle happens to constrain possible measurements everywhere in this world, the one world in which we can take measurements.

The moment they recognize our deflationary commitment to modality, some will call the seriousness of our realism into question. Such critics might contend that we overstate the difference between our view and constructive empiricism by pretending that deflated modality is 'true' modality. We dealt with this critical thrust back in Chapter 2. But the critic might now attempt a more direct way of trying to drive a wedge between our commitments to verificationism and realism. This direct way would begin by insisting that our primary commitment to naturalism tethers us unequivocally to verificationism. Then it would press an argument based on the claims that verificationism just is empiricism, and that empiricism is (with instrumentalist constructivism and relativist scepticism) one of the three foils against which realism is identified.[1]

It may be wondered how this direct criticism could still be dangling unaccounted for so late in the book. The answer is that few philosophers since the fall of logical empiricism have supposed that naturalism implies verificationism. Thus none of the various arguments against which we have sparred—arguments coming from standard scientific realists, from constructive empiricists, from naturalistic advocates of disunity—have been based on this premise. But it is a premise we endorse. Thus it is an available premise for a critic we have not yet considered. Seeing off this hypothetical critic will complete the presentation of our view. However, before we do so, let us stress that we do not care whether we're deemed better entitled to 'empiricist' or 'realist' party cards. The other comprehensive PNC-compatible recent book of metaphysics, with which our view may usefully be compared, is that of Melnyk (2003). Melnyk takes pains to defend his position as a realist one. As we have said, our first commitment is to verificationism. Therefore, the fault line between these labels is a convenient hook for addressing the relationship between contemporary naturalistic metaphysics and standardly identified philosophical isms.

The first modern philosopher who was consistently naturalistic was Hume, and he was of course an empiricist. As van Fraassen (2002) argues, however (see 1.7) to say this at the level of beliefs (as opposed to stances) is to say something ambiguous. Melnyk (2003, 229–30), in trying to give content to the identification of his own metaphysical view as a realist one, says that a realist will 'find repugnant' the 'empiricist idea that the difference between being observable and being unobservable marks a distinction of great epistemological significance'. Let us agree that this marks one diagnostic distinction between empiricism and realism. Verificationism is usually conceived as the idea that we can know (or justifiably believe) only what we could in principle confirm (or test) by observation. Hume thought this. Hume's empiricism in so far as it went beyond his acceptance of the empirical stance and included an epistemological doctrine consisted in what we would now call verificationism. Furthermore, his naturalism is also nearly the same thing as (or is only a very few logical steps away from) his verificationism: he rejected supernatural entities and influences because he thought they fail the verificationist's requirement; indeed he more or less defined 'supernatural entities and influences' as posits of hypotheses failing that requirement.

Thus the logical tie amongst naturalism, verificationism, and empiricism as a contrasting foil for realism has solid historical roots. (While we continue

---

[1] This would be a critic who did not accept van Fraassen's account of the core contrasting principles of empiricism and (naturalistic) realism.

to endorse van Fraassen's 'stance stance' with respect to 'strong' metaphysics, we think his identification of naturalism with the 'materialist stance' forces an unduly strained history of philosophy.) Kant abandoned the whole package, but then the logical positivists showed how to remain Kantians in other main respects while taking the package back again. For them, rejection of the Humean package was an unnecessarily high price to pay for the Kantian construction of the foundations of scientific knowledge, particularly its analysis of the relationship between particular judgements and generalizations. The passage from logical positivism to logical empiricism was a slide from Kantianism back to Humeanism on those topics, which Quine, the completer of that line of thinkers, decisively ratified by pronouncing officially for naturalism (Friedman 1999a). The observable/unobservable distinction does the same work for Quine as it does for Hume.[2]

We thus endorse Melnyk's analysis of what it is to be a realist, where this analysis includes a clause explicitly denying the sort of significance to the observable/unobservable distinction historically promoted by naturalists/verificationists (and most rigorously and carefully promoted, as discussed in Chapter 2, by van Fraassen). Here is Melnyk's analysis:

The deepest commitment of those who would call themselves scientific realists and antirelativists ... consists ... in the respectful way in which they regard (certain) current scientific hypotheses. By and large, they regard these current scientific hypotheses as

(I) true or false in virtue of the way the mind-independent world is;

(II) objectively superior, in some truth-connected sense, to earlier hypotheses in the field, so that science has, in this sense, progressed;

(III) objectively superior, in the same sense, to current rival scientific hypotheses;

(IV) objectively superior, in the same sense again, to current rival hypotheses advocated by people outside the scientific establishment; and

(V) such that whether the regard for a hypothesis expressed by I–IV is appropriate is generally independent of whether or not the hypothesis postulates entities and properties that cannot be observed. (Melnyk 2003, 229)

Reproduction of this analysis out of its textual context may leave the reader unsure about how to interpret clause (III). Let us therefore note that it does not imply that philosophers, qua philosophers, should take sides amongst current rival scientific hypotheses—something we deny.[3] Rather, it says that realist philosophers recognize that scientists establish preferences among current scientific hypotheses on the basis of their evaluations of truth-related considerations. Note that it is

---

[2] This is clearest in Quine's book *The Roots of Reference* (1974).
[3] What if a current scientific hypothesis reduced the unity of science? We claim that unity of science is a scientific norm, implying that scientists themselves police this in considering hypotheses. The job of the philosopher (qua philosopher) is to articulate the unity scientists collectively foster, not to show the scientists which hypotheses best achieve it.

precisely naturalism, of the sort captured by the PNC, that makes clause (III) essential to a realist position; if clause (III) did not hold but clause (II) did, then we would have to suppose that all theory change occurred through epistemically irrational paradigm shifts. But then if the history of science consistently found the scientific community deciding after the fact that theories they had initially rejected just because they were unfashionable were in fact false—yielding a kind of 'Stalinist' picture of internal scientific historiography—this would put intolerable strain on the plausibility of the PNC itself. We conjecture that this is mainly how constructivist critics *will* aim to put pressure on the PNC; but in doing so they will need to surmount the battery of standard realist responses found in the literature, and try to undermine OSR, since OSR minimizes the extent to which the history of science should be read as a history of revolutions.

On Melnyk's other clauses, we have dealt extensively with the ways in which our view incorporates commitments (II) and (IV). Some will worry that we violate clause (I) with respect to special-science hypotheses by sliding too far towards a Kantian position on them; we addressed this concern in 5.7. But even someone left unconvinced by that discussion should at least grant that our view is realist where fundamental physics is concerned—provided we can satisfy her that we respect Melnyk's clause (V).

Our critical encounter with constructive empiricism in Chapter 2 did not directly address this issue. We argued, contra van Fraassen, that science describes (in ontological seriousness) a modally structured reality that extends beyond the scope of what can be observed. However, this does not imply clause (V). We deny, on grounds defended throughout the book, that science describes (in ontological seriousness) unobservable entities. However, we have not denied that it describes properties (in structural terms). Our claim that science describes modal structures implies that science describes unobservable instances of properties, since some of its generalizations will apply to counterfactual instantiations of them. But our endorsement of verificationism might be taken to imply that we think that science doesn't describe (in ontological seriousness) unobservable *types* of properties. And then it will be objected that if that is so we are not realists according to the part of Melnyk's criterion expressed in clause (V).

However, clause (V) needs more clarification than Melnyk provides. In fact he provides none; we take it that clause (V) is merely intended to gesture towards the 'core' idea of any version of empiricism, which might be given somewhat different exact interpretations by different empiricists. This is a reasonable procedure on Melnyk's part given the doctrines with which he means to contrast realism: classical, logical, and constructive empiricism. However, the approach requires supplementary work when what is at stake is the question of whether a view Melnyk does not have in mind—in this case, our ITSR—is an expression of empiricism or realism. The obvious way to do this supplementary work, in a way consistent with Melnyk's procedure, is to ask how closely the new view compares with the prototypes he has in mind with respect to some or all under-explicated elements of clause (V).

The element on which we focus attention is 'observation' (as this enters into the construction 'cannot be observed' in (V)). Classical, logical, and constructive empiricists all mean by 'observation of P by x' that x transduces some property or properties of P by operations of one or more of her sensory modalities, mediated by no more than minimal inference. (We say 'minimal inference' to allow for the fact that contemporary versions of classical empiricism often concede that there is no such thing as 'direct' perception. Empiricist theories informed by cognitive science sometimes gloss 'minimum' as meaning the minimum required for normal performance by the sensory modalities in question of the functions for which they were selected. 'Minimal' for Hume, and for the logical positivists, meant zero.) Then when classical and logical empiricists endorse verificationism, they incorporate into this endorsement the idea that a verifiable claim p about P is one with respect to which observation of P by x, or reliable testimony to x about someone else's observation of P, bears on x's justification for believing p.

The verifiability criterion that we endorse is based on a different relation: that of 'being informationally connected to'. x is informationally connected to P iff there exists an information channel from P to x. Following Dretske (1981), an informational channel from P to x is a set of standing conditions necessary for the receipt of information about P by x, but which itself carries no information about P, or only redundant information about P. By 'redundant information' we refer to information carried with less noise by another information channel from P to x. x is a 'perspective' as defined in 4.4. In that chapter, we identified perspectives with 'observation points'. Perspective is thus our surrogate for the philosophical idea of observation. Is this surrogate enough like the empiricist's idea of observation that, when we plug it into a statement of verificationism, it would be disingenuous for us not to consider ourselves empiricists?

There are two main differences between the relation of being informationally connected and the traditional empiricist relation of 'being an observer of [ ]'. The first is that informational connectedness invokes no bias against mediation by inference. If I receive the information that the price of a commodity in a particular market has been stable for a year, and I receive the information that the price in question is not controlled by institutional rules, then if standing conditions include my awareness of basic economics, I receive the information that there is an equilibrium of supply and demand in that market with respect to the commodity and its close substitutes and main complements. My knowledge of economics is an objective fact about the world and so can count as part of channel conditions. Chains of inference can be information channels as long as they are sound.[4] (Astrological inferences are not information channels, a

---

[4] Barwise and Seligman (1997) provide a formal theory of this.

fact we diagnose when we see that they do not support projection as defined in 4.4.)

Many philosophers, told we are going to formulate a verifiability criterion in terms of information channels, will now object that this criterion is bound to come out extraordinarily weak. If some inferential connections can stand in where traditional empiricists put observation relations, then if some scientifically dubious metaphysical speculation is the consequence of a sound inference to the best explanation our verifiability criterion will not block it.

Well, just so; this is precisely to the point of our emphasizing our distance from traditional empiricist accounts. Informational connection is a much less demanding relation than traditional empiricist observation. As we stressed in Chapter 1, what rules out the kind of metaphysics the logical positivists were right to reject should not be, for a naturalist who stands opposed to strong metaphysics, an a priori epistemic rule like a verifiability criterion. In the case of our account, what is supposed to block the kinds of inference to 'the best explanation' in which 'strong' metaphysicians engage is the PNC. The kind of 'empirical' premise that the PNC requires for a justified speculation is not picked out by reference to any philosophical analyses, but by reference to evolved (and evolving) norms of scientific practice.

In that case, our critic may wonder, why do we want to bother with any verifiability criterion at all? The answer brings us to the second major difference between our verificationism and that of classical and logical empiricists. For the latter, verifiability criteria were intended to rule out certain *propositions* as significant. Ours may do something similar—declaring some propositions to be not worth taking seriously, even if they are perfectly meaningful—but only as a by-product. What we really want a verifiability criterion to capture is the pointlessness of merely putative *domains* of inquiry.

When the classical and logical empiricists thought about an observational relation between some x and some P, they thought of x as a cognitive agent and of P as an object of predication. When we think about a relation of informational connectedness between some x and some P, we are thinking about *both* x and P as points (nodes) or regions (interconnected sets of nodes) in networks. Some relevant x's are indeed cognitive agents, but many are not. As discussed in Chapter 4, many perspectives are unoccupied, or occupied only by very stupid agents. (Increasingly many perspectives are occupied by information processors that are more powerful than humans or networks of humans; large computers open new informational channels (Humphreys 2004).) Which perspectives an agent occupies—which x she instantiates, if you like—is partly a function of the inferences available to her, which is in turn a function of both her computational capacities and her position in the network of information flow. This includes her position in spacetime, and her position with respect to testifiers with whom she is informationally connected; thus it incorporates her background knowledge (again, as in Dretske 1981).[5]

On the prototypical P's, we differ even more radically from traditional empiricists. We do not conceive of these as *objects* but as *locations*.[6] The basic sorts of elements with which perspectives are, or aren't, informationally connected are parts of the universe to which they can direct attention (perform the operation of locating) and at which they can take measurements. Then science—not philosophy—tells us that there are terminal points in chains of informational connection. A sound verifiability criterion will tell us not to waste resources (including by misleading us about what we can aim to know) trying to extend our inquiries and the coverage of our theories beyond these terminal points. *If* the Big Bang is a singularity (which is for physics to determine) then there is no point in wondering about the other side of it. Within upper and lower bounds established by inference from other cases and available specific evidence, there is no point in us wondering exactly how many hairs Napoleon sported at Waterloo, because we are collectively disconnected from that information.[7] Note that this is not because of anything that is defective about any *proposition*; it is because we have lost connection with *that region of spacetime at the relevant measurement scale*.

We have now said enough to be able to state informatively our version of a normative verifiability criterion: the science and metaphysics of a community of inquirers should remain silent about putative domains of inquiry from which the community is collectively informationally disconnected.

Let us stress that we really mean 'silent', not just 'muted'. Again, if the Big Bang is a singularity there are no grounds for regarding the other side of it as part of reality or talking about the other side of it as if that locution was functioning as a locator. With respect to locutions such as 'origins of reality'—which exemplifies a range of examples—our view indeed converges with that of the positivists (but on grounds that have nothing directly to do with empiricism). It is far from obvious that 'the origins of reality' has *any* intelligible sense unless 'reality' is

---

[5] This paragraph adds nothing to what we argued in Chapter 4; it merely synopsizes that material in such a way as to make salient its relevance to the set of issues now before us.

[6] In Chapters 4 and 5, of course, we reconstructed reference to objects in terms of dynamic relationships between reference to locations and applications of theories.

[7] Here is a good place at which to remind readers again of our pervasive fallibilism. Current evidence suggests that we are disconnected from this information. But suppose that it turns out that a person's quantity of hair at any given age is rigorously determined by a genetic algorithm, modulo chemotherapy or malnutrition or the influence of basketball fashion. Then, since we know that Napoleon did not have chemotherapy or shave his head, if someone recovered his DNA from surviving corpse fragments, we might establish a connection to the information about his exact hirsuteness at Waterloo. As far as we now know, however, the speculative possibility on which this whole fantasy depends is false.

deflated. (For example, when people consider whether God created reality, they have deflated reality so as to allow for there to be something more.) The positivists tried to get too much philosophical work out of the idea of meaninglessness. But it's true that people very often imagine themselves to be expressing coherent content when they are not, merely because the strings of words they are using are grammatical. Because the positivists' verifiability criterion of meaning was refuted, it was widely concluded that philosophers no longer have grounds for alleging that, for example, religious or neo-scholastic metaphysicians are saying nothing clear or coherent enough to evaluate. But this is a non-sequitur. We think that such people are indeed doing nothing but revealing properties of themselves and don't usually realize it.

Our verifiability criterion has an essentially different status from that of classical and logical empiricists. The content of ours is settled by science, not by an a priori theory of the cognitive role of perception. Its purpose is to describe limits identified by scientists, not to prescribe such limits. It is not part of the foundations of our view, but a consequence of the view's application to the state of scientific knowledge.

Nevertheless, we *do* have in common with naturalistic empiricists (for example, Hume and Quine) the property that our verification*ism* is epistemically foundational. This is bound to be so given our agreement that naturalism and verificationism are the same thesis, or almost the same thesis. But in so far as our verificationism is foundational, it is captured by the PNC, not by the verifiability criterion, and the PNC does not mention observability, or any surrogate notion. Therefore our view does not fall afoul of Melnyk's clause (V).

We thus conclude that what we have defended in this book, having assumed naturalism, are verificationism and realism. Since these two things have been generally thought to be incompatible, it is no wonder that a significant logical space in the metaphysics of science has gone unexplored, and some conundra have seemed insurmountable. Of all the main historical positions in philosophy, the logical positivists and logical empiricists came closest to the insights we have urged. Over-reactions to their errors have led metaphysicians over the past few decades into widespread unscientific and even anti-scientific intellectual waters. We urge them to come back and rejoin the great epistemic enterprise of modern civilization.

References

Adams, R. (1979). Primitive thisness and primitive identity. *Journal of Philosophy* 76: 5–26.
Addleson, M. (1997). *Equilibrium versus Understanding*. London: Routledge.
Adler, S. (2003). Why decoherence has not solved the measurement problem: A response to P. W. Anderson. *Studies in the History and Philosophy of Modern Physics* 34: 135–42.
Aharonov Y., Bergmann, P. G. and Lebowitz, J. L. (1964). Time symmetry in the quantum process of measurement. *Physical Review* 134B: 1410–16.
Ainslie, G. (2001). *Breakdown of Will*. Cambridge: Cambridge University Press.
Albert, D. (2000). *Time and Chance*. Cambridge, MA: Harvard University Press.
____ and Loewer, B. (1988). Interpreting the many worlds interpretation. *Synthese* 77: 195–213.
Almeder, R. (1998). *Harmless Naturalism*. La Salle: Open Court.
Alspector-Kelley, M. (2001). Should the empiricist be a constructive empiricist? *Philosophy of Science* 68: 413–31.
Ambjorn, J, Jurkiewicz, J., and Loll, R. (2004). Emergence of a 4D world from causal quantum gravity. *Physical Review Letters* 98: 131–301.
Annas, J. and Barnes, J. (eds.) (1985). *The Modes of Scepticism: Ancient Texts and Modern Interpretations*. Cambridge: Cambridge University Press.
Armstrong, D. (1983). *What is a Law of Nature?* Cambridge: Cambridge University Press.
____ (2004). Going through an open door again: Counterfactual versus singularist theories of causation. In J. Collins, N. Hall and L. A. Paul (eds.), *Causation and Counterfactuals*, pp. 445–58. Cambridge, MA: MIT Press.
Auyang, S. (1995). *How Is Quantum Field Theory Possible?* Oxford: Oxford University Press.
Baez, J. (2001). Higher-dimensional algebra and Planck scale physics. In C. Callender and N. Huggett (eds.), *Physics Meets Philosophy at the Planck Scale: Contemporary Theories in Quantum Gravity*, pp. 177–95. Cambridge: Cambridge University Press.
Baillargeon, R., Kotovsky, L., and Needham, A. (1995). The acquisition of physical knowledge in infancy. In D. Sperber, D. Premack, and A. James Premack (eds.), *Causal Cognition*, pp. 79–116. Oxford: Oxford University Press.
Bain, J. (2003). Towards structural realism. http://ls.poly.edu/~jbain/papers/SR.pdf
____ (2004). Theories of Newtonian gravity and empirical indistinguishability. *Studies in History and Philosophy of Modern Physics* 35: 345–76.
____ and Norton, J. (2001). What should philosophers of science learn from the history of the electron. In J. Z. Buchwald and A. Warwick (eds.), *Histories of the Electron*, pp. 451–65. Cambridge, MA: MIT Press.
Baldwin, J. M. (ed.) (1901). *Dictionary of Philosophy and Psychology*. London: Macmillan.
Ball, P. (2004). *Critical Mass*. London: Heinemann.
Barrett, J. (1999). *The Quantum Mechanics of Minds and Worlds*. Oxford: Oxford University Press.

Barrow, J., Davies, P., and Harper, C. (eds.) (2004). *Science and Ultimate Reality: Quantum Theory, Cosmology and Computation*. Cambridge: Cambridge University Press.
Barwise, J. and Seligman, J. (1997). *Information Flow*. Cambridge: Cambridge University Press.
Batterman, R. W. (2002). *The Devil in the Details*. Oxford: Oxford University Press.
____ (2005). Response to Belot's 'Whose devil? Which details?'. *Philosophy of Science* 72: 154–63.
Bealer, G. (1987). The philosophical limits of scientific essentialism. *Philosophical Perspectives*, vol. 1, Metaphysics: 289–365.
Beckermann, A. (1992). Supervenience, emergence, and reduction. In A. Beckermann (ed.), *Emergence or Reduction?* pp. 94–118. Berlin & New York: De Gruyter.
Bell, J. (1976). How to teach special relativity. *Progress in Scientific Culture* 1. Reprinted in his (1987).
____ (1987). *Speakable and Unspeakable in Quantum Mechanics*. Cambridge: Cambridge University Press.
Belot, G. (2005a). Dust, time and symmetry. *The British Journal for the Philosophy of Science* 56: 255–91.
____ (2005b). Whose devil? Which details? *Philosophy of Science* 72: 128–53.
____ and Earman, J. (2000). From metaphysics to physics. In J. Butterfield and C. Pagonis (eds.), *From Physics to Philosophy*, pp. 166–86. Cambridge: Cambridge University Press.
____ ____ (2001). Pre-Socratic quantum gravity. In C. Callender and N. Huggett (eds.), *Physics Meets Philosophy at the Planck Scale: Contemporary Theories in Quantum Gravity*, pp. 213–55. Cambridge: Cambridge University Press.
Benacerraf, P. (1965 [1983]). What numbers could not be. *Philosophical Review* 74: 47–73. Reprinted in P. Benacerraf and H. Putnam (eds.), *Philosophy of Mathematics: Selected Readings*. Cambridge: Cambridge University Press. 1983.
Bennett, C. (1990). How to define complexity in physics, and why. In W. Zurek (ed.), *Complexity, Entropy and the Physics of Information*, pp. 137–48. Boulder: Westview.
Berry, M. (1989). *Principles of Cosmology and Gravitation*. Bristol: Adam Hilger.
Beth, E. (1949). Towards an up-to-date philosophy of the natural sciences. *Methodos* 1: 178–85.
Bickerton, D. (1990). *Language and Species*. Chicago: University of Chicago Press.
Bickle, J. (1992). Mental anomaly and the new mind–brain reductionism. *Philosophy of Science* 59/2: 217–30.
____ (1996). New wave reductionism and the methodological caveats. *Philosophy and Phenomenological Research* 56/1: 57–78.
____ (1998). *Psychoneural Reduction: The New Wave*. Cambridge, MA: MIT Press.
Bird, A. (2007). *Nature's Metaphysics: Dispositions, Laws, and Properties*. Oxford: Oxford University Press.
Birkhoff, G. and Bennett, M. (1988). Felix Klein and his 'Erlanger Programme'. *History and Philosophy of Modern Mathematics, Minnesota Studies in Philosophy of Science, Volume XI*, pp. 145–76. Minneapolis: University of Minnesota Press.
Bishop, M. (2003). The pessimistic induction, the flight to reference and the metaphysical zoo. *International Studies in Philosophy of Science* 17: 161–78.

Bitbol, M. (forthcoming). Structuralism and the a priori (2): Kant's ungrounded relations and quantum non-supervenient relations.
Black, M. (1952). The identity of indiscernibles. *Mind* 61: 153–64.
Blackburn, S. (2002). Realism: Deconstructing the debate. *Ratio* 15: 111–33.
Block, N. (1997). Anti-reductionism slaps back. *Mind, Causation, World: Philosophical Perspectives* 11: 107–33.
Bontly, T. (2001). The supervenience argument generalizes. *Philosophical Studies* 109: 75–96.
Bookstein, F. (1983). Comment on a 'nonequilibrium' approach to evolution. *Systematic Zoology* 32: 291–300.
Born, M. (1953). Physical reality. *Philosophical Quarterly* 3: 139–49.
Bourne, C. (2004). Becoming inflated. *The British Journal for the Philosophy of Science* 55: 107–19.
Boyd, R. N. (1984). The current status of scientific realism. In J. Leplin (ed.), *Scientific Realism*, 41–82. Berkeley: University of California Press.
_____ (1985). *Lex orandi est lex credendi*. In P. M. Churchland and C. A. Hooker (eds.), *Images of Science*, pp. 3–34. Chicago: University of Chicago Press.
Braithwaite, R. B. (1940). Critical notice: *The Philosophy of Physical Science*. *Mind* 49: 455–66.
_____ (1953). *Scientific Explanation: A Study of the Function of Theory, Probability and Law in Science*. Cambridge: Cambridge University Press.
Brighouse, C. (1994). Spacetime and holes. In D. Hull, M. Forbes, and A. Fine (eds.), *PSA 1994, volume 1*, pp. 117–25. East Lansing: Philosophy of Science Association.
Brillouin, L. (1956). *Science and Information Theory*. 2nd edn., 1962. New York: Academic Press.
Brittan, G. G. (1970). Explanation and reduction. *Journal of Philosophy* 67/13: 336–457.
Broad, C. D. (1925). *The Mind and its Place in Nature*. London: Routledge and Kegan Paul.
Brooks, D. R. and Wiley, E. O. (1985). Nonequilibrium thermodynamics and evolution: Responses to Bookstein and Wicken. *Systematic Zoology* 34: 89–97.
Brooks, D. R. and Wiley, E. O. (1986). *Evolution as Entropy*. Chicago: University of Chicago Press.
Brown, H. (1993). Correspondence, invariance and heuristics in the emergence of Special Relativity. In S. French and H. Kaminga (eds.), *Correspondence, Invariance and Heuristics: Essays in Honour of Heinz Post, Boston Studies in the Philosophy of Science, Volume 148*, pp. 227–60. Dordrecht: Kluwer.
_____ (2005). *Physical Relativity: Space-Time Structure from a Dynamical Perspective*. Oxford: Oxford University Press.
Brown, H. and Pooley, O. (forthcoming). Minkowski space-time: a glorious non-entity. In Petkov (ed.), *The Ontology of Spacetime*.
Brown, H. and Wallace, D. (forthcoming). Solving the measurement problem: de Broglie-Bohm loses out to Everett. *Foundations of Physics*
Brown, H., Elby, A., and Weingard, R. (1996). Cause and effect in the pilot-wave interpretation of quantum mechanics. In J. T. Cushing et al. (eds.), *Bohmian Mechanics and Quantum Theory: An Appraisal*, pp. 309–19. Dordrecht: Kluwer.
Brown, J. (1999). *Philosophy of Mathematics*. London: Routledge.

Bub, J. (2004). Why the quantum? *Studies in History and Philosophy of Modern Physics* 35: 241–66.

Bueno, O. (1997). Empirical adequacy: A partial structures approach. *Studies in the History and Philosophy of Science* 28: 585–610.

Bueno, O., French, S., and Ladyman, J. (2002). On representing the relationship between the mathematical and the empirical. *Philosophy of Science* 69: 452–73.

Busch, J. (2003). What structures could not be. *International Studies in the Philosophy of Science* 17: 211–25.

Butterfield, J. (1989). The hole truth. *British Journal for the Philosophy of Science* 40: 1–28.

_____ (1992). Review of B. C. van Fraassen, *Quantum Mechanics: An Empiricist View*. *Studies in History and Philosophy of Modern Physics* 24: 443–76.

_____ (2002). Critical notice: Julian Barbour's *The End of Time*. *The British Journal for the Philosophy of Science* 53: 289–330.

_____ (2006). The Rotating Discs Argument Defeated. *The British Journal for the Philosophy of Science* 57: 1–45.

Byrne, R. W. and Whiten, A. (1988). *Machiavellian Intelligence: Social Expertise and the Evolution of Intellect in Monkeys, Apes and Humans*. Oxford: Oxford University Press.

_____ _____ (1997). Machiavellian intelligence. In A. Whiten and R. W. Byrne, *Machiavellian Intelligence II: Extensions and Evaluations*, pp. 1–23. Cambridge: Cambridge University Press.

Callender, C. and Huggett, N. (eds.) (2001). *Physics Meets Philosophy at the Planck Scale: Contemporary Theories in Quantum Gravity*. Cambridge: Cambridge University Press.

Camerer, C. (2003). *Behavioral Game Theory*. Princeton: Princeton University Press.

Cao, T. (1997). *Conceptual Development of 20th Century Field Theories*. Cambridge: Cambridge University Press.

_____ (2003). Structural realism and the interpretation of quantum field theory. *Synthese* 136: 3–24.

_____ and Schweber, S. (1993). The conceptual foundations and the philosophical aspects of renormalization theory. *Synthese* 97: 33–108.

Carnap, R. (1928). *The Logical Structure of the World*. Berkeley; University of California Press.

_____ (1936–7). Testability and meaning. *Philosophy of Science* 3: 419–71.

_____ (1939). *Foundations of Logic and Mathematics*. Chicago: University of Chicago Press.

_____ (1952). *The Continuum of Inductive Methods*. Chicago: University of Chicago Press.

Carroll, L. (1895). What the tortoise said to Achilles. *Mind* 4: 278–80.

Cartwright, N. (1980). Do the laws of physics state the facts? *Pacific Philosophical Quarterly* 61: 75–84.

_____ (1983). *How the Laws of Physics Lie*. Oxford: Oxford University Press.

_____ (1989). *Nature's Capacities and their Measurement*. Oxford: Oxford University Press.

_____ (1992). Aristotelian natures and the modern experimental method. In J. Earman (ed.), *Inference, Explanation and Other Frustrations in the Philosophy of Science*, pp. 44–71. Berkeley: University of California Press.

____ (1999). *The Dappled World*. Cambridge: Cambridge University Press.
____ (2002). Reply. In Book Symposium on *The Dappled World*. *Philosophical Books* 43: 271–8.
Cassirer, E. (1936 [1956]). *Determinism and Indeterminism in Modern Physics*. New Haven: Yale University Press.
____ (1944). Group concept and perception theory. *Philosophy and Phenomenological Research* 5: 1–35.
Castellani, E. (1998). Galilean particles: An example of constitution of objects. In E. Castellani (ed.), *Interpreting Bodies: Classical and Quantum Objects in Modern Physics*, pp. 181–94. Princeton: Princeton University Press.
____ (2002). Reductionism, emergence and effective field theories. *Studies in History and Philosophy of Modern Physics* 33: 251–67.
____ (ed.) (1998). *Interpreting Bodies: Classical and Quantum Objects in Modern Physics*. Princeton: Princeton University Press.
Caves, C. (1990). Entropy and information: How much information is needed to assign a probability? In W. Zurek (ed.), *Complexity, Entropy and the Physics of Information*, pp. 91–115. Boulder: Westview.
Chaitin, G. (1966). On the length of programs for computing finite binary sequences. *Journal of the Association for Computing Machinery* 13: 547–69.
____ (1987). *Algorithmic Information Theory*. Cambridge: Cambridge University Press.
Chakravartty A. (1998). Semirealism. *Studies in History and Philosophy of Modern Science* 29: 391–408.
Chiara, D., Krause, D., and Giuntini, R. (1998). Quasiset theories for microobjects: a comparison. In E. Castellani (ed.), *Interpreting Bodies: Classical and Quantum Objects in Modern Physics*, pp. 142–52. Princeton: Princeton University Press.
Clark, A. (1997). *Being There*. Cambridge, MA: MIT Press.
____ (2004). *Natural Born Cyborgs*. Oxford: Oxford University Press.
Cleland, C. (1984). Space: An abstract system of non-supervenient relations. *Philosophical Studies* 46: 19–40.
Clifton, R. and Halvorson, H. (2002). No place for particles in relativistic quantum theories? *Philosophy of Science* 69: 1–28.
Clifton, R., Bub, J., and Halvorson, H. (2003). Characterising quantum theory in terms of information-theoretic constraints. *Foundations of Physics* 33: 1561–91.
Coffa, A. (1993). *The Semantic Tradition from Kant to Carnap: To the Vienna Station*. Cambridge: Cambridge University Press.
Collier, J. (1986). Entropy in evolution. *Biology and Philosophy* 1: 5–24.
____ (1988). Supervenience and reduction in biological hierarchies. *Canadian Journal of Philosophy* Supplementary Volume 14: 209–34.
____ (1990). Two faces of Maxwell's demon reveal the nature of irreversibility. *Studies in the History and Philosophy of Science* 21: 257–68.
____ (1996). Information originates in symmetry breaking. *Science and Culture* 7: 247–56.
____ (1999). Causation is the transfer of information. In H. Sankey (ed.), *Causation, Natural Laws and Explanation*, pp. 279–331. Dordrecht: Kluwer.
____ (2000). Autonomy and process closure as the basis for functionality. In J. L. R. Chandler and G. de Vijver (eds.), *Closure: Emergent Organizations and their Dynamics*, pp. 280–91. The Annals of the New York Academy of Science vol. 901.

Collier, J. (2001). Dealing with the unexpected. *International Journal of Computing Anticipatory Systems* 10: 21–30.

—— (2003). Hierarchical dynamical information systems with a focus on biology. *Entropy* 5: 100–24.

—— (2004). Self-organization, individuation and identity. *Revue Internationale de Philosophie* 59: 151–72.

—— and Burch, M. (1998). Order from rhythmic entrainment and the origin of levels through dissipation. *Symmetry: Culture and Science* 9: 2–4.

—— and Hooker, C. (1999). Complexly organized dynamical systems. *Open System and Information Dynamics* 6: 241–302.

—— and Muller, S. (1998). The dynamical basis of emergence in natural hierarchies. In G. Farre and T. Oksala (eds.), *Emergence, Complexity, Hierarchy and Organization, Selected and Edited Papers from the ECHO III Conference, Acta Polytechnica Scandinavica, MA91.* Espoo: Finish Academy of Technology.

Connes, A. (1994). *Non-commutative Geometry.* New York: Academic Press.

Cosmides, L. (1989). The logic of social exchange: Has natural selection shaped how humans reason? Studies with the Wason selection task. *Cognitive Psychology* 31: 187–276.

Couvalis, S. G. (1997). *The Philosophy of Science: Science and Objectivity.* London: Sage.

Crane, T. (1995). Mental causation. *Proceedings of the Aristotelian Society* Supplementary vol. 69: 211–36.

Creath, R. (ed.) (1991). *Dear Carnap, Dear Van: The Quine–Carnap Correspondence and Related Work.* Berkeley: University of California Press.

Crook, S. and Gillett, C. (2001). Why physics alone cannot define the 'physical': materialism, metaphysics, and the formulation of physicalism. *Canadian Journal of Philosophy* 31: 333–60.

Cruse, P. (2005). Ramsey sentences, structural realism and trivial realization. *Studies in History and Philosophy of Science* 36: 557–76.

Curiel, E. (2001). Against the excesses of qauntum gravity: A plea for modesty. *Philosophy of Science* supplementary vol. 68: 424–41.

Cushing, J. and McMullin, E. (eds.) (1987). *Philosophical Consequences of Quantum Theory: Reflections on Bell's Theorem.* Notre Dame: University of Notre Dame Press.

Cussins, A. (1990). The Connectionist Construction of Concepts. In M. Boden (ed.), *The Philosophy of Artificial Intelligence*, pp. 368–440. Oxford: Oxford University Press.

da Costa, N. C. A. and French, S. (2003). *Science and Partial Truth: A Unitary Approach to Models and Scientific Reasoning.* Oxford: Oxford University Press.

Dahlbom, B. (ed.) (1993). *Dennett and his Critics.* Oxford: Blackwell.

Darrigol, O. (1992). *From c-Numbers to q-Numbers: The Classical Analogy in the History of Quantum Theory*, California Studies in the History of Science. Berkeley: University of California Press.

Davidson, D. (1970 [1980]). Mental events. In L. Foster and J. W. Swanson (eds.), *Experience and Theory*, pp. 79–101. Amherst: University of Massachusetts Press. Reprinted in N. Block, *Readings in Philosophy of Psychology Volume 1*, pp. 107–19. London: Methuen. 1980.

Davies, P. (1990). Why is the physical world so comprehensible? In W. Zurek (ed.), *Complexity, Entropy and the Physics of Information*, pp. 61–70. Boulder: Westview.

Dawid, R. (forthcoming). Scientific realism in the age of string theory. http://philsci-archive.pitt.edu/archive/00001240

Demopoulos, W. (forthcoming a). Carnap's philosophy of science. In R. Creath and M. Friedman (eds.), *The Cambridge Companion to Carnap*. Cambridge: Cambridge University Press.

Demopoulos, W. (forthcoming b). Review of Hallvard Lillehammer and Hugh Mellor (eds.), *Ramsey's Legacy, Notre Dame Philosophy Reviews*.

Demopoulos, W. and Friedman, M. (1985 [1989]). Critical notice: Bertrand Russell's *The Analysis of Matter*: Its historical context and contemporary interest. *Philosophy of Science* 52: 621–639. Reprinted in C. W. Savage and C. A. Anderson (eds.) (1989), *Rereading Russell: Essays on Bertrand Russell's Metaphysics and Epistemology. Minnesota Studies in the Philosophy of Science, Volume XII*. Minneapolis: University of Minnesota Press.

Dennett, D. (1971). Intentional systems. *Journal of Philosophy* 68: 87–106.

—— (1991a). Real patterns. *Journal of Philosophy* 88: 27–51.

—— (1991b). *Consciousness Explained*. Boston: Little Brown.

—— (1993). Back from the drawing board. In B. Dahlbom (ed.), *Dennett and his Critics*, pp. 203–35. Oxford: Blackwell.

—— (2000). With a little help from my friends. In D. Ross, A. Brook and D. Thompson (eds.), *Dennett's Philosophy: A Comprehensive Assessment*, pp. 327–88. Cambridge, MA: MIT Press.

—— (2005). *Sweet Dreams*. Cambridge, MA: MIT Press.

—— (2006). *Breaking the Spell*. New York: Viking.

DePaul, M and Ramsey, W. (1998). *Rethinking Intuition: The Psychology of Intuition and its Role in Philosophical Inquiry*. Oxford: Rowman and Littlefield.

Deutsch, D. (1997). *The Fabric of Reality*. London: Penguin.

—— (1999). Quantum theory of probability and decisions. *Proceedings of the Royal Society of London* A455: 3129–37.

—— (2004). It from qubit. In J. Barrow, P. Davies, and C. Harper (eds.), *Science and Ultimate Reality*, pp. 90–102. Cambridge: Cambridge University Press.

Dipert, R. R. (1997). The mathematical structure of the world: The world as graph. *Journal of Philosophy* 94: 329–58.

Dirac, P. (1930). *The Principles of Quantum Mechanics*. Oxford: Oxford University Press.

DiSalle, R. (1994). On dynamics, indiscernibility, and spacetime ontology. *The British Journal for the Philosophy of Science* 45: 265–87.

Domski, M. (1999). The epistemological foundations of structural realism: Poincaré and the structure of relations. Paper presented to the research workshop at the University of Leeds.

Dorato, M. (1999). Cao on substantivalism and the development of 20th century field theories. *Epistemologia* 22: 151–66.

—— (2000). Substantivalism, relationism and structural spacetime realism. *Foundations of Physics* 30/10: 1605–28.

Dowe, P. (1992). Wesley Salmon's process theory of causality and the conserved quantity theory. *Philosophy of Science* 59/2: 195–216.

Dowell, J. (2006). The Physical: Empirical, Not Metaphysical. *Philosophical Studies* 131: 25–60.

Dowker, F. (2003). Real time. *New Scientist* 180: 36–9.
Dretske, F. (1981). *Knowledge and the Flow of Information*. Cambridge, MA: MIT Press.
Dunbar, R. I. M. (1992). Neocortex size as a constraint on group size in primates. *Journal of Human Evolution* 20: 469–93.
Dupré, J. (1993). *The Disorder of Things*. Cambridge, MA: Harvard University Press.
_____ (1999). On the impossibility of a monistic account of species. In R. Wilson (ed.), *Species*, pp. 3–22. Cambridge, MA: MIT Press.
_____ (2001). *Human Nature and the Limits of Science*. Oxford: Oxford University Press.
_____ (2004). Review of Laporte, *Natural Kinds and Conceptual Change*. *Notre Dame Philosophical Reviews*. http://ndpr.nd.edu/review.cfm?id=1439
Durlauf, S. (1997). Statistical mechanics approaches to socioeconomic behavior. In W. Arthur, S. Durlauf and D. Lane (eds.), *The Economy as an Evolving Complex System II*, pp. 81–104. Reading, MA: Addison-Wesley.
Duwell, A. (2003). Quantum information does not exist. *Studies in History and Philosophy of Modern Physics* 34: 479–99.
Earman, J. and Norton, J. (1987). What price space-time substantivalism? The hole story. *The British Journal for the Philosophy of Science* 38: 515–25.
Earman, J. and Roberts, J. (1999). *Ceteris paribus*, there is no problem of provisos. *Synthese* 118: 439–78.
Eddington, A. (1928). *The Nature of the Physical World*. Cambridge: Cambridge University Press.
_____ (1939). *The Philosophy of Physical Science*. Cambridge: Cambridge University Press.
Elder, C. (2004). *Real Natures and Familiar Objects*. Cambridge, MA: MIT Press.
Ellis, B. (1985). What science aims to do. In P. M. Churchland and C. A. Hooker (eds.), *Images of Science*, pp. 48–74. Chicago: University of Chicago Press.
Elster, J. (1989). *Nuts and Bolts for the Social Sciences*. Cambridge: Cambridge University Press.
English, J. (1973). Underdetermination: Craig and Ramsey. *The Journal of Philosophy* 70: 453–62.
Esfeld, M. (2004). Quantum entanglement and a metaphysics of relations. *Studies in History and Philosophy of Modern Physics* 35: 601–17.
Esfeld, M. and Lam, V. (forthcoming). Moderate structural realism about space-time. *Synthese*
Feigl, H. (1950). Existential hypotheses: Realistic versus phenomenalistic interpretations. *Philosophy of Science* 17: 35–62.
Feyerabend, P. (1962). Explanation, reduction and empiricism. In H. Feigl and G. Maxwell (eds.), *Minnesota Studies in the Philosophy of Science, Volume 3*, pp. 29–97. Minneapolis: University of Minnesota Press.
Field, H. (forthcoming). Causation in a physical world. In M. Loux and D. Zimmerman (eds.), *Oxford Handbook of Metaphysics*. Oxford: Oxford University Press.
Fine, A. (1984a). The natural ontological attitude. In J. Leplin (ed.), *Scientific Realism*, pp. 83–107. Berkeley: University of California Press.
_____ (1984b). And not anti-realism either. *Noûs* 18: 51–65.
_____ (1986). *The Shaky Game*. Chicago: University of Chicago Press.

____ (1987). Do correlations need to be explained? In J. Cushing and E. McMullin (eds.), *Philosophical Consequences of Quantum Theory: Reflections on Bell's Theorem*, pp. 175–94. Notre Dame: University of Notre Dame Press.

Floridi, L. (2003). Informational realism. *Computers and Philosophy* 37: 7–12.

Fodor, J. (1974). Special sciences, or the disunity of science as a working hypothesis. *Synthese* 28: 77–115.

____ (1991). You can fool some of the people all the time, everything else being equal: Hedged laws and psychological explanation. *Mind* 100: 19–34.

____ (2004). Water's water everywhere. *London Review of Books* 26/20, 21 October.

Forrest, P. (1994). Why most of us should be scientific realists: A reply to van Fraassen. *Monist* 77: 47–70.

Forrest, S. (ed.) (1991). *Emergent Computation*. Cambridge, MA: MIT Press.

French, S. (1989). Identity and individuality in classical and quantum physics. *Australasian Journal of Philosophy* 67: 432–46.

____ (1995). The esperable uberty of quantum chromodynamics. *Studies in History and Philosophy of Modern Physics* 26: 87–105.

____ (1997). Partiality, pursuit and practice. In M. L. Dalla Chiara, K. Doets, D. Mundici and J. van Benthem (eds.), *Structures and Norms in Science: Proceedings of the 10th International Congress on Logic, Methodology and Philosophy of Science*, pp. 35–52. Dordrecht: Reidel.

____ (1998). On the withering away of physical objects. In E. Castellani (ed.), *Interpreting Bodies: Classical and Quantum Objects in Modern Physics*, pp. 93–113. Princeton: Princeton University Press.

____ (1999). Models and mathematics in physics: The role of group theory. In J. Butterfield and C. Pagonis (eds.), *From Physics to Philosophy*, pp. 187–207. Cambridge: Cambridge University Press.

____ (2000). The reasonable effectiveness of mathematics: Partial structures and the application of group theory to physics. *Synthese* 125: 103–20.

____ and Kaminga, H. (eds.) (1993). *Correspondence, Invariance and Heuristics: Essays in Honour of Heinz Post, Boston Studies in the Philosophy of Science, Volume 148*. Dordrecht: Kluwer.

____ and Krause, D. (2006). *Identity in Physics: A Historical, Philosophical and Formal Analysis*. Oxford: Oxford University Press.

____ and Ladyman, J. (1997). Superconductivity and structures: Revisiting the London account. *Studies in the History and Philosophy of Modern Physics* 28: 363–93.

____ ____ (1998). A semantic perspective on idealism in quantum mechanics. In N. Shanks (ed.), *Idealization IX: Idealization in Contemporary Physics: Poznan Studies in the Philosophy of the Sciences and the Humanities*, pp. 51–73. Amsterdam: Rodopi.

____ ____ (1999). Reinflating the semantic approach. *International Studies in the Philosophy of Science* 13: 103–21.

____ ____ (2003a). Remodelling structural realism: Quantum physics and the metaphysics of structure. *Synthese* 136: 31–56.

____ ____ (2003b). Between platonism and phenomenalism: Reply to Cao. *Synthese* 136: 73–8.

____ and Redhead, M. (1988). Quantum physics and the identity of indiscernibles. *British Journal for the Philosophy of Science* 39: 233–46.

Frieden, B. (1998). *Physics from Fisher Information*. Cambridge: Cambridge University Press.
Friedman, Milton (1953). *Essays in Positive Economics*. Chicago: University of Chicago Press.
Friedman, Michael (1974). Explanation and scientific understanding. *Journal of Philosophy* 71: 5–19.
——(1981). Theoretical explanation. In R. Healey (ed.), *Reduction, Time and Reality: Studies in the Philosophy of Natural Sciences*, pp. 1–16. Cambridge: Cambridge University Press.
——(1983). *Foundations of Spacetime Theories: Relativistic Physics and Philosophy of Science*. Princeton: Princeton University Press.
——(1992). *Kant and the Exact Sciences*. Cambridge, MA: Harvard University Press.
——(1999a). *Reconsidering Logical Positivism*. Cambridge: Cambridge University Press.
——(1999b). *Dynamics of Reason*. Stanford: CSLI.
——(2000). *A Parting of the Ways*. Chicago: Open Court.
Fuchs, C. and Peres, A. (2000). Quantum theory needs no interpretation. *Physics Today* 3: 70–1.
Garcia, J. and Koelling, R. A. (1966 [1972]). Relation of cue to consequence in avoidance learning. *Psychosomatic Science* 4: 123–4. Reprinted in Seligman and Hager (1972, 10–14).
Gasper, P. (1992). Reduction and instrumentalism in genetics. *Philosophy of Science* 594: 655–70.
Geach, P. (1972). *Logic Matters*. Oxford: Oxford University Press.
Gell-Mann, M. and Hartle, J. (1990). Quantum mechanics in the light of quantum cosmology. In W. Zurek (ed.), *Complexity, Entropy and the Physics of Information*, pp. 425–58. Boulder: Westview.
Gelman, R., Durgin, F., and Kaufman, L. (1995). Distinguishing between animates and inanimates: Not by motion alone. In D. Sperber, D. Premack, and A. James Premack (eds.), *Causal Cognition*, pp. 150–84. Oxford: Oxford University Press.
Ghemawat, P. (1997). *Games Businesses Play*. Cambridge, MA: MIT Press.
Ghiradi, G. C., Rimini, A., and Weber, T. (1986). Unified dynamics for microscopic and macroscopic systems. *Physics Review D* 34: 470–9.
Giere, R. (1985). Constructive realism. In P. M. Churchland and C. A. Hooker (eds.), *Images of Science*, pp. 75–98. Chicago: University of Chicago Press.
Giere, R. (1988a). Laws, theories and generalizations. In A. Grünbaum and W. Salmon (eds.), *The Limits of Deductivism*, pp. 37–46. Berkeley: University of California Press.
——(1988b). *Explaining Science*. Chicago: University of Chicago Press.
Glimcher, P. (2003). *Decisions, Uncertainty and the Brain*. Cambridge, MA: MIT Press.
Glymour, C. (1999). A Mind is a Terrible Thing to Waste. Critical Notice: Jaegwon Kim, *Mind in a Physical World*. *Philosophy of Science* 66: 455–71.
Gödel, K. (1949). A remark about the relationship between Relativity Theory and idealistic philosophy. In A. Schilpp (ed.), *Albert Einstein, Philosopher-Scientist*, pp. 557–62. La Salle: Open Court.
Godfrey-Smith, P. (2001). Three kinds of adaptationism. In S. Orzack and E. Sober (eds.), *Adaptationism and Optimality*, pp. 335–57. Cambridge: Cambridge University Press.

Gold, T. (1962). The arrow of time. *American Journal of Physics* 30: 403–10.
Goodman, N. (1955). *Fact, Fiction and Forecast*. Indianapolis: Bobbs-Merrill.
―― (1978). *Ways of Worldmaking*. Indianapolis: Hackett.
Gopnik, A., Glymour, C., Sobel, D. M., Schulz, L. E., Kushnir, T., and Danks, D. (2004). A theory of causal learning in children: Causal maps and Bayes nets. *Psychological Review* 111/1: 1–31.
Gould, S. J. (1999). *Rocks of Ages*. New York: Ballantine.
Gower, B. (2000). Cassirer, Schlick and 'structural' realism: The philosophy of the exact sciences in the background to early logical empiricism. *British Journal for the History of Philosophy* 8: 71–106.
Greaves, H. (2004). Understanding Deutsch's probability in a deterministic multiverse. *Studies in History and Philosophy of Modern Physics* 35: 423–56.
Greenberg, O. W. and Messiah, A. M. L. (1964). Symmetrization postulate and its experimental foundation. *Physical Review* 136B: 248–67.
Greene, B. (1999). *The Elegant Universe*. London: Jonathan Cape.
―― (2004). *The Fabric of the Cosmos: Space, Time and the Texture of Reality*. London: Allen Lane.
Guhl, A. M. (1956). The social order of chickens. *Scientific American* 194: 42–6.
Hackermüller, L., Hornberger, K., Brezger, B., Zeilinger, A., and Arndt, M. (2004). Decoherence of matter waves by thermal emission of radiation. *Nature* 427: 711–14.
Hacking, I. (1985). Do we see through a microscope? In P. M. Churchland and C. A. Hooker (eds.), *Images of Science*, pp. 132–52. Chicago: University of Chicago Press.
Hagar, A. (2005). A philosopher looks at quantum information theory. *Philosophy of Science* 70: 752–75.
Hale, S. C. (1990). Elementarity and anti-matter in contemporary physics: Comments on Michael Resnik's 'Between mathematics and physics'. *PSA 1990, Vol. 2*, pp. 379–83. East Lansing: Philosophy of Science Association.
Halvorson, H. (2004). On information-theoretic characterizations of physical theories. *Studies in History and Philosophy of Modern Physics* 35: 277–93.
Hanna. J. (2004). Contra Ladyman: What really is right with constructive empiricism. *The British Journal for the Philosophy of Science* 55: 767–77.
Hardin, C. L. and Rosenberg, A. (1982). In defence of convergent realism. *Philosophy of Science* 49: 604–15.
Hardy, L. (2003). Probability theories in general and quantum theory in particular. *Studies in History and Philosophy of Modern Physics* 34: 381–93.
―― (2005). Why is nature described by quantum theory. In J. Barrow, P. Davies, and C. Harper (eds.), *Science and Ultimate Reality*, pp. 45–71. Cambridge: Cambridge University Press.
Harris, R. (2005). *The Semantics of Science*. London: Continuum.
Harte, V. (2002). *Plato on Parts and Wholes: The Metaphysics of Structure*. Oxford: Oxford University Press.
Hartle, J. (2004). General Relativity and quantum cosmology. http://www.arxiv.org/abs/gr-qc/0403001
Hartmann, S. (2001). Effective field theories, reductionism and scientific explanation. *Studies in History and Philosophy of Modern Physics* 32: 267–304.

Haugeland, J. (1993). Pattern and being. In B. Dahlbom (ed.), *Dennett and his Critics*, pp. 53–69. Oxford: Blackwell.
Hausman, D. (1998). *Causal Asymmetries*. Cambridge: Cambridge University Press.
Hayes, P. (1979). The naïve physics manifesto. In D. Michie (ed.), *Expert Systems in the Micro-electronic Age*, pp. 242–70. Edinburgh: Edinburgh University Press.
Heil, J. (2003). *From an Ontological Point of View*. Oxford: Oxford University Press.
Heisenberg, W. (1962). *Physics and Philosophy*. New York: Harper and Row.
Hellman, G. (2001). Three varieties of mathematical structuralism. *Philosophia Mathematica* 3: 184–211.
─── (2005). Structuralism. In S. Shapiro (ed.), *The Oxford Handbook if Philosophy of Mathematics and Logic*. Oxford: Oxford University Press.
Hellman, G. P. and Thompson, F. W. (1975). Physicalism: Ontology, determination, and reduction. *Journal of Philosophy* 72: 551–64.
Hempel, C. (1950). Problems and changes in the empiricist criterion of meaning. *Revue Internationale de Philosophie* 11: 41–63.
─── (1963). Implications of Carnap's work for philosophy of science. In P. Schilpp (ed.), *The Philosophy of Rudolf Carnap*. LaSalle: Open Court.
─── (1965). *Aspects of Scientific Explanation and Other Essays in the Philosophy of Science*. New York: Free Press.
─── (1988). Provisos: A problem concerning the inferential function of scientific laws. In A. Grünbaum and W. Salmon (eds.), *The Limits of Deductivism*, pp. 19–36. Berkeley: University of California Press.
Hempel, C. and Oppenheim, P. (1953). The logic of explanation. In H. Feigl and M. Brodbeck (eds.), *Readings in the Philosophy of Science*, pp. 319–52. New York: Appleton-Century-Crofts.
Henderson, L. (2003). The Von Neumann entropy: A reply to Shenker. *British Journal for the Philosophy of Science* 54: 291–6.
Hesse, M. (1966). *Models and Analogies in Science*. Oxford: Oxford University Press.
Hochberg, H. (1994). Causal connections, universals and Russell's hypothetico-scientific realism. *Monist* 77: 71–92.
Hoefer, C. (1996). The metaphysics of space-time substantivalism. *Journal of Philosophy* 93: 5–27.
Hogarth, M. L. (1992). Does General Relativity allow an observer to view an eternity in a finite time? *Foundations of Physics Letters* 5: 173–81.
─── (1994). Non-Turing computers and non-Turing computability. In D. Hull, M. Forbens and R. M. Burian (eds.), *PSA 1994*, pp. 126–38. East Lansing: Philosophy of Science Association.
─── (2001). *Predictability, computability, and spacetime*. PhD Dissertation, University of Cambridge.
─── (2004). Deciding arithmetic using SAD computers. *The British Journal for the Philosophy of Science* 55/4: 681–91.
Hoover, K. (2001). *Causality in Macroeconomics*. Cambridge: Cambridge University Press.
Hopf, F. (1988). Entropy and evolution: Sorting through the confusion. In B. Weber, D. Depew, and J. Smith (eds.), *Entropy, Information and Evolution*, pp. 263–74. Cambridge, MA: MIT Press.

Horgan, T. (1982) Supervenience and Microphysics, *Pacific Philosophical Quarterly* 63: 29–43.
Horgan, T. and Potrc, M. (2000). Blobjectivism and indirect correspondence. *Facta Philosophica* 2: 249–70.
—— (2002). Addressing questions for blobjectivism. *Facta Philosophica* 4: 311–22.
Howson, C. (2000). *Hume's Problem: Induction and the Justification of Belief*. Oxford: Oxford University Press.
Huggett, N. (1997). Identity, quantum mechanics and common sense. *The Monist* 80: 118–30.
—— (1999). Atomic metaphysics. *Journal of Philosophy* 96: 5–24.
Hull, D. (1976). Are species really individuals? *Systematic Zoology* 25: 174–91.
Hume, D. (1975 [1777]). *Enquiries Concerning Human Understanding and Concerning the Principles of Morals*, (3rd edn.), ed. L. A. Selby-Bigge, rev. P. H. Nidditch. Oxford: Oxford University Press.
—— (1978 [1740]) *A Treatise of Human Nature*, ed. L. A. Selby-Bigge. Oxford: Oxford University Press.
Humphreys, P. (2004). *Extending Ourselves*. Oxford: Oxford University Press.
Hüttemann, A. (2004). *What's Wrong with Microphysicalism?* London: Routledge.
—— and Papineau, D. (2005) Physicalism decomposed. *Analysis*, 65: 33–9.
Jackson, F. (1986) What Mary didn't know. *Journal of Philosophy* 83: 291–5.
—— (1998). *From Metaphysics to Ethics: A Defence of Conceptual Analysis*. Oxford: Oxford University Press.
Jauch, J. M. (1968). *Foundations of Quantum Mechanics*. Reading, MA: Addison-Wesley.
Jaynes, E. (1957a). Information theory in statistical mechanics I. *Physical Review* 106: 620–30.
—— (1957b). Information theory in statistical mechanics II. *Physical Review* 108: 171–90.
Johnson, S. (2001). *Emergence*. London: Penguin.
Jones, R. (1991). Realism about what? *Philosophy of Science* 58: 185–202.
Jozsa, R. (1998). Quantum information and its properties. In H.-K. Lo, S. Popescu, and T. Spiller (eds.), *Introduction to Quantum Computation and Information*, pp. 49–75. Singapore: World Scientific.
Kant, I. (1933) *Critique of Pure Reason*. Trans. N. Kemp Smith. London: Macmillan.
Kantorovich, A. (2003). The priority of internal symmetries in particle physics. *Studies in History and Philosophy of Modern Physics* 34: 651–75.
Kauffman, S. (1995). *At Home in the Universe*. London: Penguin.
Kelso, S. (1995). *Dynamic Patterns*. Cambridge, MA: MIT Press.
Kemeny, J. C. and Oppenheim, P. (1956). On reduction. *Philosophical Studies* 7: 6–19.
Keränen, J. (2001). The identity problem for realist structuralism. *Philosophia Mathematica* 3: 308–30.
Ketland, J. (2004). Empirical adequacy and ramsification. *The British Journal for the Philosophy of Science* 55: 409–24.
Kim, J. (1993). *Mind and Supervenience*. Cambridge: Cambridge University Press.
—— (1996). *Philosophy of Mind*. Boulder: Westview Press.
—— (1998). *Mind in a Physical World*. Cambridge, MA: MIT Press.
—— (1999). Making sense of emergence. *Philosophical Studies* 95: 3–36.

Kim, J. (2005). *Physicalism or Something near Enough*. Princeton: Princeton University Press.
Kincaid, H. (1990). Defending laws in the social sciences. *Philosophy of the Social Sciences* 20: 56–83.
―― (1996). *Philosophical Foundations of the Social Sciences*. Cambridge: Cambridge University Press.
―― (1997). *Individualism and the Unity of Science*. Lanham, MD: Rowman and Littlefield.
―― (2004). There are laws in the social sciences. In C. Hitchcock (ed.), *Contemporary Debates in Philosophy of Science*, pp. 168–85. Oxford: Blackwell.
Kitcher, P. (1976). Explanation, conjunction and unification. *Journal of Philosophy* 73: 207–12.
―― (1981). Explanatory unification. *Philosophy of Science* 48: 507–31.
―― (1982). *Abusing Science*. Cambridge, MA: MIT Press.
―― (1984). 1953 and all that: A tale of two sciences. *Philosophical Review* 93: 335–73.
―― (1989). Explanatory unification and the causal structure of the world. In P. Kitcher and W. Salmon (eds.), *Scientific Explanation*, pp. 410–505. Minneapolis: University of Minnesota Press.
―― (1993). *The Advancement of Science*. Oxford: Oxford University Press.
Krause, D. (1992). On a quasi-set theory. *Notre Dame Journal of Formal Logic* 33: 402–11.
Kripke, S. (1973). *Naming and Necessity*. Cambridge, MA: Harvard University Press.
Kuhn, T. (1962). *The Structure of Scientific Revolutions*. Chicago: Chicago University Press.
Kukla, A. (1993). Laudan, Leplin, empirical equivalence, and underdetermination. *Analysis* 53: 1–7.
―― (1994). Non-empirical theoretical virtues and the argument from underdetermination. *Erkenntnis* 41: 157–70.
―― (1995). Scientific realism and theoretical unification. *Analysis* 55: 230–8.
―― (1996a). Does every theory have empirically equivalent rivals? *Erkenntnis* 44: 137–66.
―― (1996b). Scientific realism, scientific practice and the natural ontological attitude. *The British Journal for the Philosophy of Science* 45: 955–75.
―― (1998). *Studies in Scientific Realism*. Oxford: Oxford University Press.
Kydland, F and Prescott, E. (1982). Time to build and aggregate fluctuations. *Econometrica* 50: 345–70.
Ladyman, J. (1998). What is structural realism? *Studies in History and Philosophy of Science* 29: 409–24.
―― (1999). Review of *A Novel Defense of Scientific Realism* by J. Leplin. *The British Journal for the Philosophy of Science* 50: 181–8.
―― (2000). What's really wrong with constructive empiricism? Van Fraassen and the metaphysics of modality. *The British Journal for the Philosophy of Science* 51: 837–56.
―― (2002a), Science, metaphysics and structural realism. *Philosophica* 67: 57–76.
―― (2002b). *Understanding Philosophy of Science*. London: Routledge.

____ (2004). Modality and constructive empiricism: A reply to van Fraassen. *The British Journal for the Philosophy of Science* 55: 755–65.

____ (2005). Mathematical structuralism and the identity of indiscernibles. *Analysis* 65: 218–21.

____ (forthcoming a). The Epistemology of Constructive Empiricism. In B. Monton (ed.), *Van Fraassen's Philosophy of Science*. Oxford: Oxford University Press.

____ (forthcoming b). On the Identity and Diversity of Individuals. In *The Proceedings of the Aristotelian Society*.

____, Douven, I., Horsten, L., and van Fraassen, B. C. (1997). In defence of van Fraassen's critique of abductive reasoning: A reply to Psillos. *Philosophical Quarterly* 47: 305–21.

____, Presnell, S., Short, A., and Groisman, B. (2007). The relationship between logical and thermodynamic reversibility. *Studies in History and Philosophy of Modern Physics*.

Lakoff, G. (1987). *Women, Fire and Dangerous Things*. Chicago: University of Chicago Press.

____ and Johnson, M. (1980). *Metaphors We Live By*. Chicago: University of Chicago Press.

Landauer, R. (1961). Irreversibility and heat generation in the computing process. *IBM Journal of Research and Development* 5: 183–91.

Lange, M. (1993). Natural laws and the problem of provisos. *Erkenntnis* 38: 233–48.

____ (2002). Baseball, pessimistic inductions and the turnover fallacy. *Analysis* 62: 281–5.

Langton, R. (1998), *Kantian Humility: Our Ignorance of Things in Themselves*. Oxford: Oxford University Press.

____ and Lewis, D. (1998). Defining 'intrinsic'. In D. Lewis, *Papers in Metaphysics and Epistemology*, pp. 116–32. Cambridge: Cambridge University Press.

Laporte, J. (2004). *Natural Kinds and Conceptual Change*. Cambridge: Cambridge University Press

Laudan, L. (1981). A confutation of convergent realism. *Philosophy of Science* 48: 19–49.

____ (1984). Discussion: Realism without the real. *Philosophy of Science* 51: 156–62.

____ and Leplin, J. (1991). Empirical equivalence and underdetermination. *Journal of Philosophy* 88: 269–85.

____ ____ (1993). Determination underdeterred. *Analysis* 53: 8–15.

Laurence, S. and Margolis, E. (2003). Concepts and conceptual analysis. *Philosophy and Phenomenological Research* 67: 253–82.

Layzer, D. (1982). Quantum mechanics, thermodynamics and the strong cosmological principle. In A. Shimony and H. Freshbach (eds.), *Physics as Natural Philosophy*, pp. 240–62. Cambridge, MA: MIT Press.

____ (1988). Growth of order in the universe. In B. Weber, D. Depew, and J. Smith (eds.), *Entropy, Information and Evolution*, pp. 23–39. Cambridge, MA: MIT Press.

____ (1990). *Cosmogenesis: The Growth of Order in the Universe*. Oxford: Oxford University Press.

Leggett, A. (1995). Time's arrow and the quantum measurement problem. In S. Savitt (ed.), *Time's Arrow Today: Recent Physical and Philosophical Work on the Direction of Time*, pp. 97–106. Cambridge: Cambridge University Press.

Leitgeb, H. and Ladyman, J. (forthcoming). Criteria of identity and structuralist ontology. *Philosophica Mathematica*.
Leplin, J. (1997). *A Novel Defense of Scientific Realism*. Oxford: Oxford University Press.
Lewis, D. (1970). How to define theoretical terms. *Journal of Philosophy* 67: 427–46.
―― (1983). Extrinsic properties. *Philosophical Studies* 44: 197–200.
―― (1986). *On the Plurality of Worlds*. Oxford: Blackwell.
―― (1991). *Parts of Classes*. Oxford: Blackwell.
―― (1999). *Papers in Metaphysics and Epistemology*. Cambridge: Cambridge University Press.
―― (2004). How many lives has Schrödinger's Cat. *Australasian Journal of Philosophy* 82: 3–22.
―― (forthcoming). Ramseyan humility. In D. Braddon-Mitchell, R. Nola, and D. Lewis (eds.), *The Canberra Programme*. Oxford: Oxford University Press.
Lewis, P. (2001). Why the pessimistic induction is a fallacy. *Synthese* 129: 371–80.
Lipton, P. (1991). *Inference to the Best Explanation*. London: Routledge.
―― (2001). Is explanation a guide to inference? In G. Hon and S. Rackover (eds.), *Explanation: Theoretical Approaches*, pp. 93–120. Dordrecht: Kluwer.
Litt, A., Eliasmith, C., Kroon, F., Weinstein, S., and Thagard, P. (forthcoming). Is the brain a quantum computer? *Cognitive Science*.
Lockwood, M. (1989). *Mind, Brain and the Quantum*. Oxford: Oxford University Press.
Loewer, B. (2001). From physics to physicalism. In C. Gillett and B. Loewer (eds.), *Physicalism and its Discontents*, pp. 37–56. Cambridge: Cambridge University Press
Loux, M. and Zimmerman, D. (2003). *The Oxford Handbook of Metaphysics*. Oxford: Oxford University Press.
Lovtrup, S. (1983). Victims of ambition: Comments on the Wiley and Brooks approach to evolution. *Systematic Zoology* 32: 90–6.
Lowe, E. J. (2002). *A Survey of Metaphysics*. Oxford: Oxford University Press.
―― (2003a). In defense of moderate-sized specimens of dry goods. *Philosophy and Phenomenological Research* 67: 704–10.
―― (2003b). Individuation. In Loux and Zimmerman (2003, 75–95).
―― (2006). *The Four-Category Ontology: A Metaphysical Foundation for Natural Science*. Oxford: Oxford University Press.
Lucas, J. and Hodgson, P. (1990). *Spacetime and Electromagnetism*. Oxford: Oxford University Press.
Lyons, T. (2006). Scientific realism and the *stratagema de divide et impera*. *The British Journal for the Philosophy of Science* 57: 537–60.
Lyre, H. (2004). Holism and structuralism in U(1) gauge theory. *Studies in History and Philosophy of Modern Physics* 35: 643–70.
MacBride, F. (2004). Introduction. *The Philosophical Quarterly* 54: 1–15.
―― (2005). Structuralism reconsidered. In S. Shapiro (ed.), *The Oxford Handbook of Logic and Mathematics*, pp. 563–89. Oxford: Oxford University Press.
McDowell, J. (1994). *Mind and World*. Cambridge, MA: Harvard University Press.
McMullin, E. (1984). The goals of natural science. *Proceedings of the American Philosophical Association* 58: 37–64.
―― (1985). Galilean idealization. *Studies in History and Philosophy of Science* 16: 247–73.

_____ (1990). Comment: Duhem's middle way. *Synthese* 83: 421–30.
McLaughlin, B. and Bennett, K. (2005). Supervenience. In *The Stanford Encyclopedia of Philosophy*. http://plato.stanford.edu/entries/supervenience
McLeish, C. (2005). Scientific realism bit by bit: Part I. Kitcher on reference. *Studies in History and Philosophy of Science* 36: 667–85.
_____ (2006). Scientific realism bit by bit: Part II. Disjunctive partial reference. *Studies in History and Philosophy of Science* 37: 171–90.
Maddy, P. (1990). *Realism in Mathematics*. Oxford: Oxford University Press.
Magnus, P. D. and Callender, C. (2004). Realist ennui and the base rate fallacy. *Philosophy of Science* 71: 320–38.
Mainzer, K. (1996). *Symmetries of Nature: A Handbook for Philosophy of Nature and Science*. Berlin: de Gruyter.
Mäki, U. (1992). Friedman and realism. In W. Samuels and J. Biddle (eds.), *Research in the History of Economic Thought and Methodology, Volume 10*, pp. 171–95. Greenwich, CT: JAI Press.
Malament, D. (1996). In defense of dogma: Why there cannot be a relativistic quantum mechanical theory of (localizable) particles. In R. Clifton (ed.), *Perspectives on Quantum Reality*, pp. 1–10. Dordrecht: Kluwer.
_____ (2006). Classical General Relativity. In J. Butterfield and J. Earman (eds.), *Handbook of the Philosophy of Physics*. North Holland: Elsevier.
Margolis, E. and Laurence, S. (2003). Should we trust our intuitions? Deflationary accounts of the analytic data. *Proceedings of the Aristotelian Society* 103/3: 299–323.
Markosian, N. (2000). What are Physical Objects? *Philosophy and Phenomenological Research* 61: 375–95.
_____ (2005). Against ontological fundamentalism. *Facta Philosophia* 7: 69–84.
Marras, A. (2000). Critical notice of Kim's *Mind in a Physical World*. *Canadian Journal of Philosophy* 30: 137–60.
_____ (2002). Kim on reduction. *Erkenntnis*, 57: 231–57.
_____ (2005). Consciousness and reduction. *British Journal for the Philosophy of Science* 56: 335–61.
Maudlin, T. (1990). Substances and spacetimes: What Aristotle would have said to Einstein. *Studies in History and Philosophy of Science* 21: 531–61.
_____ (1994). *Quantum Non-Locality and Relativity*. Oxford: Blackwell.
_____ (1998). Part and whole in quantum mechanics. In E. Castellani (ed.) (1998), 46–60.
_____ (2002a). *Quantum Non-Locality and Relativity*, 2nd edn. Oxford: Blackwell.
_____ (2002b). Remarks on the passing of time. *Proceedings of the Aristotelian Society* 102: 237–52.
_____ (2007). Non-local correlations in quantum theory: some ways the trick might be done. In Q. Smith and W. L. Craig (eds.), *Einstein, Relativity, and Absolute Simultaneity*. London: Routledge.
Maxwell, G. (1962). The ontological status of theoretical entities. In H. Feigl and G. Maxwell (eds.), *Minnesota Studies in the Philosophy of Science, Volume 3*, pp. 3–14. Minneapolis: University of Minnesota Press.
_____ (1970a). Structural realism and the meaning of theoretical terms. In S. Winokur and M. Radner (eds.), *Analyses of Theories and Methods of Physics and Psychology:*

Minnesota Studies in the Philosophy of Science, Volume 4, pp. 181–92. Minneapolis: University of Minnesota Press.

―― (1970b). Theories, perception and structural realism. In R. Colodny (ed.), *The Nature and Function of Scientific Theories: University of Pittsburgh Series in the Philosophy of Science, Volume 4*, pp. 3–34. Pittsburgh: University of Pittsburgh Press.

―― (1972). Scientific methodology and the causal theory of perception. In H. Feigl, H. Sellars, and K. Lehrer (eds.), *New Readings in Philosophical Analysis*, pp. 148–77. New York: Appleton-Century Crofts.

Melia, J. (1999). Holes, haecceitism and two conceptions of determinism. *The British Journal of Philosophy of Science* 50: 639–64.

―― and Saatsi, J. (2006). Ramsification and theoretical content. *The British Journal for the Philosophy of Science* 57: 561–85.

Melnyk, A. (2003). *A Physicalist Manifesto*. Cambridge: Cambridge University Press.

Mermin, N. (1998). What is quantum mechanics trying to tell us? *American Journal of Physics* 66: 753–67.

Merricks, T. (2001). *Objects and Persons*. Oxford: Oxford University Press.

―― (2003a). Précis of *Objects and Persons*. *Philosophy and Phenomenological Research* 67: 700–3.

―― (2003b). Replies. *Philosophy and Phenomenological Research* 67: 727–44.

Meyering, T. (2000). Physicalism and downward causation in psychology and the special sciences. *Inquiry* 43: 181–202.

Mill, J. S. (1974). *A System of Logic Ratiocinative and Deductive, Being a Connected View of the Principles of Evidence and the Methods of Scientific Investigation*, vol. 1, ed. J. M. Robson; vol. VII, J. M. Robson (general ed.), *The Collected Works of John Stuart Mill*. Toronto: University of Toronto Press.

Møller, A. P. (1994). Directional selection on directional asymmetry: Testes size and secondary sexual characters in birds. *Proceedings of the Royal Society of London*, Series B258: 147–51.

Montero, B. (2006). Physicalism in an Infinitely Decomposable World, *Erkenntnis* 64: 177–91.

Monton, B. (forthcoming). Presentism and quantum gravity.

Monton, B. and van Fraassen, B. C. (2003). Constructive empiricism and modal nominalism. *The British Journal for the Philosophy of Science* 54: 405–22.

Morgan, M., Morrison, M., Skinner, Q., Tulley, J., Daston, L., and Ross, D. (1999). *Models as Mediators*. Cambridge: Cambridge University Press.

Morganti, M. (2004). On the preferability of epistemic structural realism. *Synthese* 142: 81–107.

Morowitz, H. (1986). Entropy and nonsense: A review of Daniel R. Brooks and E. O. Wiley. *Evolution as Entropy, Biology and Philosophy* 1: 473–6.

Morris, M. W., Nisbett, R. E., and Peng, K. (1995). Causal attribution across domains and cultures. In D. Sperber, D. Premack, and A. James Premack (eds.), *Causal Cognition*, pp. 577–612. Oxford: Oxford University Press.

Muller, F. (2005). The deep black sea: Observability and modality afloat. *The British Journal for the Philosophy of Science* 56: 61–99.

Mumford, S. (2004). *Laws in Nature. Routledge Studies in Twentieth-Century Philosophy*. London: Routledge.

Musgrave, A. (1988). The ultimate argument for scientific realism. In R. Nola (ed.), *Relativism and Realism in Sciences*, pp. 229–52. Dordrecht: Kluwer.
Myrvold, W. (2002). On peaceful coexistence: Is the collapse postulate incompatible with relativity? *Studies in History and Philosophy of Modern Physics* 33: 435–66.
Nagel, E. (1949). The meaning of reduction in the natural sciences. In R. C. Stouffer (ed.), *Science and Civilization*, pp. 99–135. Madison: University of Wisconsin Press.
____ (1961). *The Structure of Science*. New York: Harcourt, Brace, and World.
Nagel, J. (2000). The empiricist conception of experience. *Philosophy* 75: 345–76.
Needham, P. (2002). The discovery that water is H_2O. *International Studies in the Philosophy of Science* 16: 205–26.
Newman, M. H. A. (1928). Mr Russell's causal theory of perception. *Mind* 37: 137–48.
Ney, A. (forthcoming a). Physicalism as an attitude. *Philosophical Studies*.
____ (forthcoming b). Physicalism and our knowledge of intrinsic properties. *Australasian Journal of Philosophy*.
Nickles, T. (1973), Two concepts of intertheoretic reduction. *The Journal of Philosophy* 70: 181–201.
Nielsen, M. and Chuang, I. (2000). *Quantum Computation and Quantum Information*. Cambridge: Cambridge University Press.
Nolan, D. (2004). Classes, worlds and hypergunk. *The Monist* 87: 3–21.
Norton, J. (2003). Causation as folk science. PhilSci archive. www.philsci-archive.pitt.edu/archive/00001214
Nottale, L. (1993). *Fractal Space-Time and Microphysics*. Singapore: World Scientific.
Oppenheim, P. and Putnam, H. (1958). Unity of science as a working hypothesis. In H. Feigl, M. Scriven, and G. Maxwell (eds.), *Minnesota Studies in the Philosophy of Science, Volume II*, pp. 3–36. Minneapolis: University of Minnesota Press.
Orzack, S., and Sober, E. (eds.) (2001). *Adaptationism and Optimality*. Cambridge: Cambridge University Press.
Pais, A. (1991). *Neils Bohr's Times, in Physics, Philosophy, and Polity*. Oxford: Oxford University Press.
Papineau, D. (1993). *Philosophical Naturalism*. Oxford: Blackwell.
____ (1996). Theory-dependent terms. *Philosophy of Science* 63: 1–20.
____ (2001). The rise of physicalism. In C. Gillet and B. Loewer, *Physicalism and its Discontents*, pp. 3–37. Cambridge: Cambridge University Press.
____ and Hüttemann, A. (2005). Physicalism decomposed. *Analysis* 65: 33–9.
Parsons, C. (1990). The structuralist view of mathematical objects. *Synthese* 84: 303–46.
Paul, L. A. (2004). The context of essence. *Australasian Journal of Philosophy* 82: 170–84.
Peirce, C. S. (1960–6). *Collected Papers*, ed. C. Hartshorne and P. Weiss. Cambridge, MA: Belknap Press of Harvard University Press.
Pennock, R. (2000). *Tower of Babel: The Evidence against the New Creationism*. Cambridge, MA: MIT Press.
Penrose, R. (2001). On gravity's role in quantum state reduction. In C. Callender and N. Huggett (eds.), *Physics Meets Philosophy at the Planck Scale*, pp. 290–304. Cambridge: Cambridge University Press.
____ (2004). *The Road to Reality: A Complete Guide to the Laws of the Universe*. London: Jonathan Cape.
Peres, A. (1990). Thermodynamic constraints on quantum axioms. In W. Zurek (ed.), *Complexity, Entropy and the Physics of Information*, pp. 345–56. Boulder: Westview.

Perry, J. (1970). The same 'F'. *Philosophical Review* 79: 181–200.
Pettit, P. (1993). A definition of physicalism. *Analysis* 53: 213–23.
Pierce, J. (1980). *An Introduction to Information Theory: Symbols, Signals and Noise.* 2nd, rev. edn. New York: Dover.
Pietroski, P. and Rey, G. (1995). When other things aren't equal: Saving *ceteris paribus* laws from vacuity. *British Journal for the Philosophy of Science* 46: 81–110.
Poincaré, H. (1898). On the foundations of geometry. *The Monist* 9: 1–43.
____ (1905 [1952]). *Science and Hypothesis.* New York: Dover.
____ (1906 [1914] [1958]). *The Value of Science.* Trans. G. B. Halsted, 1914. Repr. 1958. New York: Dover.
Poland, J. (1994). *Physicalism: The Philosophical Foundations.* Oxford: Oxford University Press.
Ponce, V. (2003). Rethinking Natural Kinds. Doctoral dissertation, Duke University.
Pooley, O. (2006). Points, particles and structural realism. In D. Rickles, S. French and J. Saatsi (eds.) pp. 83–120.
Post, H. R. (1971 [1993]). Correspondence, invariance and heuristics. *Studies in History and Philosophy of Science* 2: 213–55. Reprinted in S. French and H. Kamminga (eds.), *Correspondence, Invariance and Heuristics: Essays in Honour of Heinz Post*, Boston Studies in the Philosophy of Science, vol. 148, pp. 1–44. Dordrecht: Kluwer Academic Press. 1993.
Poundstone, W. (1985). *The Recursive Universe.* New York: Morrow.
Prigogine, I (1980). *From Being to Becoming: Time and Complexity in Physical Science.* San Francisco: Freeman.
Psillos, S. (1994). A philosophical study of the transition from the caloric theory of heat to thermodynamics: Resisting the pessimistic meta-induction. *Studies in the History and Philosophy of Science* 25: 159–90.
____ (1995). Is structural realism the best of both worlds? *Dialectica* 49: 15–46.
____ (1996). On van Fraassen's critique of abductive reasoning. *Philosophical Quarterly* 46: 31–47.
____ (1997). How not to defend constructive empiricism: A rejoinder. *Philosophical Quarterly* 47: 369–72.
____ (1999). *Scientific Realism: How Science Tracks Truth.* London: Routledge.
____ (2000). Carnap, the Ramsey-sentence and realistic empiricism. *Erkenntnis* 52: 253–79.
____ (2001). Is structural realism possible? *Philosophy of Science* 68 (supplementary volume): S13–S24.
____ (forthcoming). The structure, the whole structure and nothing but the structure? In *Philosophy of Science: Supplementary Volume.*
Putnam, H. (1962). What theories are not. In H. Putnam, *Mathematics, Matter and Method: Philosophical Papers Volume 1.* Cambridge: Cambridge University Press.
____ (1967). Time and physical geometry. *Journal of Philosophy* 64: 240–7.
____ (1975). *Mathematics, Matter and Method.* Cambridge: Cambridge University Press.
____ (1975a). Philosophy and our mental life. In H. Putnam *Mind, Language, and Reality*, pp. 291–303. New York: Cambridge University Press.
____ (1975b). *Mind, Language and Reality.* Cambridge: Cambridge University Press.

_____ (1995). Pragmatism. *Proceedings of the Aristotelian Society*, 95: 291–306.
Quine, W. V. (1951). Two dogmas of empiricism. *The Philosophical Review*, 60/1: 20–43.
_____ (1953 [1966]). Mr. Strawson on logical theory. *Mind*, 62/248: 433–51. Repr. in *The Ways of Paradox and Other Essays*. New York: Random House. 1966.
_____ (1953). *From a Logical Point of View*. Cambridge, MA: Harvard University Press.
_____ (1960). *Word and Object*. Cambridge, MA: MIT Press.
_____ (1969). *Ontological Relativity and Other Essays*. New York: Columbia University Press.
_____ (1974). *The Roots of Reference*. La Salle: Open Court.
_____ (1976 [1981]). Grades of discriminability. *Journal of Philosophy* 73: 113–16. Reprinted in *Theories and Things*, 1981, Cambridge, MA: Harvard University Press.
_____ (1978). A postscript on metaphor. *Critical Inquiry* 5: 161–2.
Ramsey, F. P. (1929). Theories. In R. B. Braithwaite (ed.), *The Foundations of Mathematics and Other Logical Essays*, pp. 212–36. Paterson, NJ: Littlefield and Adams.
Rea, M. (2002). *World without Design: The Ontological Consequences of Naturalism*. Oxford: Oxford University Press.
Reck, E. and Price, M. (2000). Structures and structuralism in contemporary philosophy of mathematics. *Synthese* 125: 341–87.
Redhead, M. L. G. (1980). Models in physics. *British Journal for the Philosophy of Science* 31: 145–63.
_____ (1987). *Incompleteness, Nonlocality and Realism: A Prolegomenon to the Philosophy of Quantum Mechanics*. Cambridge, MA: Cambridge University Press.
_____ (1990). Explanation. In D. Knowles (ed.), *Explanation and its Limits*, pp. 135–54. Cambridge: Cambridge University Press.
_____ (1995). *From Physics to Metaphysics*. Cambridge: Cambridge University Press.
_____ (1999). Quantum field theory and the philosopher. In T. Cao (ed.), *Conceptual Foundations of Quantum Field Theory*. Cambridge: Cambridge University Press.
_____ (2004). Asymptotic reasoning. *Studies in History and Philosophy of Modern Physics* 35: 527–30.
Reichenbach, H. (1956). *The Direction of Time*. Berkeley: University of California Press.
_____ (1957). *The Philosophy of Space and Time*. New York: Dover.
Resnik, M. (1997). *Mathematics as a Science of Patterns*. Oxford: Oxford University Press.
_____ (1990) Between mathematics and physics. *PSA 1990, Vol. 2*, pp. 369–78. East Lansing: Philosophy of Science Association.
Rickles, D. (2005). A new spin on the hole argument. *Studies in History and Philosophy of Modern Physics* 36: 415–34.
_____, French, S., and Saatsi, J. (eds.) (2006). *Structural Foundations of Quantum Gravity*. Oxford: Oxford University Press
Rickles, D., French, S., and Saatsi, J. (forthcoming). *The Structural Foundations of Quantum Gravity*. Oxford: Oxford University Press.
Rorty, R. (1993). Holism, intrinsicality and the ambition of transcendence. In B. Dahllbom (ed.), *Dennett and His Critics*, pp. 184–202. Oxford: Blackwell.
Rosen, G. (1984). What is constructive empiricism? *Philosophical Studies* 74: 143–78.
Rosenberg, A. (1992). *Economics: Mathematical Politics or Science of Diminishing Returns?* Chicago: University of Chicago Press.

Rosenberg, A. (1994). *Instrumental Biology or the Disunity of Science*. Chicago: University of Chicago Press.
Ross, D. (2000). Rainforest realism: A Dennettian theory of existence. In D. Ross, A. Brook, and D. Thompson (eds.), *Dennett's Philosophy: A Comprehensive Assessment*, pp. 147–68. Cambridge, MA: MIT Press.
—— (2004). Metalinguistic signalling for coordination amongst social animals. *Language Sciences* 26: 621–42.
—— (2005). *Economic Theory and Cognitive Science: Microexplanation*. Cambridge, MA: MIT Press.
—— (forthcoming a). The economic and evolutionary basis of selves. *Journal of Cognitive Systems Research*.
—— (forthcoming b). *H. sapiens* as ecologically special: what does language contribute? *Language Sciences*.
—— and Spurrett, D. (2004). What to say to a skeptical metaphysician: A defense manual for cognitive and behavioral scientists. *Behavioral and Brain Sciences*, 27/5: 603–27.
Rovelli, C. (1997). Relational quantum mechanics. *International Journal of Theoretical Physics* 35: 1637–78.
—— (2004). *Quantum Gravity*. Cambridge: Cambridge University Press.
Rowlands, M. (1999). *The Body in Mind*. Cambridge: Cambridge University Press.
Russell, B. (1903). *The Principles of Mathematics*. Cambridge: Cambridge University Press.
—— (1913 [1917]). On the notion of cause. In B. Russell, *Mysticism and Logic*, pp. 173–99. London: Unwin.
—— (1925). *ABC of Relativity*. London: Allen and Unwin.
—— (1927). *The Analysis of Matter*. London: Routledge Kegan Paul.
—— (1948). *Human Knowledge*. New York: Simon and Schuster.
Ruttkamp, E. (2002). *A Model-Theoretic Realist Interpretation of Science*. Dordrecht: Kluwer.
Ryckman, T. (2005). *The Reign of Relativity: Philosophy in Physics 1915–1925*. Oxford: Oxford University Press.
Saatsi, J. (2005). Reconsidering the Fresnel-Maxwell theory shift: how the realist can have her cake and EAT it too. *Studies in History and Philosophy of Science* 36: 509–38.
Salmon, W. (1984). *Scientific Explanation and the Causal Structure of the World*. Princeton: Princeton University Press.
—— (1989). Four decades of scientific explanation. In P. Kitcher and W. C. Salmon (eds.), *Scientific Explanation: Minnesota Studies in the Philosophy of Science, Volume 13*, pp. 3–219. Minneapolis: University of Minnesota Press,
—— (1990). Scientific explanation: Causation *and* unification. *Critica Revista Hispanoamericana de Filosofia* 22: 3–21.
Saunders, S. (1993a). To what physics corresponds. In S. French and H. Kamminga (eds.), *Correspondence, Invariance and Heuristics: Essays in Honour of Heinz Post, Boston Studies in the Philosophy of Science, Vol 148*, pp. 295–325. Dordrecht: Kluwer.
—— (1993b). Decoherence, relative states and evolutionary adaptation. *Foundations of Physics* 23: 1553–85.
—— (1995). Time, decoherence and quantum mechanics. *Synthese* 102: 235–66.
—— (2003a). Indiscernibles, general covariance, and other symmetries. In A. Ashtekar, D. Howard, J. Renn, S. Sarkar and A. Shimony, eds., *Revisiting the Foundations of Relativistic Physics: Festschrift in Honour of John Stachel*. Dordrecht: Kluwer.

—— (2003b). Physics and Leibniz's principles. In K. Brading and E. Castellani, eds., *Symmetries in Physics: Philosophical Reflections*, pp. 289–307. Cambridge: Cambridge University Press.
—— (2003c). Structural realism again. *Synthese* 136: 127–33.
—— (2003d). Critical notice: Cao's 'The Conceptual Development of 20th Century Field Theories'. *Synthese* 136: 79–105.
—— (2006). Are quantum particles objects? *Analysis* 66: 52–63.
Savitt, S. (ed.) (1995). *Time's Arrow Today: Recent Physical and Philosophical Work on the Direction of Time*. Cambridge: Cambridge University Press.
Scerri, E. R. and McIntyre, L. (1997). The case for the philosophy of chemistry. *Synthese* 111: 213–32.
Schaffer, J. (2003). Is there a fundamental level? *Noûs* 37: 498–517.
Schrödinger, E. (1935–6). Discussion of probability relations between separated systems. *Proceedings of the Cambridge Philosophical Society* 31 (1935): 555–63; 32 (1936): 446–51.
—— (1944). *What is Life?* Cambridge: Cambridge University Press.
Seife, C. (2006). *Decoding the Universe*. New York: Viking.
Sellars, W. (1962). Philosophy and the Scientific Image of Man. In R. Colodny (ed.), *Frontiers of Science and Philosophy*, pp. 35–78. Pittsburgh: University of Pittsburgh Press.
Shalizi, C. (2004). Functionalism, emergence, and collective coordinates: A statistical physics perspective on 'What to say to a sceptical metaphysician'. *Behavioral and Brain Sciences* 27: 635–6.
—— and Moore, C. (2003). What is a macrostate? Subjective observations and objective dynamics. Philsci Archive, University of Pittsburgh. http://philsci-archive.pitt.edu/archive/00001119/
Shannon, C. and Weaver, W. (1949). *The Mathematical Theory of Communication*. Urbana: University of Illinois Press.
Shapiro, S. (1997). *Philosophy of Mathematics: Structure and Ontology*. Oxford: Oxford University Press.
Shenker, O. (1999). Is -kTr({rho} ln {rho}) the entropy in quantum mechanics? *British Journal for the Philosophy of Science* 50: 33–48.
Shoemaker, S. (1980). Causality and properties. In P. van Inwagen (ed.), *Time and Cause*, pp. 109–36. Dordrecht: Reidel.
Sider, T. (1993). Van Inwagen and the possibility of gunk. *Analysis* 53: 285–9.
—— (2001). *Four-Dimensionalism: An Ontology of Persistence and Time*. Oxford: Oxford University Press.
Skinner, B. F. (1971). *Beyond Freedom and Dignity*. New York: Knopf.
Sklar, L. (1967) Types of inter-theoretic reduction. *British Journal for the Philosophy of Science* 18/2: 109–24.
—— (1974). *Space, Time and Spacetime*. Berkeley: University of California Press.
—— (1993). *Physics and Chance: Philosophical Issues in the Foundations of Statistical Mechanics*. Cambridge: Cambridge University Press.
Smart, J. (1963). *Philosophy and Scientific Realism*. London: Routledge.
—— (1989). *Our Place in the Universe: A Metaphysical Discussion*. Oxford: Basil Blackwell.

Smolin, L. (1997). *The Life of the Cosmos*. Oxford: Oxford University Press.
____ (2000). *Three Roads to Quantum Gravity*. London: Weidenfeld and Nicolson.
Smolin, L. (2006). The case for background independence. In D. Rickles, S. French, and J. Saatsi, *Structural Foundations of Quantum Gravity*, pp. 196–239. Oxford: Oxford University Press.
Sober, E. (2001). Venetian sea levels, British bread prices, and the principle of the common cause. *British Journal for the Philosophy of Science* 52: 331–46.
Sorkin, R. (1995). A specimen of theory construction from quantum gravity. In J. Leplin (ed.), *The Creation of Ideas in Physics*, pp. 167–79. Dordrecht: Kluwer.
Sorrell, T. (1991). *Scientism*. London: Routledge.
Spelke, E. S., Phillips, A., and Woodward, A. L. (1995). Infants' knowledge of object motion and human action. In D. Sperber, D. Premack, and A. James Premack (eds.), *Causal Cognition*, pp. 44–78. Oxford: Oxford University Press.
Spurrett, D. and Papineau, D. (1999). A note on the completeness of 'physics'. *Analysis* 59/1: 25–9.
Stachel, J. (2002). 'The relations between things' versus 'the things between relations': The deeper meaning of the hole argument. In D. Malament (ed.), *Reading Natural Philosophy: Essays in the History and Philosophy of Science and Mathematics*, 231–66. Chicago and LaSalle, IL: Open Court.
____ (2006). Structure, individuality and quantum gravity. In D. Rickles, S. French, and J. Saatsi (eds.), *Structural Foundations of Quantum Gravity*, pp. 53–82. Oxford: Clarendon Press.
Stamp, P. (1995). Time, decoherence and 'reversible' measurements. In S. Savitt (ed.), *Time's Arrow Today: Recent Physical and Philosophical Work on the Direction of Time*, pp. 107–54. Cambridge: Cambridge University Press.
Stanford, P. K. (2000). An antirealist explanation of the success of science. *Philosophy of Science* 67: 266–84.
____ (2003). Pyrrhic victories for scientific realism. *Journal of Philosophy* 11: 551–72.
Stapp, H. (1987). Quantum nonlocality and the description of nature. In J. Cushing and E. McMullin (eds.), *Philosophical Consequences of Quantum Theory: Reflections on Bell's Theorem*, pp. 154–74. Notre Dame: University of Notre Dame Press.
Stein, H. (1968). On Einstein-Minkowski space-time. *Journal of Philosophy* 65: 5–23.
____ (1989). Yes, but, ... Some skeptical remarks on realism and antirealism. *Dialectica* 43: 47–65.
____ (1991). On relativity and openness of the future. *Philosophy of Science* 58: 147–67.
Stent, G. (1971). An ode to objectivity: Does God play dice? *Atlantic*, November: 128.
Strawson, P. (1959). *Individuals*. London: Methuen.
Suppe, F. (1977). *The Structure of Scientific Theories*. Urbana: University of Illinois Press.
____ (1989). *Scientific Realism and Semantic Conceptions of Theories*. Urbana: University of Illinois Press.
____ (1961). A comparison of the meaning and use of models in mathematics and the empirical sciences. In P. Suppes, *Studies in the Methodology and Foundations of Science* pp. 10–23. Dordrecht: Reidel.
____ (1962). Models of data. In E. Nagel, P. Suppes and A. Tarksy (eds.), *Logic, Methodology and Philosophy of Science*, pp. 252–61. Stanford: Stanford University Press.

_____ (1969). *Studies in the methodology and Foundations of Science*. Dordrecht: Reidel.
Swoyer, C. (1983). Realism and Explanation. *Philosophical Inquiry* 5: 14–28.
Teller, P. (1989). Relativity, relational holism, and the Bell inequalities. In J. T. Cushing and E. McMullin (eds.), *Philosophical Consequences of Quantum Theory: Reflections on Bell's Theorem*, pp. 208–21. Notre Dame: University of Notre Dame Press.
_____ (1990). Prolegomenon to a proper interpretation of quantum field theory. *Philosophy of Science* 57: 594–618.
Thagard, P. (1992). *Conceptual Revolutions*. Princeton: Princeton University Press.
Thornhill, R. and Gangestad, S. W. (1999). The scent of symmetry: A human sex pheromone that signals fitness? *Evolution and Human Behavior* 20: 175–201.
Timpson, C. (2004). *Quantum Information Theory and the Foundations of Quantum Mechanics*. Unpublished doctoral thesis.
_____ (2006). The grammar of teleportation. *The British Journal for the Philosophy of Science* 57: 587–621.
_____ and Brown, H. (forthcoming). Relativity and entanglement. In R. Lupacchini and V. Fano (eds.), *Understanding Physical Knowledge* (Preprint no. 24, Department of Philosophy, University of Bologna, CLUEB, 2002).
Tooley, M. (1997). *Time, Tense and Causation*. Oxford: Oxford University Press.
Tversky, A. and Kahneman, D. (1983). Extensional versus intuitive reasoning: The conjunction fallacy in probability judgement. *Psychological Review* 90: 293–315.
Twardy, C. R. and Bingham, G. P. (2002). Causation, causal perception, and conservation laws. *Perception & Psychophysics* 64/6: 956–68.
Tymoczko, T. (1991). Mathematics, science and ontology. *Synthese* 88: 201–28.
Unger, P. (2001). Why there are no people. *Midwest Studies in Philosophy* 4: 177–222.
Vaidman, L. (1994). On the paradoxical aspects of new quantum experiments. In D. Hull, M. Forbes, and R. Burian (eds.), *PSA 1994, volume 1*, pp. 211–17. East Lansing: Philosophy of Science Association.
Valentini, A. (2002). Signal locality and subquantum information in deterministic hidden variable theories. In J. Butterfield and T. Placek (eds.), *Non-locality and Modality*, vol. 64 of *Nato Science Series: II*, pp. 81–103. Dordrecht: Kluwer.
van Brakel, J. (1986). The chemistry of substances and the philosophy of mass terms. *Synthese* 69: 291–324.
van der Waerden, B. L. (ed.) (1968). *Sources of Quantum Mechanics*. New York: Dover.
van Dyck, M. (forthcoming). Constructive empiricism and the argument from underdetermination. In B. Monton (ed.), *Van Fraassen's Philosophy of Science, Mind Occasional Series*. Oxford: Oxford University Press.
van Fraassen, B. C. (1977). The only necessity is verbal necessity. *Journal of Philosophy* 74: 71–85.
_____ (1978). Essence and existence. In N. Rescher (ed.), *Studies in Ontology*, pp. 1–25. Oxford: Blackwell.
_____ (1979). Russell's philosophical account of probability. In G. W. Roberts (ed.), *Bertrand Russell Memorial Volume*. London: George Allen and Unwin.
_____ (1980). *The Scientific Image*. Oxford: Oxford University Press.

van Fraassen, B. C. (1981). Essences and the laws of nature. In R. Healey (ed.), *Reduction, Time and Reality: Studies in the Philosophy of the Natural Sciences*, pp. 189–200. Cambridge: Cambridge University Press.

―― (1985). Empiricism and the philosophy of science. In P. Churchland and C. Hooker (eds.), *Images of Science*, pp. 245–308. Chicago: Chicago University Press.

―― (1989). *Laws and Symmetry*. Oxford: Oxford University Press.

―― (1991). *Quantum Mechanics: An Empiricist View*. Oxford: Oxford University Press.

―― (1994). Against transcendental empiricism. In T. J. Stapleton (ed.), *The Question of Hermeneutics*, pp. 309–35. Dordrecht: Kluwer.

―― (1995). Against naturalised epistemology. In P. Leonardi and M. Santambrogio (eds.), *On Quine: New Essays*, pp. 68–88. Cambridge: Cambridge University Press.

―― (1997). Sola experimentia? Feyerabend's refutation of classical empiricism. *Philosophy of Science* 64 Supplementary Volume: 385–95.

―― (1997). Structure and perspective: Philosophical perplexity and paradox. In M. L. Dalla Chiara, K. Doets, D. Mundici, and J. van Benthem (eds.), *Logic and Scientific Methods*, pp. 511–30. Dordrecht: Kluwer.

―― (2002). *The Empirical Stance*. New Haven: Yale University Press.

―― (2006). Structure: Its shadow and substance. *The British Journal for the Philosophy of Science* 57: 275–307.

―― (2008). *Scientific Representation: Paradoxes of Perspective*. Oxford: Oxford University Press.

van Inwagen, P. (1990). *Material Beings*. Ithaca, NY: Cornell University Press.

von Foerster, H. (1960 [2003]). On self-organising systems and their environments. Reprinted in *Understanding Understanding*, pp. 1–19. New York: Springer-Verlag.

―― (1981). *Observing Systems: Selected Papers of Heinz von Foerster*. Seaside, CA: Intersystems.

Vorbeij, G. (2005). *Nature: An Economic History*. Princeton: Princeton University Press.

Votsis, I. (forthcoming). The upward path to structural realism. *Philosophy of Science* supplementary volume.

Wald, R. M. (2004). *Quantum Field Theory in Curved Spacetime and Blackhole Thermodynamics*. Chicago: Chicago University Press.

Wallace, D. (2000). The quantization of gravity: An introduction. Unpublished manuscript. Available at http://www.arXiv:gr-qc/0004005 v1.

―― (2001). Implications of quantum theory in the foundations of statistical mechanics. Unpublished manuscript. Available at http://philsci-archive.pitt.edu

―― (2002). Worlds in the Everett interpretation. *Studies in History and Philosophy of Modern Physics* 33: 637–61.

―― (2003a). Everett and structure. *Studies in History and Philosophy of Modern Physics* 34: 87–105.

―― (2003b). Everettian rationality: Defending Deutsch's approach to probability in the Everett interpretation. *Studies in History and Philosophy of Modern Physics* 34: 415–39.

―― (2004). Protecting cognitive science from quantum theory. *Behavioral and Brain Sciences* 27: 636–7.

____ (2006). In Defence of Naiveté: The Conceptual Status of Lagrangian Quantum Field Theory. *Synthese* 151: 33–80.

____ (2006). Epistemology quantised: Circumstances in which we should come to believe in the Everett interpretation. *The British Journal for the Philosophy of Science* 57: 655–89.

Walter, S. and Heckmann, H.-D. (eds.) (2003). *Physicalism and Mental Causation*. Bowling Green: Academic Imprint.

Weatherson, B. (2003). What Good are Counterexamples? *Philosophical Studies* 115: 1–31.

Weyl, H. (1931, 1950). *The Theory of Groups and Quantum Mechanics* (trans. H. P. Robertson). New York: Dover.

Wheeler, J. A. (1990). Information, physics, quantum: the search for links. In W. Zurek (ed.), *Complexity, Entropy and the Physics of Information*, pp. 3–28. Boulder: Westview.

Wheeler, Q. and Meier, R. (eds.) (2000). *Species Concepts and Phylogenetic Theory*. New York: Columbia University Press.

Whiten, A. and Byrne, R. W. (1997). *Machiavellian Intelligence II: Extensions and Evaluations*. Cambridge: Cambridge University Press.

Wicken, J. (1983). Entropy, information, and nonequilibrium evolution. *Systematic Zoology* 32: 438–43.

____ (1988). Thermodynamics, evolution and emergence: ingredients for a new synthesis. In B. Weber, D. Depew, and J. Smith (eds.), *Entropy, Information and Evolution*, pp. 139–69. Cambridge, MA: MIT Press.

Wiley, E. O. (1981). *Phylogenetics: The Theory and Practice of Phylogenetic Systematics*. New York: Wiley.

____ and Brooks, D. R. (1983). Nonequilibrium thermodynamics and evolution: A response to Lovtrup. *Systematic Zoology* 32: 209–19.

Wilson, J. (2005). Supervenience-based formulations of physicalism. *Noûs* 39: 426–59.

____ (2006). On characterizing the physical. *Philosophical Studies* 131: 61–99.

Wilson, R. (ed.) (1999). *Species*. Cambridge, MA: MIT Press.

Wolpert, L. (1992). *The Unnatural Nature of Science*. Cambridge, MA: Harvard University Press.

Worrall, J. (1984). An unreal image: Review article of van Fraassen. *British Journal for the Philosophy of Science* 35: 65–79.

____ (1989). Structural realism: The best of both worlds? *Dialectica* 43: 99–124.

____ (1994). How to remain (reasonably) optimistic: Scientific realism and the 'luminiferous ether'. In D. Hull, M. Forbes, and R. M. Burian (eds.), *PSA 1994, Volume 1*, pp. 334–42. East Lansing: Philosophy of Science Association.

____ (1996). Structural realism: The best of both worlds? In D. Papineau (ed.), *The Philosophy of Science*, pp. 139–65. Oxford: Oxford University Press (originally published in *Dialectica* 43 (1989): 99–124).

____ Worrall, J. and Zahar, E. (2001). Ramsification and structural realism. Appendix in E. Zahar, *Poincaré's Philosophy: From Conventionalism to Phenomenology*, pp. 236–51. Chicago, IL: Open Court Publishing Co.

Yi, S. W. (2003). Reduction of thermodynamics. *Philosophy of Science* 70: 1028–38.

Zahar, E. (1994). Poincaré's structural realism and his logic of discovery. In G. Heinzmann et al. (eds.), *Henri Poincaré*, pp. 43–68. Nancy: Akten Des Internationale Kongresses.

Zeh, H. (1990). Quantum measurements and entropy. In W. Zurek (ed.), *Complexity, Entropy and the Physics of Information*, pp. 405–20. Boulder: Westview.

Zeilinger, A. (2004). Why the quantum? 'It' from 'bit'? A participatory universe? Three far-reaching challenges from John Archibald Wheeler and their relation to experiment. In J. Barrow, P. Davies, and C. Harper (eds.), *Science and Ultimate Reality: Quantum Theory, Cosmology and Computation*, pp. 201–20. Cambridge: Cambridge University Press.

Zurek, W. (1990b). Algorithmic information content, Church-Turing thesis, physical entropy, and Maxwell's demon. In W. Zurek (ed.), *Complexity, Entropy and the Physics of Information*, pp. 73–89. Boulder: Westview.

_____ (ed.) (1990a). *Complexity, Entropy and the Physics of Information*. Boulder: Westview.

Index

abstract/concrete distinction 159–60, 186
Addleson, M. 7
Ainslie, G. 45
Albert, D. 166, 177 n. 80, 194 n. 5
Anaxagoras 20
Aristotle 3, 61, 65, 86, 148, 251
Armstrong, D. 4 n. 1, 255, 268, 271
 bundle theory 144
 metaphysical naturalism 23–4
atomism 20, 234
 mereological 17–22, 50, 54–5, 178 n. 83, 261
Auyang, S. 140 n. 20, 146 n. 30, 158

background independence 169–70
Baillargeon, R. 267
Bain, J. 82–3, 94, 132 n. 1, 145 n. 27
Baldwin, J. M. 266 n. 9
Barbour's relationism 171
Barwise, J. 224, 307 n.
Batterman, R. W. 49, 95 n. 23 193, 204–5, 237–8,
 emergence 264–5
 universality classes 204, 290
Belot, G. 143, 164 n. 63, 170, 204 n.
Bealer, G. 17 n. 19
Benacerraf, P. 138
Bennett, C. 220
Bennett, K. 41
Bennett, M. 147 n. 34
Bergson, H. 260
Berkeley, G. 227
Bickle, J. 48, 49, 197
biology 21, 214–15, 216–17
 biological thermodynamics 217–18
 species 222–3
Bishop, M. 92
Blackburn, S. 76
Black, M. 137
Block universe 172–3, 175–80, 215, 216, 219–20
 information transfer 211–12
Bohm, D.
 pilot wave theory 136 n., 165, 173, 181, 185 n. 93
Bohr, N. 43, 90, 95, 173–4
Bookstein, F. 217 n.
Born, M. 146 n. 30, 147, 151 n. 40
 Born rule 175–6, 182
Bontly, T. 192, 197

boson 135 n. 13, 136, 145
Boyd, R. N. 69, 70, 73, 88
Boyle, R. 11, 20, 48
Braithwaite, R. B. 75, 156 n. 52
Brittan, G. G. 48, 49 n. 47
Broad, C. D. 42–3, 56
de Broglie, L. V. 261, 275
Brooks, D. R. 217, 250 n. 65
Brown, J. 237
Bueno, O. 97 n., 116, 117 n. 42
Burch, M. 217
Butterfield, J. 9 n. 9, 143, 149, 171 n. 76
Byrne, R. W. 2

Callender, C. 74, 85 n. 14
caloric 84, 88–93
Carnap, R. 8, 29, 75, 123, 125
 definition of theoretical terms 107 n. 36, 111, 113–14
Cartwright, N. 5, 6, 44, 56, 71, 76, 226 n. 45, 238, 262, 283
 auxiliary assumptions 115
 ontological plurality 118 n. 44, 260–1
Cassirer, E. 123, 132, 140, 145, 147
causation 1, 15, 16, 31 n., 38–9, 160
 causal asymmetry 163, 274, 277 see also time, temporal asymmetry
 causal capacities 56
 causal process theories 263, 265–6, 273, 277–8, 280 see also Salmon
 downward 45, 57
 efficient 3–4 see also containment metaphor
 eliminativism 4, 259, 258–66, 275
 essentialism 156 n. 50
 folk concept 266–80
 functional interdependence 22, 262
 and information 210–11
 in science 22, 44, 269–74
 and structural realism 158–9, 191–4
cellular automata 201–2
Chaitin, G. 183, 213, 214
Chakravartty, A. 90 n. 18, 154
Chuang, I. 183
cladistics 223 n.
Clark A. 2, 209 n. 19
Cleland, C. 150 n. 39
cohesion 240–57, 294–7
Collier, J. 211, 215, 216–18, 263 n. 6
 bound information 186, 227
 cohesion 240, 246–7

Index

common sense realism 106 *see also* metaphysics, intuitions and common sense
complexity 197–8, 214–15, 223
composition relation 11
  in economic models 21
  in the physical sciences 21
computation 214, 216–18 *see also* complexity; compressibility; projectibility
  physical limitations 208–10
compressibility 202–7, 213–15, 220–1, 250 n. 64, 251
constructive empiricism 95–111 *see also* van Fraassen, B.
  critcism of 97, 101–11, 306–7
  modality 107–11
containment metaphor 3–4, 269 n. 12 *see also* metaphysics, habitual
  microbangings 4, 23
  spacetime substantivilism 142
Conway, J. 201
Cosmides, L. 268
Couvalis, S. G. 107 n. 34
Crook, S. 39 n. 37

Dahlbom, B. 228 n. 28
Darwin, C. 28, 73, 78–9, 203
Davidson D. 22, 49, 51
  Mental Events 41
Davies, P. 221
Dawid, R. 175 n.
demarcation 7, 27–8, 32–4 *see also* (PNC) principle of naturalistic closure; positivism
  anthropocentric investigations 36–7
  institutional factors 33–8
Democritus 20
Demopoulos, W. 113 n., 126, 127 n. 52
Dennett, D. 64 n., 225, 231, 232, 242, 254, 256, 272, 298–300
  dynamic patterns 200–2
  information 210, 213 n. 24
  instrumentalism 199, 233
  intuitions in metaphysics 14–15
  notional worlds 244–6, 247–9, 252, 300
  real patterns 36, 196–210, 218, 220–38
DePaul, M. 10 n. 10
derivatives, financial 92 n.
Descartes, R. 10, 11, 20
determinism 142, 173
Deutsch, D. 176
Dipert, R. R. 138 n., 148
Dirac, P. 146
discernibility 134
  of quantum particles 153
  relative 137
  strong 137
  weak 137, 152
distinguishability *see* discernibility
disunity hypotheses 190–6
domestication 149 n. 36, 189
  and causation 3, 265
  God 29–30
  of scientific discoveries 1–6, 17, 24, 29–30, 52
Domski, M. 123, 155 n. 47
Dorato, M. 144–5
Dowe, P. 263 n. 6
Dowell, J. 39 n. 37
Dretske, F. 307, 309
Dunbar, R. I. M. 50
Dupré, J. 7 n. 4, 40, 44, 196 n., 223, 226 n. 45, 292
  ontological plurality 5–6, 71, 194–5
Durlauf, S. 250 n. 65
dynamic patterns 200–2

Earman, J. 142–3, 170, 256–7, 281–8
Eddington, A. 132, 145, 156 n. 52, 252–3, 274
Einstein, A. 28, 77, 90, 141, 151 n. 40, 153, 261
Elder, C. 23
Elster, J. 271
emergentism 20, 45, 56, 192–3, 204, 215, 264–5 *see also* scale relativity
  of ontology
entropy 165–7, 212–13, 214, 216–17, 219–20
(ESR) epistemic structural realism 124–9, 130–1, 144 n. 26, 158
Esfeld, M. 67 n. 1, 132 n. 1, 145 n. 27, 150 n. 39, 152–4, 188 n. 101
ether 84, 85, 88–93
extrinsic properties 134 n. 11, 135, 148

Faraday, M. 42
Feigl, H. 114
Feyerabend, P. 47, 48, 93
fermions 135 n. 13, 136, 145
Field, H. 262 n.
Fine, A. 75, 187, 200, 253
Floridi, L. 121 n. 45, 190–1, 203
Fodor, J. 9 n. 8, 22, 49, 50, 51, 281, 284
formal mode of discourse 119, 120
Forrest, S. 45
Frege, G. 100
French, S. 9, 44, 67, 116, 135, 97 n.
  semantic approach and isomorphism 117
  ontic structural realism 94 n. 21, 131
  identity of particles 132
Fresnel, A.
  wave optics 76–7, 88–9, 93

Friedman, M. 27, 126, 198, 261, 293, 305
  argument for realism 71–2
fundamentalism 178 *see also* reductionism;
  metaphysics, levels

Galilean transformations 95
Game theory 222–3
Garcia, J. 267
Gassendi, P. 11, 20
Geach, P. 17
Gell-Mann, M. 219, 225
Gelman, R. 267
(GR) General Relativity 11, 57, 76–7, 141–5
  equivalence principle 167
  general covariance 141–4
  stress-energy tensor 141, 283
  time 164
geometry 155
Giere, R. 67, 99, 111, 284
  semantic approach 117–18
Gillett, C. 39 n. 37
Glimcher, P. 53
Glymour, C. 14 n. 14, 18 n. 22, 22 n. 27, 265 n.
Gödel, K. 163 n. 60, 235
Godfrey-Smith, P. 203
Goodman, N. 70 n. 4, 80, 194
Gopnik, A. 267, 268
Gould, S. J. 28 n.
Gower, B. 123, 132, 157
graph theory 138 n., 148
gravitational entropy 166
Greene, B. 277
group theory 145–7

Hacking, I 69 n. 3
haecceity 134, 136 n., 142–3, 175 *see also* discernibility
Hanson, R. 70 n. 4
Hardin, C. L. 85
Harris, R. 7 n. 4
Hartle, J. 173, 178 n. 82, 219, 220 n., 225
Haugeland, J. 199
Hayes, P. 14, 269
Hegel, G. 62
Heidegger, M. 5
Heil, J. 54 n.
Heisenberg, W. 90, 146, 165
Hellman, G. 38, 39, 161 n. 56
Helmholtz, H. 42
Hempel, C. 8, 32, 39 n. 37, 46 n., 284–5, 291
Henderson, L. 213 n. 23
hermeneutic economics 7
Hilbert, D. 100
Hoefer, C. 143
Hogarth, M. L. 235

Holbach, P. D. 62
hole problem 142–3, 170
Hooker, C. 218
Hoover, K. 203
Hopf, F. 217–18
Horgan, T. 40, 253–4
Howson, C. 74, 79
Hull, D. 296
Hume, D. 8, 26–7, 304–5, 310
  causation 262–3, 266–7
  Humean supervenience 19–22, 54, 148–51, 175
  verificationism 64
Humphreys P. 209 n. 20, 300, 308
Hüttemann, A. 24, 25, 41, 55–7, 203

identity 131, 132–40
  (PII) identity of indiscernibles 135–8, 151–3
  synchronic identity 135 n. 14
individualism 191–2, 196
intelligent design theory 7
idealism 194–5
impenetrability 134
indexical redundancy 295–7
indispensible terms 86–9 *also see* scientific realism
indistinguishability postulate 133 *see also* identity
individuality 132–40, 242–5 *see also* discernibility
induction 255
inductive libertarianism 103–11
(IBE) inference to the best explanation 17 n. 18, 102
  defense of realism 69, 75, 98
information 220–1
  bound information 186, 227
  and causation 210–11
  (ITSR) Information-Theoretic Structural Realism 238–57, 279–80
  information theory 183–9, 210–20 *see also* quantum information theory
  information transfer *see* block universe
  informational connection *see* verificationism
instrumentalism 196–210
intrinsic properties 130–7, 148–51, 152 n. 44
irreversible processes
  thermodynamics 214–15, 217
isomorphism
  between models and structures 117
  underdetermination 100

Jackson, F. 16, 19, 39 n. 38, 40, 127 n. 53, 154

Johnson, S. 3, 3 n. 3, 197
Jones, R. 82
Joule, J. P. 42

Kahneman, D. 268
Kant, E. 266, 299–300, 305
Kantorovich, A. 147
Kauffman, S.
  complexity theory 45
Kelso, S. 201
Kemeny, J. C. 46
Ketland, J. 127 n. 52
Kim, J. 39 n. 37, 40, 51, 197, 252, 258, 268, 272
  causal exclusion 191–4, 265
  emergentism 199, 204
  Mind in a Physical World 18
  reductionism 19, 26, 261
Kincaid, H. 191, 200 n. 10, 261
  causation 270–2, 275, 277, 279
  special sciences 256–7, 281–3, 285–9
Kitcher, P. 7 n. 5, 11, 27, 46, 71 n., 86, 87, 90, 265, 274, 291
  explanatory argument patterns 30–2, 261, 270, 273, 279, 293
Koelling, R. A. 267
Krause, D. 10 n., 139 n.
Kripke, S. 91
  Naming and Necessity 9, 9 n. 8
Kuhn, T. 9, 47, 70 n. 4
  Kuhn-losses 157
Kukla, A. 72, 81 n.
Kydland, F. 92 n.

Ladyman, J. 44, 77 n, 99, 116, 125, 138 n.
  (PII) identity of indiscernibles 151
  Landauer's Principle 189 n. 102, 208 n. 18
  ontic structural realism 67, 131–2
  semantic approach and isomorphism 117
Lakoff, G. 3
Landauer, R. 218
  Landauer's Principle 189 n. 102, 208 n. 18, 214 n. 26
Lange, P. 85 n. 14, 281, 284
Laplace, P. S. 62
Laporte, J. 291–2, 295, 296
Laudan, L. 75, 81 n., 84, 85, 86, 87
laws (in physics and the special sciences) 281–90
Layzer, D. 214 n. 27, 215, 217–18
Leggett, A. 165 n. 64, 219 n.
Leibniz, G. W. 20, 141 n. 22, 148
  equivalence 143
  identity of indiscernibles 134
Leitgeb, H. 138 n., 161 n. 56
  (PII) identity of indiscernibles 151

Leplin, J. 77, 78, 81 n., 86
Lewis, D. 9, 13 n. 19, 41, 62, 134 n. 11, 142, 153, 154, 188 n.100
  cost–benefit analysis 12
  gunk 13 n. 12
  Humean supervenience 19, 54, 148–51
  internal/external relations 20, 149–51
Lewis, P. 85 n. 14
Lipton, P. 6, 74, 99 n.
Litt, A. 208 n. 17
locators 121–2, 158, 221–3, 239, 244, 295
Locke, J. 20, 101–2, 241
  primary qualities 11
Lockwood, M. 177 n. 80
Lowe, J. 22, 151 n. 41, 139 n., 151 n. 41,
  against naturalized metaphysics 6, 7, 13–14, 15–16,
Loewer, B. 39 n. 37, 177 n. 80, 194 n. 5
logical depth 218–20, 231–2, 240
logical empiricism 63
  fall of 7–8, 290–1
logical positivism 8–9 see also logical empiricism
  demarcation problem 7
  and verificationism 8, 27, 30, 63
Loux, M. 12
Lovtrup, S. 217 n.
Lyre, H. 94, 128, 132 n. 1, 140, 147, 154
Lyell, C. 78, 203

Magnus, P. D. 74, 79, 85 n. 14
Malebranche, N. 227 n. 47
Mäki U. 198
many minds theory 177 n. 80 also see quantum mechanics, Everett interpretation
Markosian, N. 17, 18, 39 n. 37, 178
Marras, A. 18 n. 21, 48, 49, 192, 197
material mode of discourse 119, 120
materialism 60, 62
Maudlin, T. 142, 151 n. 40, 153, 162 n. 58, 165, 188
Maxwell, J. C.
  Maxwell's Demon 217
  electrodynamics 85, 88–9, 90 n. 18, 91, 93–4
Maxwell, G. 125–8
McIntyre, L. 43 n. 42
McLaughlin, B. 41 n. 41
Melnyk, A. 39, 55 n. 53, 197 n. 8, 266 n. 10, 304, 305, 306, 310
  core-sense reductionism 52–3, 233
  realization physicalism 41–2
Merricks, T. 5, 15, 22–3, 252, 253
meta-induction
  theory change 83–93
metaphysics

and contemporary science 2
habitual 2–5
individuation 11, 13–14, 44, 132–40
intuitions and common sense 10–15, 132, 247–9, 252–4, 260, 268
levels 53–7, 178–80, 193, 197, 229 *see also* scale relativity of ontology
matter 3, 20–3
methodology 60–2 *see also* a priori metaphysics
naturalistic 1–7, 23–4, 53, 59–60
  failure to follow 17, 22
  as philosophy of science 24
  neo-scholastic 7–27, 61, 194, 196
  a priori 15–17, 57
  pseudo-scientific 17–27
realism 290–7, 305
relations 148–54
revival of 9–10
unification of 6, 16, 45, 65, 234–8, 260–1, 272
value of 27, 58–60
mere patterns 231
Mermin, D. 187–8
metric essentialism 142
Michelson Morley experiment 85
microgovernance 55–6, 203
Mill, J. S. 43, 255, 261, 266
modality 132, 130, 153, 186–8
Møller, A. P. 50
Montero, B. 39 n. 37
Monton, B. 99, 107 n. 33, 111, 173 n.
Moore, C. 203
Morowitz, H. 217 n.
Morris, M. W. 10, 268
Minkowski metric 141
Muller, F. 107 n. 33
Muller, S. 240
Mustgrave, A. 76
Myrvold, W. 165 n. 65

Nagel, E. 46–53, 111
  bridge laws 49–50
  Nagelian reductions 47–53
Nagel, J. 98 n. 30
natural kinds 294–7
naturalism 5–7, 9, 44, 62, 288
  constraints on metaphysics 30–5, 300–2, 303–10
  and the primacy of physics 40–2
  pseudo-naturalist philosophy 7, 27
Needham, P. 191
Newman, M. H. A.
  cardinality objection 126, 127 n. 52, 128, 147 n. 33
Newton, I. 25, 42, 49, 59

Ney, A. 18, 39 n. 37, 41 n. 40, 127 n. 53
Nickles, T. 48, 49 n. 47
Nielsen, M. 183
Nolan, D. 20 n., 33 n.
nominalism 159–61
no-miracles argument 69–76, 84–5, 298
  circularity 75
Norton, J. 94, 142, 143, 191, 265, 269 n. 11, 272, 274, 276, 277, 278 n.
Nottale, L. 204

observability 107–11
Ontological Physicalism *see* physicalism
(OSR) ontic structural realism 67, 151–4, 251–2
  objections 154–9
  ontological implications 228–9
  and philosophy of physics 130–89
  quantum mechanics 175–80
open set 155
Oppenheim, P. 50, 197, 207, 237
  physicalism 40, 41, 46, 47, 49
Orzack, S. 249

Pais, A. 43
Papineau, D. 38, 39, 41, 44, 56, 75, 156
particularism 148–9
Past Hypothesis 166
Paul, L. A. 13, 21 n. 25
Peirce, C. S. 27, 28, 29, 63, 255
  modality 129
Penrose, R. 166, 171 n. 73, 173
Peres, A. 184, 212–13, 214, 216, 227
Perry, J. 248
perspective 220–1, 224–6, 308
Pettit, P. 40, 41
philosophy of mind 15, 18
physicalism
  naturalist defence 40
  objections to 39–40
  part–whole 41
physics 275–9, 281–90
  Big-Bang 29–30
  classical 134
  ideal gas laws 48
  information theory 210–20
  naïve *see also* metaphysics, intuitions and common sense
  obsolete 19–20, 44
  special status of 42
  unification of 42
Pietroski, P. 281
Plato 10, 61, 152
Platonism 236–7
Poincaré, H. 93, 100, 123–4, 125, 155
Poisson bracket 95 n. 24

Index

Ponce, V. 191
Pooley, O. 141 n. 24, 143–4, 161 n. 57
Potrc, M. 253–4
Poundstone, W. 201
positivism 7–10, 26–32
  neo-positivism 303–10
pragmatism 27–8
Prescott, E 92 n.
(PPC) Primacy of Physics Constraint 38–45, 191–2, 275
  arguments for 43
  formulation of 44
(PNC) principle of naturalistic closure 27–38, 39, 66, 90, 121
  constraints on metaphysics 190–6, 301
  stipulations of 37
projectibility 220–1, 223, 227–32, 236
Psillos, S. 67 n., 68 n., 69, 75, 76, 86–91
  caloric theory 91
  'upward path' to structural realism 125 n.
psychology 208–10
Putnam, H. 163 n. 60, 241
  against anti-realism 70
  demarcation 27–8
  metaphysical realism 290–1
  no-miracles argument 69
  reference relations 9, 91, 92–3
  unity of science 40, 41, 45–53, 197, 207, 237

quantum gravity 167–75, 219
  causal set theory 170–1
  causal triangulation 171
  loop quantum gravity 170
  problem of time 167–70
quantum mechanics 9, 95, 145
  arrow of time 165, 219
  Bell's theorem 26, 73, 149 n. 36, 175, 186–8
  collapse of the wave function 165, 175–80, 181–3, 218–19
  decoherence theory 176–8, 219 n.
  entanglement 19, 25–6, 41 n. 40, 57, 73, 135–7, 149–51, 175, 183–8, 196, 204
  Everett interpretation 131–2, 165, 173, 175–80, 181–3, 194–5, 219
  preferred basis problem 176
  hidden variables 165, 213
  individuality 132–40
  locality 26
  measurement problem 26, 53, 177, 180–3, 200
  ontic structural realism 175–80
  permutation of particles 132–6
  pilot wave model *see* Bohm, D.

quantum field theory 138–40, 171
  Fock space 158
  Schrödinger time evolution 180–1
  singlet state 137, 149–51, 187–8
  superposition 180–3, 203–4
quantum information theory 183–9
  (CBH) Clifton–Bub–Halvorson theorem 184–5
quantum logic 236 n.
Quine, W. V. 29, 62, 207, 234, 237, 255, 310
  destruction of positivism 8
  naturalization of epistemology 2, 305

(RR) Rainforest Realism 220–38, 251–2, 256
  causation 279–80
  verificationism 230–4
Ramsey, F. P. 157
  Ramsey sentence 123, 125–8
Ramsey sentence 125–7
Rea, M. 301–2
real patterns 118–22, 177–9, 218, 220–38
  compressibility 220–1, 226
  Dennett, D. 196–210
  natural kinds 290–7
  pattern-reality 36, 226–9, 233, 236 *see also* projectibility
Redhead, M. 90 n. 19, 131–2, 139, 144 n. 26, 154, 186, 191, 204 n.
  causation in physics 211, 264, 274, 276, 277, 283, 289
  identity of particles 132, 134 n. 10
reductionism 19, 26, 196–210
  core-sense reductionism 52–3, 233
  micro-reduction 46–54
  multiple realization argument 49–51
  Nagelian reductions 47–53
  thermodynamics to statistical mechanics 48
  type reductionism 49–54
  and the unity of science 45–53
reference 85–8, 92, 290–2
relationalism 141–5
representation 242–5
Resnik, M. 159–60, 161 n. 56
Rey, G. 281
Reichenbach, H. 26, 111, 163, 211, 219, 263
  illata and abstracta 178, 207–8, 228
  (PCC) principle of common cause 72–3, 163
Rickles, D. 170, 175
Roberts, J. 256–7, 281–8
Rorty, R. 198, 200 n. 10
Rosenberg, A. 85, 198
Ross, D. 2, 6, 8 n. 6, 8 n. 7, 31, 53, 200, 203, 207, 228 n. 48, 239, 249, 265
  causal glue 262–3
  individualism 191, 192, 197

pattern reality 226–7, 231, 233 *see also* Rainforest Realism
Rowlands M. 209 n. 19
Rovelli, C. 172, 188 n. 99
Russell, B. 8, 22, 24, 26, 27, 100, 123, 127, 137–8
 causal theory of perception 125–6, 132 n. 2
 eliminativism on causation 44 n., 191, 211, 258–66
Rutherford, E. 43
Ruttkamp, E. 118 n. 44

Saatsi, J. 90 n. 18, 127 n. 52
Salmon, W. 71 n., 211, 273, 274, 291
 causal process theory 263, 265–6, 273, 277–8, 280
Santa Fe Institute 56, 211–12, 216, 237
Saunders, S. 94, 131, 158 n. 54, 159, 188, 200, 220 n., 228 n. 28
 Everettian physics 175–80, 182
 (PII) principle of identity of indiscernibles 137–8, 151–3
 scale relativity of ontology 193–6, 199–204, 234, 237–8, 252
Scerri, E. R. 43 n. 42
Schaffer, J. 54–7, 178
Schlick, M. 123
Schrödinger, E. 19, 132, 146, 211
 time evolution 180–1, 213
science
 epistemic reliability 37
 error-filter 58
 intuition 15–16 *see also* metaphysics, intuitions and common sense
 non-actual 18–19, 24–5
 scientific progress 63
 semantic approach 111–18, 159
 syntactic view 111–18
 unanswerable questions in 29–30
scientism 303
 defense of 1–65
scientific realism 67–83, 95–6
 novel prediction 76–9
 theory conjunction 70–1
Second Law of Thermodynamics 165–7, 211–18, 250, 252
Seife, C. 213 n. 24
self-organizing systems 214–15, 217
Seligman, J. 224, 307 n.
Sellars, W. 1
semantic approach to scientific theories 111–18, 159
sense datum 8, 29
Shalizi, C. 203
Shannon, C. 220
 Shannon entropy 183, 213, 214, 216–17

Shenker, O. 213 n. 23
Sider, T. 13, 16
Skinner, Q. 12
Sklar, L. 49 n. 47, 141 n. 23, 142 n., 164 n. 61, 212 n.
Smart, J. 52 n., 69
Smolin, L. 133 n. 5, 166–73, 275–7
 string theory 168–9
Sober, E. 228, 249
Sorrell, T. 9
(SR) Special Relativity 9, 17, 94–5
 Lorentz transformations 94
 time 163–4, 175
 violation of the action–reaction principle 141
special sciences 190–6, 238–57, 298–300
 definition of 195
 laws 256–7, 281–90, 285–9
Spelke, E. S. 11, 267
Spurrett, D. 8 n. 6, 18 n. 21, 39 n. 37, 44 n. 43, 192, 197, 265
 causal glue 262–3
Stachel, J. 132 n. 1, 140 n. 21, 143, 151–4, 170 n., 175, 228 n. 28
Stamp, P. 165 n. 64, 219 n.
Stein, H. 124, 143, 147, 163 n. 60
Stent, G. 264 n.
Strawson, P. 9, 229
string theory 167, 168
 M-theory 169
structural empiricism 95–100
structuralism *see also* OSR; ESR; ITSR
 mathematical 159–61
 physical 159–61
 structural realism 67, 92–129, 122–9, 148
 continuity of structure 93–4
Suarez, F. 117 n. 43, 134 n. 9
substantivalism 141–5
supervenience 38, 41–2, 54–5
 Humean 19, 54, 148–51, 175
Suppe, F. 114–17
Swoyer, C. 17 n. 18
symmetry 145–7
syntactic view of scientific theories 111–18

Teller, P. 99 n., 132 n. 4, 139, 148, 149
tensed universe 172, 215, 219–20
Thagard, P. 27, 32, 46
Thales 3
Thompson, F. W. 38–9
theory change 83–93 *also see* meta-induction
thermodynamics
 arrow of time 166, 212–13
 biological thermodynamics 217–18
 irreversibility 214–15, 217 *see also* Landauer's Principle

thermodynamics (*cont.*)
   reduction to statistical mechanics 48
   Second Law of Thermodynamics 165–7, 211–18, 250, 252
   thermodynamic depth 218–20, 240
time 162–7, 172–3
   problem of time 167–70 *see also* quantum gravity
   temporal asymmetry 165, 212–213, 219, 268, 274, 277 *see also* causal asymmetry
   thermodynamic arrow 166, 212–13
Timpson, C. 165, 183 n. 87, 184 n. 91, 185
Tooley, M. 17 n. 20, 33, 35
transcendent individuality 134, 135
transduction 225–7, 307
Tversky, A. 268

Underdetermination 79–83, 100
   strong form 81
   weak form 79–80
universals 152

Valentini, A. 185 n. 93
Van Brakel, J. 21 n. 26
van Dyck, M. 82 n.
van Fraassen, B. 10, 81–2, 116, 123, 124 n., 157–8, 253, 306
   against no-miracles argument 72–4
   constructive empiricism 95–107, 198, 227
   Darwinian explanation of predictive success 79
   empirically surplus factors 136
   naturalism and metaphysics 58–65, 304–5
   observability 107–11
   scientific realism 67–74, 95–6
van Inwagen, P. 19, 20, 22, 253

verificationism 61, 112, 130–1, 303–4
   Hume 64
   informational connection 307–10
   logical positivism 8, 27, 30, 63
   (PNC) principle of naturalistic closure 29–38
   (RR) Rainforest Realism 230–4
Vienna Circle 26
von Neumann, J. 146, 183, 213, 216–17
Vorbeij, G. 239
Votsis, I. 67 n., 125 n.

Wallace, D. 25 n., 167 n., 174 n., 199, 203, 228 n. 28, 253
   Everett interpretation 131, 175–9, 181 n., 182
Ward, J. 260
Weaver, W. 213, 214, 216–17, 220
Weyl, H. 100, 132, 145–6, 147 n. 34, 151 n. 42
Wheeler, J. A. 186, 188 n. 100, 212, 227
   Wheeler–DeWitt equation 170, 219
Whiten, A. 2
Wicken, J. 214–15, 216–18
Wiley, E. O. 217, 240, 250 n. 65
Witten, E. 169 n. 70
Wittgenstein, L. 9
Wolpert, L. 11, 12
Worrall, J. 76, 77, 89, 123, 124
   structural realism 67, 93–129

Zahar, E. 123, 124, 125 n., 127 n. 52, 128
Zeh, H. 218–19
Zimmerman, D. 12
Zurek, W. 211, 227
   physical entropy 213–17